# Springer Series in Electrophysics
# Volume 10

Edited by Walter Engl

## Springer Series in Electrophysics

Editors: Günter Ecker Walter Engl Leopold B. Felsen

---

Volume 1 **Structural Pattern Recognition**
By T. Pavlidis

Volume 2 **Noise in Physical Systems**
Editor: D. Wolf

Volume 3 **The Boundary-Layer Method in Diffraction Problems**
By V. M. Babič, N. Y. Kirpičnikova

Volume 4 **Cavitation and Inhomogeneities**
in Underwater Acoustics
Editor: W. Lauterborn

Volume 5 **Very Large Scale Integration (VLSI)**
Fundamentals and Applications
Editor: D. F. Barbe

Volume 6 **Parametric Electronics**
An Introduction
By K.-H. Löcherer, C. D. Brandt

Volume 7 **Insulating Films on Semiconductors**
Editors: M. Schulz, G. Pensl

Volume 8 **Theoretical Fundamentals of the Ocean Acoustics**
By L. Brekhovskikh, Y. P. Lysanov

Volume 9 **Principles of Plasma Electrodynamics**
By A. F. Alexandrov, L. S. Bogdankevich, A. A. Rukhadze

Volume 10 **Ion Implantation Techniques**
Editors: H. Ryssel, H. Glawischnig

# Ion Implantation Techniques

Lectures given at the Ion Implantation School
in Connection with the Fourth International Conference
on Ion Implantation: Equipment and Techniques

Berchtesgaden, Fed. Rep. of Germany, September 13–15, 1982

Editors: H. Ryssel and H. Glawischnig

With 245 Figures

Springer-Verlag Berlin Heidelberg New York 1982

Dr. Heiner Ryssel
Frauenhofer - Institut für Festkörpertechnologie, Paul-Gerhardt-Allee 42
D-8000 München 60, Fed. Rep. of Germany

Dr. Hans Glawischnig
Siemens AG, Balanstraße 73
D-8000 München 80, Fed. Rep. of Germany

*Series Editors:*

Professor Dr. Günter Ecker
Ruhr-Universität Bochum, Theoretische Physik, Lehrstuhl I,
Universitätsstrasse 150, D-4630 Bochum-Querenburg, Fed. Rep. of Germany

Professor Dr. Walter Engl
Institut für Theoretische Elektrotechnik,
Rhein.-Westf. Technische Hochschule, Templergraben 55,
D-5100 Aachen, Fed. Rep. of Germany

Professor Leopold B. Felsen Ph.D.
Polytechnic Institute of New York, 333 Jay Street, Brooklyn, NY 11201, USA

ISBN 3-540-11878-0 Springer-Verlag Berlin Heidelberg New York
ISBN 0-387-11878-0 Springer-Verlag New York Heidelberg Berlin


Printed in Germany

Offset printing: Beltz Offsetdruck, 6944 Hemsbach/Bergstr.
Bookbinding: J. Schäffer OHG, 6718 Grünstadt.
2153/3130-543210

# Preface

In recent years, ion implantation has developed into the major doping technique for integrated circuits. Several series of conferences have dealt with the application of ion implantation to semiconductors and other materials (Thousand Oaks 1970, Garmisch-Partenkirchen 1971, Osaka 1974, Warwick 1975, Boulder 1976, Budapest 1978, and Albany 1980). Another series of conferences was devoted more to implantation equipment and techniques (Salford 1977, Trento 1978, and Kingston 1980). In connection with the Third International Conference on Ion Implantation: Equipment and Techniques, held at Queen's University, Kingston, Ontario, Canada, July 8-11, 1980, a two-day instructional program was organized parallel to an implantation conference for the first time. This implantation school concentrated on aspects of implantation-equipment design.

This book contains all lectures presented at the International Ion Implantation School organized in connection with the Fourth International Conference on Ion Implantation: Equipment and Techniques, held at the Convention Center, Berchtesgaden, Germany, September 13-17, 1982. In contrast to the first school, the main emphasis in this school was placed on practical aspects of implanter operation and application. In three chapters, various machine aspects of ion implantation (general concepts, ion sources, safety, calibration, dosimetry), range distributions (stopping power, range profiles), and measuring techniques (electrical and nonelectrical measuring techniques, annealing) are discussed. In the appendix, a review of the state of the art in modern implantation equipment is given.

We would like to thank all of the authors who made this book possible. The help of Miss Forster, Miss Schmiedt, Miss Traumüller, Mrs. Podstowka and Mrs. David in connection with the editing and typing of the manuscript is greatly appreciated.

Munich *Heiner Ryssel · Hans Glawischnig*

# Contents

*Part I* Machine Aspects of Ion Implantation

## *Part II* Ion Ranges in Solids

# Part I

# Machine Aspects of Ion Implantation

# Ion Implantation System Concepts

Hans Glawischnig

Siemens AG
D-8000 München, Fed. Rep. of Germany

Abstract

Following a description of the principal ion implantation systems, some criteria for the ideal production implanter are given, after which a review of the presently available commercial equipment is presented. The system-limiting aspects, such as throughput, uniformity, wafer heating, energy range, wafer size, charge-up, and contamination, are reported in detail. Finally, some remarks on the operation and control of an ion implanter are made.

## 1. Implanter Concepts

An ion implanter consists in general of an ion source which ionizes solids, liquids, or gases, an electrostatic extraction field to extract these ions, an acceleration system, an analyzing system where these ions are separated according to their mass, and a scanning system to distribute the ions uniformly over a target. Depending on the arrangement of the acceleration apertures, the magnets, and their respective grounding, there are three principal kinds of configurations in use, which are depicted in Fig.1.

### 1.1 Post-Analysis

In this case, the ions are accelerated to their full energy before mass separation. This has the advantage of having only the source on high voltage and remote control, whereas all the other sections are grounded and directly accessible. Disadvantages are the rather large magnet needed to handle the high ion energies as well as the necessity of changing the magnet current and the focusing conditions with changing energies. The usable ion beam depends greatly on the acceleration voltage, resulting in higher currents only at higher energies.

### 1.2 Pre-Analysis

With this configuration, the ions are extracted with typical energies of 15 to 40 kV, are then analyzed, and finally accelerated to their desired energy. The magnet can be kept small; energy variations (for example, to implant controlled profiles) are easy to carry out by adjustment of the final acceleration voltage only. The ion current is not very sensitive to energy variations. A disadvantage is the fact that electric power, cooling, vacuum, and data control of the source and the analyzing section have to be electrically isolated. Since these problems can be solved satis-

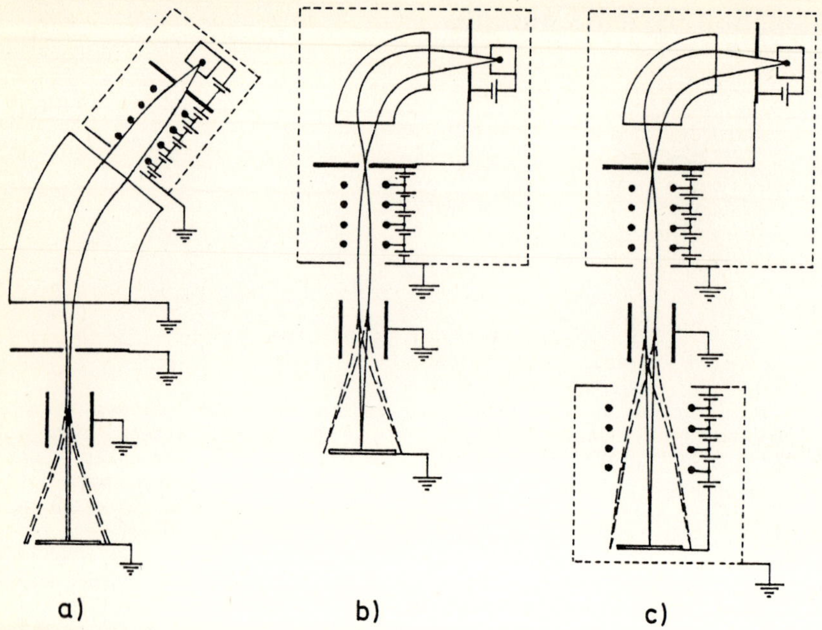


Fig. 1. Principal system configurations: a) post-analysis, b) pre-analysis, c) additional post-acceleration

factorily by modern techniques, especially by data communication via optical links, this configuration has become the favored implanter type.

### 1.3 Post-Analysis with Post-Acceleration

In this case, for an additional increase in the energy range, a second acceleration stage is provided by placing the target on high voltage. However, this system is rather inconvenient for wafer handling and dosimetry. Only a few systems of this kind have been reported [1].

## 2. Criteria for the Ideal Semiconductor-Manufacturing Implanter

The principal implanter characteristics in the ideal case are: High throughput at low cost, with sufficient uniformity to handle any presently available wafer size over the dose range from $10^{10}cm^{-2}$ to $10^{16}cm^{-2}$ at energies from several keV's to several MeV's, for a variety of ion species; automatic control, wafer handling, and documentation; high reliability, including self-diagnosis or remote diagnosis by the manufacturer via modems; perfect safety. Since these partly contradictory demands cannot be realized at present in only one single type of implanter, different machine types are in use for various applications, as follows:

- Low-current or medium-current implanters (MCI) with total currents up to 3 mA and maximum energies of 100 to 500 keV, mainly with an electrostatic scan system.

- High-current implanters (HCI) with maximum currents up to 15 mA and top energy ranges up to 200 keV, preferably with a double-mechanical or hybrid scan system.
- High-energy implanters up to several MeV. These operate according to the Van-de-Graaff principle. Due to problems in handling, HV stability, and thermal heating, use of these machines is still restricted to research.

The trend to higher usable beam currents in combination with improved wafer-cooling systems in MCI, as well as the development of HCI with higher energies and use of double-charged ions, tend to close the gap between these two types. Isolated production facilities, which have only a minimum number of implanters and therefore no back-up possibilities, could favor acquisition of only two MCI or two HCI for both high-current and low-current applications, in order to diminish the risk of a complete production stop due to the failure of one type of machine.

## 3. Low-Current and Medium-Current Implanter Concepts

Nearly all of the presently available commercial MCI [2-6] are built according to the same concept, which is shown in Fig.2.

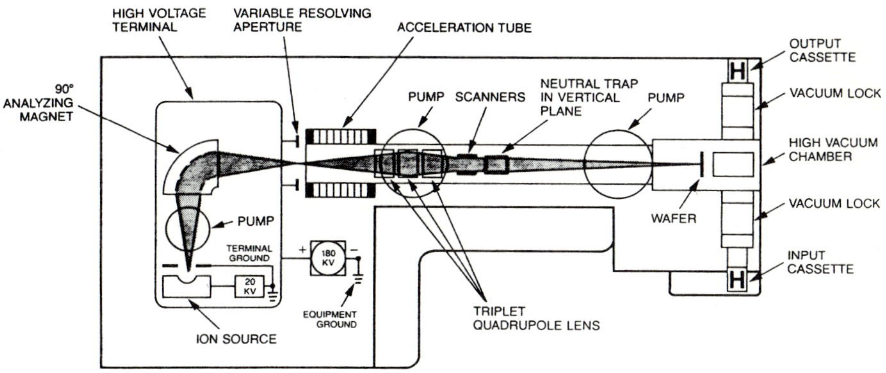


Fig. 2. Typical commercial medium-current implanter

The system configuration is of the pre-analysis type, with source and magnet at high voltage.

LCI use Penning sources [7]; MCI mainly use slit sources of the Freeman type [8]. The preferred feed gases for standard applications are $BF_3$, $PH_3$, $AsH_3$, or $AsF_3$ and $SiF_4$. For ions produced from solids, like Sb or Zn, a vaporizer oven is necessary. Solid As is sometimes used to avoid the highly toxic $AsH_3$ gas. More details on feed materials are given in the chapter by Aitken.

To extract ions from the source, and to accelerate and collimate them for subsequent focusing into the desired beam shape, sophisticated designed sets of extraction electrodes are needed. MCI's typically apply single gaps with an additional deceleration electrode, for focusing as well as to prevent the entrance of electrons into the source, which drastically reduces the available beam current. Often, these electrodes are mechanically adjustable in two or three axes, in order to improve the beam transport. The extraction voltage is typically fixed at 25 kV; in some cases it is adjustable up to 35 kV to yield higher beam currents at increased energies. The deceleration voltage is usually kept at -2 kV. The extraction power supplies have to be extremely stable with less than 0.01 % ripple, in order to achieve a stable beam position at the mass slits as well as a constant noise-free beam current.

The extracted ion beam is a mixture of different fractions of molecules and isotopes of the source-feed material. $BF_3$ gas, for example, will dissociate into $B^{++}$, $B^+$, $BF^+$, and $BF_2^+$, for both the boron mass-10 and mass-11 isotopes. In addition, there exists a certain amount of ions which is created by sputtering from the walls of the source or by ionization of residual gases. The separation of the required dopant in modern implanters is done with an analyzing magnet, typically with 90-degree deflection. A special shape of the pole pieces is used to provide additional beam focusing. When a charged particle passes through the field of the analyzing magnet, it is deflected into a circular trajectory, whose radius is given by:

$$R = \frac{143.95}{H} \sqrt{\frac{M\,U}{n}} \qquad (1)$$

where R is the radius (cm), M is the ion mass (a.m.u.), U is the accelerating voltage (volts), H is the magnetic field (gauss), and n is the charge state of the ion.

To check the quality of the separation, and to identify unknown ion species, mass spectra of the implanter for the desired source-feed materials are taken. On an x-y chart recorder as a variable in x, the magnitude of the magnetic field or the momentary current of the magnetic coils is recorded, whereas the y direction indicates the maximum beam current of the different ions. Figure 3 shows a $BF_3$ spectrum of a MCI. Since the

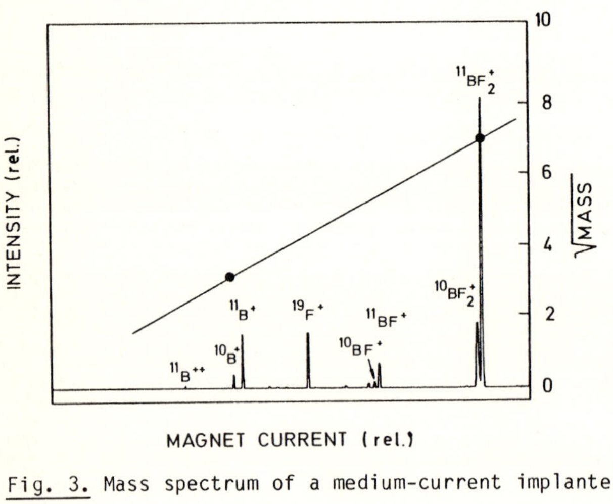


Fig. 3. Mass spectrum of a medium-current implanter

deflection is proportional to the square root of the ion mass, the following procedure can be used for identification. The x axis is labeled linearly in arbitrary units of the analyzer setting, while the y axis is labeled with the square root of the mass. By marking two well-known isotopes with dots, a straight line can be drawn through these two points, and from this line, all the other peaks can be identified. Note that double-charged particles appear at a position which is half of their mass, while molecules such as $As_2^+$ appear at twice the mass of $As^+$. In production implanters, the resolution is sufficient to separate two neighboring masses over the required mass range.

This is especially important for neighbors such as $^{11}BF^+$ and $^{31}P^+$ after a change of the ion species. The highest resolvable mass is typically $^{75}As^+$ or $^{121}Sb^+$.

Since the high-voltage terminal in a pre-analysis implanter is on high potential, all support systems have to be isolated. The supply voltage for source, extraction, and magnet is separated by an isolation transformer, housed in an oil tank. The vacuum pumps are fixed via isolating cylinders. For the cooling cycle, freon is very frequently used. The remote controls are operated by means of glass rods or light pipes. The control meters are usually observed directly.

The post-acceleration voltage is between 175 and 400 kV. Low-capacitance sparking-resistant RF power supplies represent the current state of the art. The acceleration occurs in multi-gap graded tubes. To improve the transfer of the beam over the entire energy range, parts of the graded tube are shortened automatically to ground at lower energies. The last gap is usually biased at -2 kV, for the suppression of secondary electrons. A failure of this voltage would increase the production of x rays dramatically.

The final focusing is done by electrostatic lenses. The simple grid einzel lenses previously in use are nowadays replaced by quadrupole triplet lenses, which need little maintenance. The electrostatic scanning is done by two or three pairs of deflection plates, including a neutral trap. This trap is absolutely necessary for electrostatic scanning, since a total of 1 % neutrals, distributed over a small area, leads to large nonuniformities [9]. In machines with currents up to 3 mA, the lenses and scanner-deflector plates, which previously were made of highly polished aluminum, are now made of carbon to reduce sputtering and therefore decrease maintenance.

The target chambers are generally built for single-wafer implantations with a maximum wafer size of 5". For batch processing, however, carousel end stations for simultaneous implantation of whole batches are also available, resulting in reduced wafer heating, especially when resist masking is done. For correct dose and uniformity control, deep Faraday and corner Faraday cups with appropriate bias rings are inserted. The bias voltage is usually set above -300 volts. A failure in the supply voltage will be indicated by the rise of the measured beam current, due to the loss of secondary electrons which contribute to the total current and lead to high dose errors. For faster wafer cycling, all target chambers are equipped with vac-lock systems. Vac-locks with cassettes for whole batches are now mainly replaced by single-wafer locks with automatic cassette-to-cassette loading.

The vacuum system on the source side is equipped with diffusion pumps because of the large quantity of doping gases, whereas the beam line and target chamber are either supplied with fast diffusion pumps or with cryo-pumps. The set-up and control over the entire vacuum system are automatic.

The tuning of the source, the selection of the ion species and the beam-focusing adjustments still have to be done manually. Wafer cycling, dose, and uniformity control, however, are done automatically by hardware logic or microcomputer control.

## 4. High-Current Implanter Concepts

In contrast to the MCI, quite a variety of HCI-system concepts exists. Machines of the pre-analysis concept, with source and magnet on high potential using an extraction voltage of 20 to 50 kV and a total energy of 200 kV, as well as predeposition machines of the post-analysis type with the magnet on ground potential and a maximum total energy up to 100 kV, are built.

All systems use hot-cathode slit sources. The preferred feed gases are $BF_3$, $PH_3$, and $AsH_3$. For solids, such as Sb, one or two vaporizer ovens are installed. Due to the high amount of toxic gases which can be present in a HCI, some systems favor the production of As and P from their untoxic solids as well. In all systems, the arc voltage of the source is adjustable. To increase the dissociation of $BF_3$ for a higher yield of $B^+$ ions, up to 150 V are applied, while voltages as low as 40 V are used for $As^+$ or $Sb^+$ to reduce sputtering on the filament by the heavy ions, and to gain lifetime under these conditions. The source lifetime is typically between 20 and 40 hours, at maximum current ratings.

For pre-analysis machines, the extraction of the ions is done using a single gap with an additional deceleration electrode, working typically at -2 kV. The final acceleration follows in a second step after mass analysis, using single or multiple gap columns.

In the case of the pre-deposition implanter, the entire acceleration occurs at the extraction gap. Therefore, a very careful design of the electrodes is necessary. For energies above 60 kV, double gaps are in use to improve the high-voltage stability. An additional advantage of double gaps is the possibility of operating these gaps in a deceleration mode for lower energies. The first gap is held at maximum energy, i.e. +60 kV, while the second is set to a negative potential, for example -40 kV, to give a total energy of 20 kV. This results in much higher currents than when running the machine at only 20 kV extraction at the first gap. In the acceleration mode, an additional deceleration aperture is used with voltages above -10 kV. The current through this gap has to be kept at a minimum, and is an extremely sensitive indication of the proper operation of the system. The electrodes are mechanically adjustable in three directions, which is the only possibility for beam steering, since there exist no electrostatic beam-manipulation possibilities at such high currents. To reduce beam blow-up in the deceleration mode, gas neutralization of the beam can be used by introducing nitrogen in the beam line.

The mass analysis is done by fairly large magnets with a 60- or 90-degree deflection. The resolution is sufficient to separate all neighboring ions. Even $^{121}Sb^+$ is usually separated from $^{123}Sb^+$, an unwanted effect, since both isotopes could be used simultaneously. The magnets are equipped with rotatable pole pieces for additional focusing purposes.

For the isolation of the supply voltage of the source system, large oil-tank transformers, or the more economical motor-generator concept, are in use. Freon as the cooling liquid is now mainly replaced by deionized water using appropriate filters, since the latter has a better cooling behavior by a factor of 3. The use of deionized water involves special care in the selection of plumbing material and welding, in order to avoid local galvanic elements and corrosion. The use of a resistivity monitor, which usually is not supplied by the manufacturer, is highly recommended. Poor isolation can lead to all kinds of high-voltage instabilities. For remote control, glass rods and direct view meters, as well as parallel multiplex or microprocessor-controlled serial data transfer via optical links, are in use.

HCI is, for thermal reasons, a batch process, using large target chambers. Figure 4 presents diagrams of the different scan and target-chamber systems. For currents up to 3 mA, a hybrid scan system [10] is in

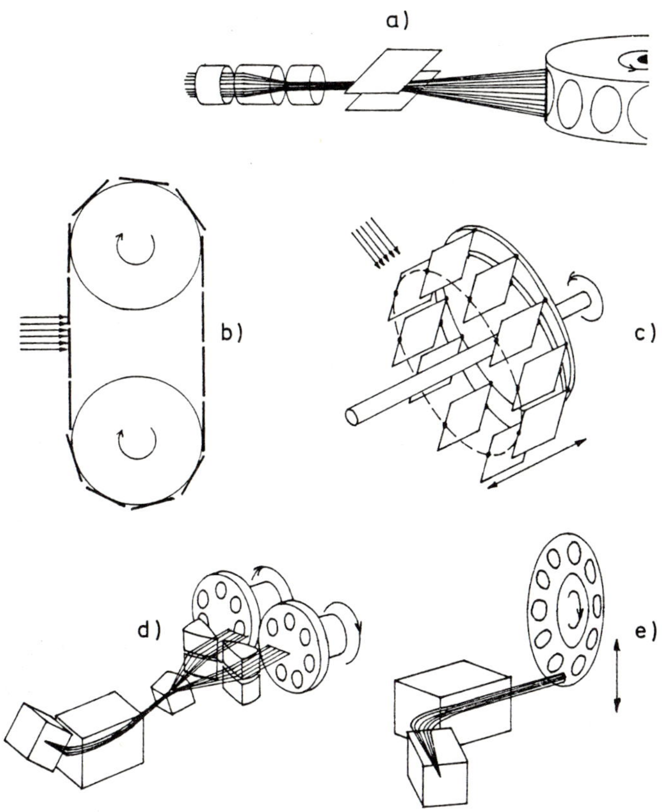


Fig. 4. High-current implanter system concepts: a) hybrid scan, b) chain wheel, c) ferris wheel, d) spinning disk with magnetic scan, e) spinning disk with double-mechanical scan

use. The beam is focused by magnetic quadrupole lenses, and is scanned electrostatically in one direction over a rapidly rotating carousel with 25 5" wafers. A vac-lock system reduces the loading time. The absolute limits in electrostatic scanning of high-current beams are not yet clear. Short scanner systems with appropriate biasing electrodes to restrict the space-charged unneutralized region to the immediate vicinity of the scanner and to avoid beam blow-up are among the possible improvements [11].

The chain-wheel implanter [12] uses double-mechanical scanning. The carousel, which carries the wafers, is so designed that the wafers pass through the beam horizontally in a straight line and at constant speed. The beam current, with some limitations, is controlled by a stepper-motor-driven vane unit, which also compensates for momentary speed deviations. The wafers are transported in a spiral form across the beam, with a fixed horizontal-to-vertical transport ratio, via a ferrofluidic vacuum feedthrough. The wafers are mounted on nine wafer plates, each of which can hold six 3" wafers or three 4" wafers. Recently, a new type of carousel has been introduced. The wafers are mounted on eight larger wafer plates, thus providing increased capacity. Furthermore, this carousel is designed with two independent drives for improved speed control. The application of large diffusion pumps or cryopumps reduces the wafer cycling.

The ferris-wheel system [13] rotates vertically, whereas the translation is arranged horizontally. In this approach, a rotary drive is used for the fast scan motion, and a parallelogram linkage is attached to each wafer plate to produce a ferris-wheel motion in such a way that the angle of incidence is held constant as the wafer traverses the beam. The horizontal scan motion is provided by a linear drive, using a stepper motor. The transport speed is calculated via a microcomputer, and is proportional to the momentary beam current. For increased throughput, a vac-lock system is employed. During the process of implantation, a cassette with the next set of 13 wafer plates, each one holding two 4" wafers, is pumped down, and the interchange of these plates occurs automatically under vacuum after the implantation of the first set is finished.

Double-disk implanters employ a hybrid system. The wafers are clamped on a disk, which typically rotates at 1000 rpm. The standard disk can hold twenty-five 4" wafers. The horizontal scanning is done magnetically, with a scan speed of less than one cycle per second. The scan control is provided by a separate microcomputer, which compensates for the radial and additional geometrical deviations. To obtain a parallel beam incidence on the wafers, additional parallelizing magnets are needed. By changing the polarity of the scan magnet, the beam is switched from one target chamber to the other. Therefore, one chamber can be loaded while the other is under implantation. An automatic cassette-to-cassette loading system is available.

Several disk implanters use a double-mechanical scan technique [14-18]. The rotary drive is provided through ferrofluidically sealed rotary shafts with over 800 rpm. For the linear drive, several different transport and sealing systems have been developed. In a commercially available system, the linear motion of the disk chamber is transmitted into the vacuum by means of a sliding-seal arrangement, which consists of four concentric seals in combination with three intermediate stages of differential pumping, thus eliminating all bearings from the vacuum. The standard disk holds thirteen 4" wafers. For faster throughput, a second disk is loaded during the implantation of the first disk, and is exchanged automatically under atmospheric pressure after the implantation is finished. In all disk systems, the linear drive needs some compensation, as the dose rate increases towards the center of the disk. This is accomplished through the online control of a microcomputer.

The vacuum system uses fast diffusion pumps on the source side, whereas the beam line and the target chamber are generally outfitted with cryopumps.

The set-up of the closed-loop-regulated source controls is accomplished manually. Some machines are entirely microprocessor-controlled. This could lead to a completely software-controlled operation in the future, assuming the use of appropriate software.

## 5. System-Limiting Aspects

### 5.1 Throughput

Handling limitations. Nearly all MCI now use vac-locks with cassette-to-cassette loading. The locks are pumped out during the implantation of the preceding wafer. The wafer-handling time is kept between 2 and 8 sec, and is therefore lower than the typical minimum implantation time of 10 sec. HCI with vac-locks or fast cryosystems also offer, at present, a handling time below 20 sec per wafer.

Uniformity limitations. To obtain sufficient uniformity, a minimum implantation time is necessary. This time is limited at first by outgassing, especially in the first seconds of an implantation when neutrals are created, which falsify the absolute dose and additionally cause nonuniformities across the wafer when electrostatic scanning is used. Secondly, a minimum time is needed to form a sufficiently uniform, dense scan pattern, which is limited by the maximum available scan frequencies. Since a minimum overscan over the implanted area of 3 $\sigma$ beam diameter is needed for 1% uniformity [19] and since the overscanned beam is useless, good focusing is a major requirement. On the other hand, precise focusing leads to a very small beam-spot size, and with low-dose, high-energy beams, cross-sections of less than 0.5 mm for 1 $\sigma$ have been measured [20, 21].

Theoretical calculations show that the overlap of the individual Gaussian beam traces has to be better than 1.5 $\sigma$ for 1% uniformity. On the other hand, a minimum of 100 identical scan patterns is needed, and a 15-second implantation therefore requires 150 msec per scan pattern.

Using electrostatic scanning with quartz-controlled scan frequencies, one can calculate the optimum scan pattern for a given maximum scan speed. Figure 5a shows an optimum pattern for a typical MCI, yielding 0.7 lines per mm. Under worst-case conditions, this pattern leads to a line-to-line nonuniformity of over 8 %. An increase in the scan speed brings about the improved picture of Fig. 5b with 2 lines per mm. The uniformity pattern in Fig. 6, across a 2" wafer obtained by a varicap measuring technique, with a measurement distance of 0.1 mm, shows that this pattern produces a line-to-line uniformity better than ± 0.5% at the same above-mentioned beam diameter. To reduce the implantation time, an additional increase of the scan frequency would be necessary. Thus, scan frequency can become the limiting factor in wafer throughput.

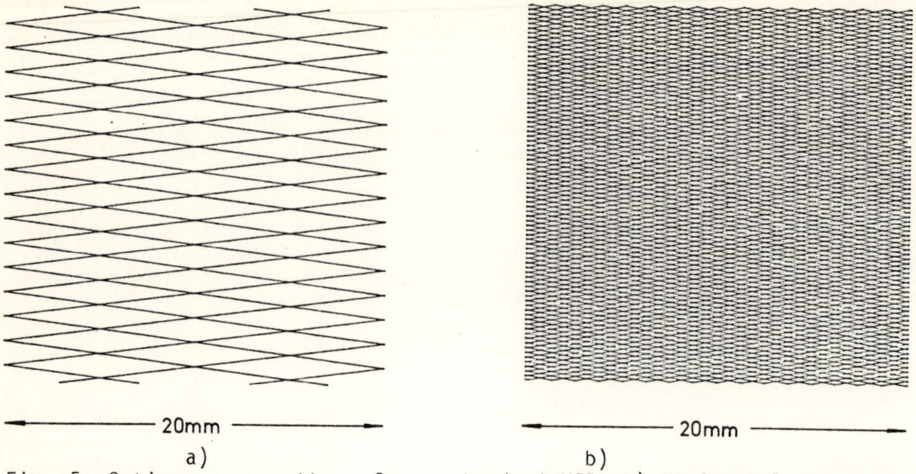


Fig. 5. Optimum scan pattern for a standard MCI. a) Horizontal frequency, 64.935 Hz, vertical frequency 500 Hz, time for one complete scan pattern 0.154 sec, 0.7 lines/mm. b) With increased frequencies, horizontal frequency 500 Hz, vertical frequency 1234.57 Hz, time for one complete scan pattern 0.162 sec, 2 lines/mm

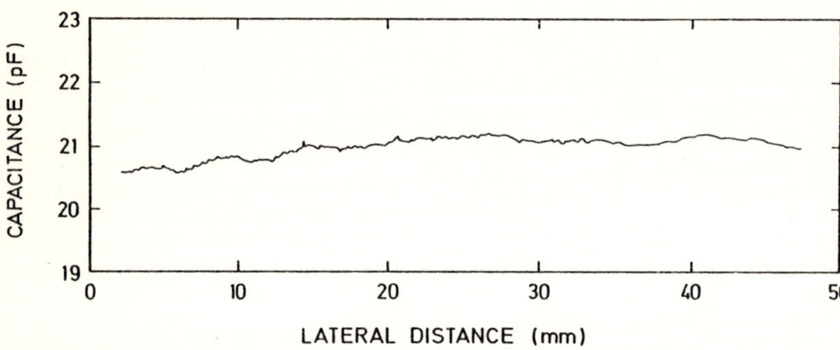


Fig. 6. Uniformity pattern across a 2" wafer obtained by varicap measurement technique; line-to-line uniformity better than ± 0.5%

For HCI with double-mechanical scanning, similar considerations can be applied. For highly focused low-dose beams, spiral-like nonuniformities could be observed in low-speed-rotating systems. The addition of a further fast electrostatic X scan, vertical to the rotation axis, can solve these problems for low currents.

Beam-current limitations. The available beam current becomes the throughput-limiting factor for high-dose implantations. Figure 7 shows the correlation between dose and necessary beam current for 400-4" wafers per hour, in a double-disk implanter, which requires no extra handling time. For example, 20 mA would be necessary for $8 \cdot 10^{15}/cm^2$, which is twice the specified and available maximum current. Such currents promise to be available in the near future.

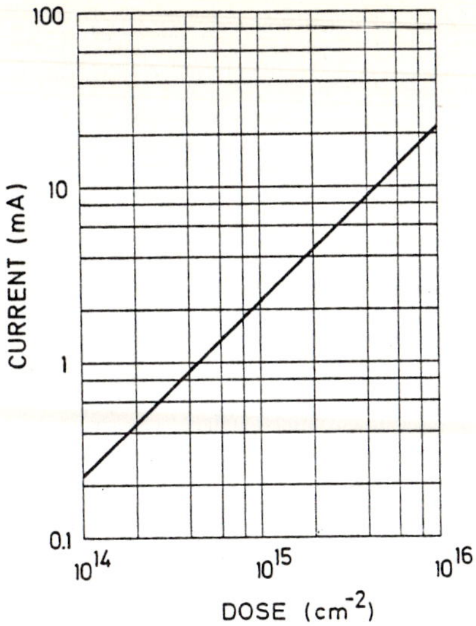


Fig. 7. Correlation between dose and necessary beam current for a 400-4" wafer-per-hour throughput in a double-disk implanter

5.2 Wafer Heating

In present applications, there are two distinctive limiting temperature points. The first one is roughly at 120°C, where positive photoresists, when used for masking, become cracked and deteriorated. The second is above 300°C, where coloration bands, due to self-annealing of the amorphous layer during high-dose implantations, occur across the wafer [22]. To obtain maximum wafer temperatures, theoretical and practical estimations have been carried out. A numerical solution of the heat-transfer equation, using the thermal diffusivity data of [23], is given in Fig. 8. The extreme beam power of 200 $W/cm^2$ already leads to a constant temperature difference between the front and the rear sides of the wafer after 0.5 msec that does not exceed 6 K. For this reason, constant temperature across the wafer can be assumed.

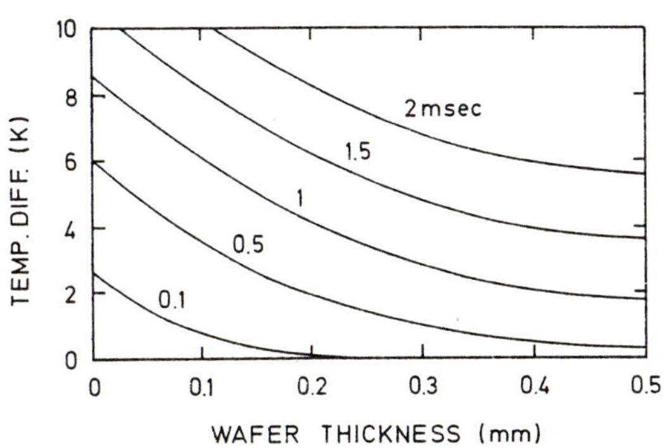


Fig. 8. Temperature distribution across a silicon wafer. Parameter is the implantation time; beam power: 200 $W/cm^2$

Wafer cooling takes place through radiation cooling and conductive cooling from the rear side of the wafer. Radiation cooling is influenced by the emissivity of the wafer, by the wafer holder, and by the surroundings, as well as by their temperatures and geometrical aspects. The spectral ε varies with wavelength, oxide thickness, percentage of oxide coverage, and doping [24-26].

For doped, partly oxide-covered wafers, a maximum of ε = 0.7 was found. ε = 0.1 was measured for the wafer holder; for the surroundings, ε = 1 was assumed. There is no rear-side cooling for unclamped wafers, but the use of silicon rubber or gas cooling, directing a cold low-pressure gas stream against the rear side of the wafer, can result in thermal resistivities below 0.02 mW/°C $cm^2$.

Figure 9 presents a comparative study of theoretical temperature rise during implantation using the above-mentioned data, with a dose of $8 \times 10^{15} cm^{-2}$ at 80 kV, which is typical for high-dose applications. The computations were made for the maximum available beam current of the respective system. Figure 9a represents a MCI with a 0.5-mA beam current. The lower curve is based on the use of a waycool system. Figure 9b presents data for a hybrid HCI with a 2-mA beam. Figure 9c shows temperature rise in a ferris-wheel implanter with a 4-mA beam. At the beginning of each scan cycle, which is equivalent to a dose of $6.5 \times 10^{14} cm^{-2}$, the temperature rises in short pulses, as the wafers are both rotated and "screwed" through the beam; it then decreases in the following cooling period. The maximum temperature is reached after four cycles, and remains constant in all additional cycles. A similar curve is seen in Fig. 9d for

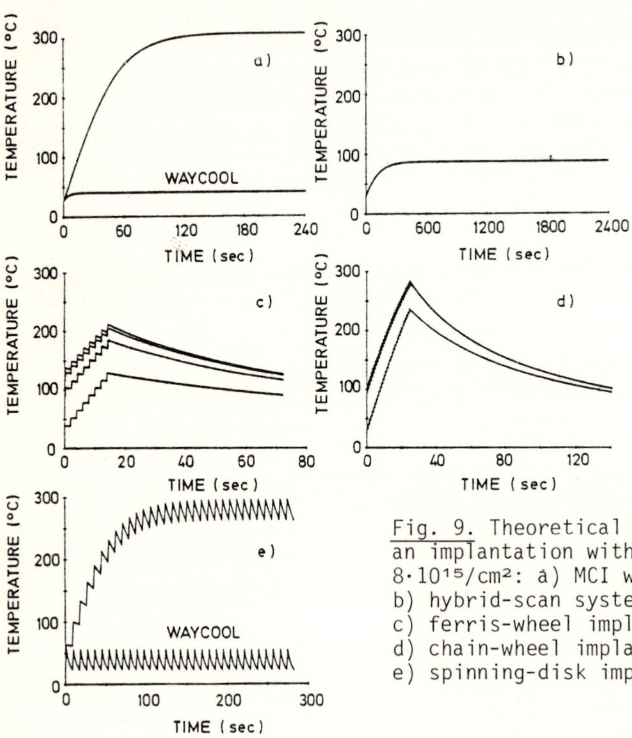


Fig. 9. Theoretical temperature rise during an implantation with 80 kV and a dose of $8 \cdot 10^{15}/cm^2$: a) MCI with 0.5 mA beam, b) hybrid-scan system with 2 mA beam, c) ferris-wheel implanter with 4 mA beam, d) chain-wheel implanter with 8 mA beam, e) spinning-disk implanter with 15 mA beam

the chain-wheel implanter with an 8-mA beam. Each scan cycle is equivalent to a dose of $3.2 \times 10^{14} cm^{-2}$. The maximum temperature is already reached after two scan cycles. Figure 9e represents a spinning-disk HCI with a 15-mA beam. The lower curve is for the use of a waycool system. Figure 9 in summary shows that, except for the hybrid system, all implanters result in approximately the same maximum temperature, when running with maximum beam current. Therefore, compared with MCI, additional thermal problems should not occur with HCI.

## 5.3 Energy Range

The low-energy range in a production implanter is limited by the rapid decrease of beam current which occurs with decreasing extraction voltage.

A better yield is achieved through an accel-decel arrangement, where the ions are extracted with the full extraction voltage of 25 kV, and decelerated by the reversely biased acceleration voltage to a total of 5 to 25 kV. Another possibility is the use of charged molecules, for example $BF_2^+$, which corresponds to only 0.22 times the energy of $B^+$, or $As_2^+$, as well as $P_2^+$, with 0.5 times the energy of $As^+$ and $P^+$, respectively. The high-energy range is extendable through use of doubly or triply charged particles.

Due to the charge exchange, an Aston band such as the splitting of $P_2^+$ into $P^0$ and $P^+$ after extraction, thus poisoning the $P^{++}$, is frequently created [27]. This effect is serious with cold-cathode sources, due to the high source pressure. It is minimized with hot-cathode sources, using solids as a gas feed. In addition, velocity filters [28], or electrostatic deflectors for HCI, can be used successfully as beam-purifying arrangements.

## 5.4 Wafer Size and Wafer Tilting

With MCI, a change in the wafer size requires an adjustment of the bumpers or guidance lines, which is done manually after venting the target chamber. A faster change is obtained by appropriate feed-throughs for the adjustment. A change in the wafer tilting is easily accomplished through alignment of a lead screw. Some systems offer automatic wafer orientation.

HCI with exchangeable wafer holders for different wafer sizes require no changes at the implanter. Disk machines usually involve changing of the entire disk for the appropriate wafer size. The wafer tilting for HCI is normally fixed at 7° or 10°. This might become problematical when wafers with different orientations or pre-tiltings are run on the same implanter. Fig. 10 shows stereographic projections of the diamond lattice for (111) and (100) orientations, with their low-indexed directions which are critical for channeling, as well as graphs showing the typical flat and pre-tilting directions of the corresponding wafers. Especially for (111) wafers, the combination of a 10° tilting with the use of misoriented wafers can lead to channeling in (221). To avoid this, correct wafer orientation, under consideration of related handling problems, is necessary.

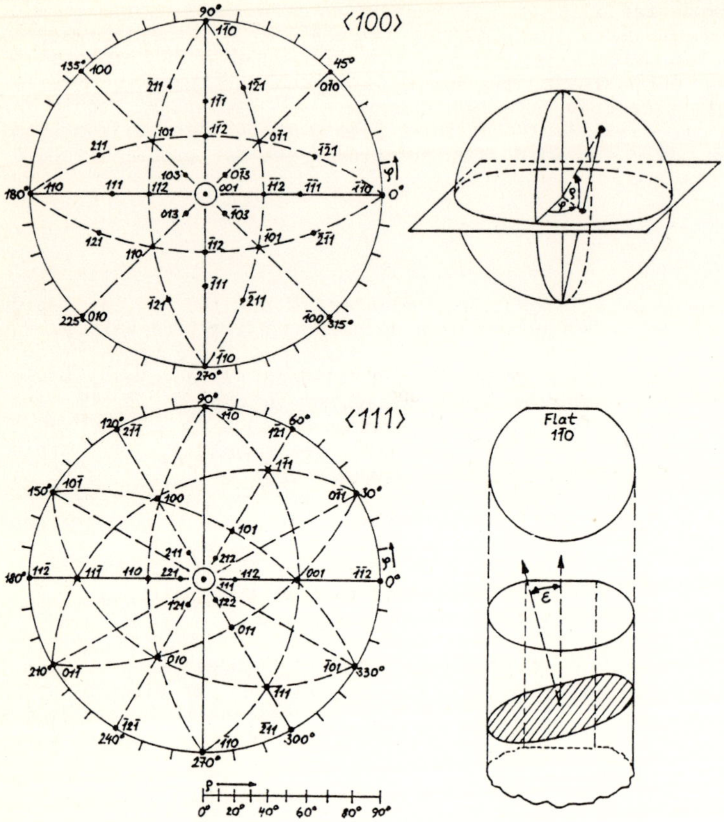


Fig. 10. Stereographic projection of the diamond lattice, for (100) and (11$\bar{1}$) orientation

## 5.5 Charge-up Phenomena

The implantation of entirely isolated islands on a wafer can lead to a charge-up, followed by a destruction of the insulating material. This was first observed as a real problem in connection with the use of silicon on sapphire wafers. Charge-up and breakdown of isolated polysilicon islands on oxide can become a problem in connection with the high current densities reached in a HCI. Special design considerations, such as oxide-free silicon areas around the chip which produce sufficient secondary electrons [29] to compensate for the charge-up of the islands, can help. New HCI's also offer the possibility of electron guns to neutralize the ion beam itself.

## 5.6 Contamination

Ion implantation, compared with other doping techniques, is assumed to be a very clean process. The use of sensitive tests, however, has revealed the presence of four typical categories of contamination, which are as follows:

Heavy metals. In the first HCI's, through use of neutron-activation analysis, large amounts (up to 2 %) of Fe, Cr, and Ni were detected in implanted wafers [30]. The composition was the same as in stainless steel, which was used in some parts of the implanter. The contamination was dependent on the dose as well as the ion species, and was a result of sputtering from the apertures. Replacing them by aluminum apertures decreased the contamination to 0.02 %. New machines now use either graphite apertures and liners, or are built entirely of copper-free aluminum. With MCI, the contamination never exceeded 0.002 % of the doping element. This is due to the use of a 7° neutral trap, which highly restrains any contamination.

Cross-contamination. All implanters show some "memory effects". After changing the ion species, some amount of the previous species can still be observed. With HCI, concentrations of up to 2 % are found. Therefore, ions which could act as lifetime killers, such as Au or Cu, should never be used in an implanter. Figure 11 shows the concentration profile of an $As^+$ implantation, made in a HCI with steel apertures. The depth profile shows that the contaminants Fe and Cr, as well as the cross-contaminants, are found in a very shallow layer amounting to only 100 Ångstroms. Therefore, the use of thin screening oxides, which are etched off before annealing, could prevent nearly all contamination. In MCI, due to the neutral trap, no cross-contamination above 0.02 % was found.

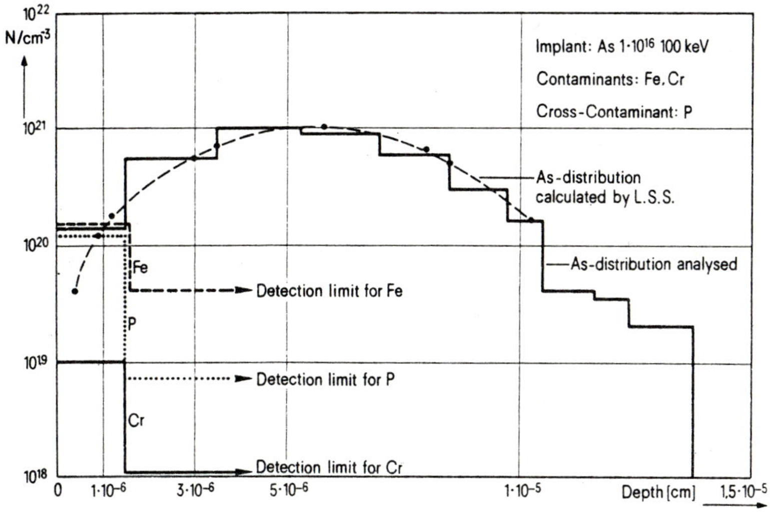


Fig. 11. Concentration profile of an As implantation and the resulting contaminants and cross-contaminants. Profile results through activation analysis and anodic oxidation

Hydrocarbons. HCI, when equipped with diffusion pumps, may form films of polymerized hydrocarbons during high-dose implantations, which are not removable by chemical cleaning. The films are only removable by oxygen plasma etching. However, the use of perfluorinated diffusion pump oils, which do not crack, can avoid this effect almost completely [31]. This effect does not occur in MCI, because of the short time the wafer is exposed to the vacuum during a single wafer implantation.

Sodium. Sodium is a severe contaminant in MOS technology, and an amount of 10 ppm for an $8x10^{15}$ $cm^{-2}$ source-drain implantation is easily detected. For HCI, the concentrations, which are a function of the implantation time, can be remarkably high. Very careful cleaning of the entire beam line, from which the Na is sputtered or evaporated, as well as the choice of appropriate materials and the use of screening oxides on the wafer, are necessary to keep this effect low. No sodium is found with MCI, because of the neutral trap and the short wafer-exposure time.

## 6. Human Engineering

### 6.1 Operation of an Implanter

The set-up of an implanter requires the adjustment of roughly a dozen mutually dependent parameters. A fully automatic set-up would therefore require an extreme amount of computerized control electronics. For this reason, all present machines are still set up manually. The human ability to learn and to execute analog adjustments makes it possible for even quite unskilled people to perform the start-up procedure with the aid of simple checklists. The operation of MCI is almost entirely automatic with cassette-to-cassette loading; and besides loading the cassettes, the operator has sufficient time to watch the system or even to run a second implanter. Set-up and operation should preferably be done by the same person.

### 6.2 Automatic Implantation Control

With a well-trained crew and carefully calibrated machines, a yield of over 99.8 % correctly implanted wafers is possible. Since the major source of failures is due to incorrect parameter adjustment, the installation of a control computer is one of the keys to improved yield. For the terminal data, optical links or infrared transmitters have to be constructed. For control of dose, uniformity, and energy, matching interfaces to the corresponding units have to be made.

Especially in production facilities, where implantation is organized as a central service for various production lines, complete implantation data for all the different device or product types should be stored in the control computer, in order to avoid false implantation. At the beginning of an implantation, the operator keys the batch number, the number of wafers, and all desired implantation data. First, a plausibility test is done, and all senseless data, such as 1,200 keV or $5x10^{18}$ $cm^{-2}$, are rejected. Second, the computer searches for the appropriate device type in its memory, the typed-in as well as the stored data are displayed for correction. To avoid daily upgrading of the data file, the following procedure can be used: If a device type is not found, or has to be corrected, the computer first asks whether this is a single test implantation, or a new type of device for regular production. In the latter case, as a cross-check, the data have to be typed in once again. If identical with the first input, they are stored in the memory in alphanumerical order. Of course, for all various device categories, regular printouts of stored data are made to control the current status. Through serial data transmission, the results can be entered in a host computer.

In case of a correct input, the computer displays the necessary adjustment data, such as analyzing magnet current for a selected extraction voltage and implantation current for a given implantation time. The operator then sets up all these data, and starts the implantation. A permanent on-line control compares ion species, calculations for extraction voltage and analzing magnet current, energy, dose, uniformity, and implantation time. Any deviation beyond a preselected level results in the implanter being placed into "hold" within a few tenths of a second, and a display of the desired and the false values. At the end of an implantation a data printout and a record for documentation are made. If the following implantation is done for the same device type, only the batch number has to be typed, using special key functions. Control computers, which to some extent fulfil the desired specifications, are now being built by some implantation-equipment manufacturers.

Besides computer control, systematic machine recalibrations are also necessary. The beam purity is checked by regular recording of the mass spectra. Leaking gas valves of an undesired gas species, as well as leaks in the source region, are easily detected. The extraction and post-acceleration voltages must be tested at regular intervals with a high-voltage probe. There are many products which are highly sensitive to implantation energy variations. Microcomputerized current integrators and uniformity monitors usually have built-in self-test mechanisms. A more reliable test, however, is obtained by the use of a precision current source, which indicates any leakage currents, especially at contaminated insulators in the Faraday-cup region.

The on-line controls are, of course, accompanied by off-line controls, using sets of appropriate test wafers. Figure 12 shows the uniformity pattern of a 3" wafer, using a Van der Pauw pattern. The wafers have to be preamorphized or covered with a screening oxide. The clearly visible nonuniformity of ± 1.7 % for two standard deviations is typical for MCI's, and is caused by the 7° neutral-trap bend, the 7° wafer tilting, and the mechanical misalignment of the entire implanter. Using such test structures and careful machine calibrations, even a dozen machines of various types can yield absolute sheet-resistivity deviations from machine to machine, as well as over the wafer, of less than ± 5 % during an entire year.

The actual usable implantation time, for a machine operated with frequent changes of ion species, energies, and doses, can decrease to as

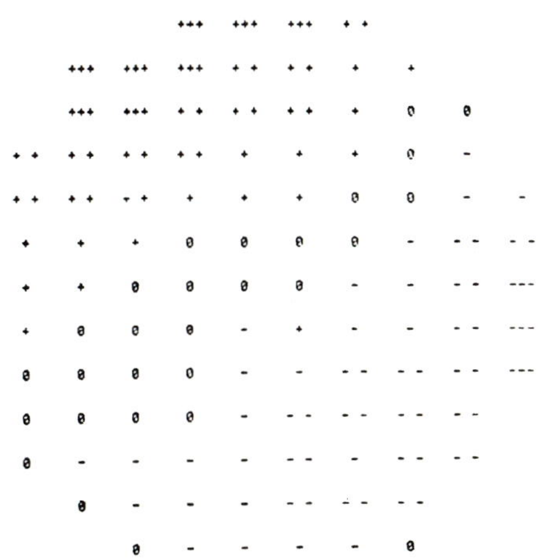

Fig. 12. Uniformity pattern of a 3" wafer obtained using Van der Pauw technique

little as 50 %. The keys to high "up-time", especially outside the USA with the increased travel, transportation and customs problems, are sufficient spare parts, a very experienced operating crew, and a local service center provided by the equipment supplier. Production implanters can have a typical rate of two or three failures per week, roughly equally divided among mechanical, electronic, and vacuum failures. With the necessary spare parts and the available service, the total repair time can usually be cut down to only a few hours.

## 6.3 Safety

Ion implanters involve the possibility of nearly all imaginable hazards which a production tool might have. First, they are run with lethal gases such as $PH_3$ and $AsH_3$, which require a carefully interlocked exhaust system. Special exhaust and respiration precautions are also needed when cleaning the sources and the source housing, as well as during the oil change of the roughing pumps. Prior to disassembly of the evacuated beam-transport system, a repeated flushing with wet nitrogen can reduce the development of poisonous gases through chemical binding. Dangerous high voltages must be kept away from personnel by at least two systems, including keys, door switches, and grounding bars.

The occurrence of X rays outside the implanter should be kept to zero, due to the general aversion which exists to such radiation, even when the values are kept within a tolerable range. This can be achieved through the use of sufficient quantities of lead, or lead oxide, for shielding purposes at all necessary positions. Further details on safety may be found in the chapter by Bustin and Rose.

## References


1. H.H. Chang, C.E. Han, W.H. Wang, and J.Y. Chen, Radiation Effects 44, 181 (1979)
2. Veeco, Series 2100: Specifications, Austin, Texas
3. Caspar, Challenger 200 MCA: Specifications, Sunnyvale, California
4. R. Allison, Jr., D. Church, H. Glavish, A. Noeth, and D. Schechter, In Proc. Symp. Electron and Ion Beam Science and Technol, ed. R. Bakish, (Electrochemical Soc., 1976) p.482
5. G. Ryding, A.B. Wittkower, and P.H. Rose, J.Vac.Sci.Technol. 13, 1030 (1976)
6. G. Dearnaley, J.H. Freeman, R.S. Nelson, and J. Stephen, Ion Implantation, North-Holland Publishing Co., London (1973)
7. J.R.J. Bennett, in: Proc. Int. Conf. on Ion Sources, I.N.S.T.N., Saclay (1969), p. 571
8. J.H. Freeman, Nucl. Inst. and Meth. 22, 306 (1963)
9. D.S. Perloff, F.E. Wall, and J.T. Kerr, In Proc. Symp. Electron and Ion Beam Science and Technol., ed. R. Bakish (Electrochemical Soc., 1976) p.464
10. J. Camplan, J. Chaumont, R. Meunier, R. Graber, J.C. Rouge, R. Stokker, and L. Wegmann, Nucl. Instr. and Meth. (1980)
11. J.H. Keller, D.K. Coultas, W.W. Hicks, and J.R. Winnard, Nucl. Instr. and Meth. 139, 41 (1976)
12. D. Aitken, Nucl. Instr. and Meth. 139, 125 (1976)
13. A. Wittkower, P. Rose, and G. Ryding: Solid State Technol. 18, 41 (Dez. 1975)
14. N.N., Solid State Technol. 22, 31 (Nov. 1979)

15. J.R. Kranik, Radiation Effects 44, 81 (1979)
16. Nova Model NV-10: Specifications, Beverly, Mass.
17. N. Sakudo, K. Tokiguchi, H. Koike, J. Kanomata, I.H. Wilson, and K.G. Stephens, Low Energy Ion Beams, Inst. Phys. Conf. Ser. 54, 36 (1980)
18. J.H. Jackson, H.M. Bird, J.P. Flemming, G.J. Hofer, J.G. McCallum, P.J. Mostek, G.I. Robertson, A.F. Rodde, B. Weissman, and N. Williams, Radiation Effects 44, 59 (1979)
19. H. Ryssel and I. Ruge, Ionenimplantation (B.G. Teubner, Stuttgart 1978) p.124
20. H. Glawischnig, in: Fourth European Conference on Electronics, EUROCON 80, Stuttgart, ed. by W. Kaiser and W.E. Proebster (North-Holland Publ.) p. 75
21. H. Glawischnig, in: Third Int. Conf. on Ion Implantation Equipment and Techniques, Ion Implantation School, Kingston (1980), unpublished
22. D.G. Beanland and D.J. Chivers, J. Electrochem. Soc. 125, 1331 (1978)
23. H.R. Shanks, P.D. Maycock, P.H. Sidles, and G.C. Danielson. Phys. Rev. 130, 1743 (1963)
24. W.R. Runyan, Silicon Semiconductor Technology, Texas Instr. Electronics Series (McGraw-Hill, New York 1965) p.200
25. P.L.A. Van der Meer, L.J. Giling, and S.G. Kroon, J. Appl.Phys. 47, 652 (1976)
26. P.D. Parry, J. Vac. Sci. Technol. 13, 622 (1976)
27. J.H. Freeman, P.J. Chivers, and G.A. Gard, Nucl. Instr. and Meth. 143, 99 (1977)
28. R.L. Seliger, J. Appl.Phys. 43, 2352 (1972)
29. G. Ferla, Sec.Int.Conf.on Ion Implantation Equipment, Trento (1978)
30. E.W. Haas, H. Glawischnig, G. Lichti, and A. Bleicher, J. Electronic Materials 7, 525 (1978)
31. M.Y. Tsai et al., J. Electrochem. Soc. 126, 98 (1979)

# Ion Sources

D. Aitken

Applied Implant Technology Inc.
Horsham, Sussex, England

## 1. Introduction

Ion implantation is becoming an increasingly important technique for introducing one material into the near surface regions of another material. It has the advantage that both the uniformity and depth distribution can be accurately controlled and the final result can be quite independent of factors such as surface chemistry and solid solubility.

The commercial exploitation of this technique started in the semiconductor industry with the realisation that threshold shifting in MOS devices could be very accurately controlled by ion implantation. The requirement in this process was for doses of approximately $10^{11}$ ions $cm^{-2}$ and the dopants were usually boron and phosphorus. This led to an early generation of machines based on the techniques used in physics-oriented departments in universities and research institutes. They were generally machines capable of producing microampere beams of ions, sometimes tens of microamperes, which used cold cathode ion sources and long beam lines with a number of focussing and deflection elements to obtain the required optics, analysis and target scanning functions. The major suppliers of commercial machines of this type for the semiconductor application were Accelerators Inc., and Extrion in the U.S.A.

Quite independently a separate technology was developing in some other laboratories and military establishments. Here the interest was not ion beam physics or ion implantation but isotope separation. The requirements were quite different. The primary requirements were high current, high ionisation efficiency and a simple optical system optimised for the most efficient separation of isotopes. The major driving force behind the development of these machines was the military requirement for the separation of uranium - 235 from uranium - 238, which led to the development of the large Calutron machines in the U.S.A. The application of this technology to the semiconductor requirement followed some years after the initial threshold-shifting implantation when the requirement for doses in the $10^{14}$ to $10^{15}$ ions $cm^{-2}$ range for sources and drains started to become significant.

Recently we have seen the beginning of the development of another ion implantation technology. This is due to the growing interest in the use of ion implantation for the surface treatment of metals. Here the requirements are quite different to the requirements of semiconductor technology.

Beam purity is, in general, not a particular concern and consequently there is not the need for magnetic analysis. This removes many of the constraints which limit the types of high-current source which can be used for isotope separation or high-dose ion implantation of semiconductors. The basic requirements in this application are simple long-life sources that can produce a range of ions of gaseous and metallic elements at very high current levels. The required doses are generally in excess of $10^{17}$ ions $cm^{-2}$.

## 2. Ion Implantation Requirements from the Ion Source

The major requirements from an implanter ion source and its associated extraction system are:

i) The production of a beam of ions, usually with a single positive charge (i.e. one electron removed) at the required current level for the required species.

ii) A beam of sufficient ion optical quality to be efficiently transmitted through the accelerator system. The "emittance" of the beam is the measure of this quality.

iii) Simplicity.

iv) Ease of operation.

v) Stability.

vi) Longest possible operational lifetime.

vii) Ease of change of species.

viii) Reliability.

ix) Serviceability.

In addition, for the case of high-current sources, there is the added requirement that the beam current extracted from the source shall be free from high-frequency modulation in order to maintain space-charge neutralisation in the beam. The concepts of beam quality and space charge will be discussed later.

## 3. The Principle of Operation of an Ion Source

The ion sources used in both research and commercial ion implanters almost universally use sources which produce ions by means of a confined electrical discharge which is wholly or partially sustained by the gas or vapour of the material to be ionised. In these sources the dominant mechanism of ionisation is electron impact and the details of the various types of source are mainly determined by the way these electrons are produced and confined.

There is no simple universal solution to the problem of heavy-ion production for ion implantation. The wide variety of ion species required, the wide variety of feed material suitable for producing the required ion, the range of ion optical requirements and the current levels from microamperes to tens of milliamperes result in a range of ion source types and sizes. Despite the range of requirements the arc discharge type of source is very versatile and all elements can be ionised in this way. The relatively small number of sources which work on a totally different principle are usually only applicable to particular species.

The principle of operation of these sources will be illustrated by particular reference to the hot cathode source, by far the most commonly found source in modern ion implanters.

In general the most important features of the source are:

1. the cathode
2. the anode
3. the extraction aperture
4. the magnetic field
5. the plasma
6. the source feed system.

## 3.1 The Cathode

The cathode supplies the electrons which will ionise the gas or vapour that is introduced into the ion source arc chamber. In the hot cathode source the cathode is at a sufficiently high temperature to thermionically emit electrons. The maximum current density which can be obtained from a metal surface of work function $\emptyset$ heated to a temperature T is:

$$j_e = AT^2 \exp\left[\frac{-e\emptyset}{kT}\right] \qquad (1)$$

where A and k are constants and $j_e$ is the electron current per cm².

The most commonly used material for the cathode is tungsten. It has the advantage of being one of the few materials that has significant mechanical strength at the required electron-emitting temperature. The work function is 4.5V, the constant A is 70 amps/cm² and kT/e is given by the conversion 11,600 K = 1eV. This gives a value of $j_e$ of 0.27 amps/cm² at 2500°K.

Cathodes are generally directly heated i.e. they are heated by passing current directly through the cathode material and are usually in the form of a coiled wire filament or a simpler geometry rod form. Tungsten is usually the preferred material but corrosion, surface poisoning and ductility problems sometimes lead to the use of alternatives such as tantalum, molybdenum and rhenium.

One of the major problems in all arc discharge sources is the fact that cathodes not only emit electrons but also attract positive ions which subsequently erode the cathode, thus limiting its useful life. In highly reactive plasmas containing halogens, for example, there can also be a significant corrosion problem to add to the erosion problem caused by sputtering. As the erosion rate due to sputtering will depend on the ion flux, which will vary over the surface of the filament, there is the additional problem of localised thinning of the filament, which is a self-perpetuating process due to the overheating of the filament in the regions of smallest cross section. The high temperature coefficient of resistance of tungsten adds to this problem.

One solution of the cathode erosion problem is to use a large block of tungsten as the cathode and to indirectly heat it by electron bombardment. Another is to remove the cathode from the arc chamber and use it as an external source of electrons, the electron beam being guided through collimating slits into the arc chamber using a magnetic field. When this technique is used another problem arises. The maximum current which can be extracted from a cathode is certainly given by equation (1) but that maximum is often not achievable because of space charge. The electrostatic mutual repulsion effect of electrons leaving the cathode surface screens the extracting electric field and inhibits the emission to a value known as the space-charge limited current. The value of this current for a parallel cathode and anode separated by a distance d with a voltage V between them is given by Langmuir's well-known formula

$$j = \frac{V^{3/2}}{9\ d^2} \left[ \frac{2e}{M} \right]^{1/2} \qquad (2)$$

for electrons:

$$j_e = 2.3 \times 10^{-6} \frac{V^{3/2}}{d^2} \quad \text{A cm}^{-2} \quad . \qquad (3)$$

If, however, the cathode is inside the arc chamber the positive ions which are attracted to the cathode tend to neutralise the negative space charge of the electrons and thus allow a significantly higher electron current to leave the cathode.

In a cold cathode source there is no thermionic emission and the discharge-maintaining electrons are produced by secondary electron emission from the cathode resulting from the positive ion bombardment.

## 3.2 The Anode

In the majority of ion sources the arc chamber body is the anode, but in a few sources it is structurally more convenient to have a cathodic arc chamber and an internal anode. An example of the latter type is the Penning source with its two cathodes being the ends of the arc chamber with a cylindrical anode between them.

As the anode usually is the large-area electrode and is generally the structure of the arc chamber, the most important considerations are geometrical (determining the distribution of the plasma) and structural (corrosion, erosion, outgassing, flake formation, etc).

## 3.3 The Extraction Aperture

Having created an arc discharge between the cathode and the anode it is now required to extract ions from the plasma. The beam can be extracted from an aperture or series of apertures in either electrode. These apertures can be circular or rectangular depending on the nature of the accelerator lens system and the required current.

The plasma in the majority of sources takes the form of a relatively long magnetically constrained column. If the extraction aperture is situated at the end of this column, the source is called an axial extraction source. If the extraction is from the side of the column then the source is referred to as a lateral extraction source. For obvious geometrical reasons there is a tendency for axial sources to have circular apertures and lateral sources to have long rectangular apertures but this need not necessarily be the case.

When there is no requirement for mass analysis and large beam currents are required then close-packed arrays of circular apertures can be used to create a large plasma surface area from which to extract the ion beam. As will be discussed in a later section the size of an individual aperture is limited by the requirement for a stable plasma meniscus from which the ion beam is to be extracted. The shape and stability of this meniscus is an important factor controlling the efficiency of transmission of an ion beam through an ion implanter mass analysis system.

## 3.4 The Magnetic Field

The majority of ion sources use a magnetic field to influence the path of the electrons from the cathode in a way which maximises their ionising efficiency. For example it is undesirable for an electron to travel in a direct straight line from the cathode to the anode, as the probability of an ionising collision would be small for such a short path length.

The magnetic field is used to control the electron distribution in the arc chamber in a number of ways:

i) collimation

ii) path lengthening

iii) focussing.

Fig.1a is an example of a source geometry where the magnetic field and electron-accelerating electric field are parallel and the electrons enter essentially a field-free arc chamber. The magnetic field prevents an excessive fraction of the electrons being attracted to the sides of the collimating slot and contrains the discharge in the arc chamber to a narrow column. Electrons attempting to travel across the lines of force are forced into a spiral path:

$$R_e = 3.37 \frac{V^{1/2}}{H} \quad . \qquad (4)$$

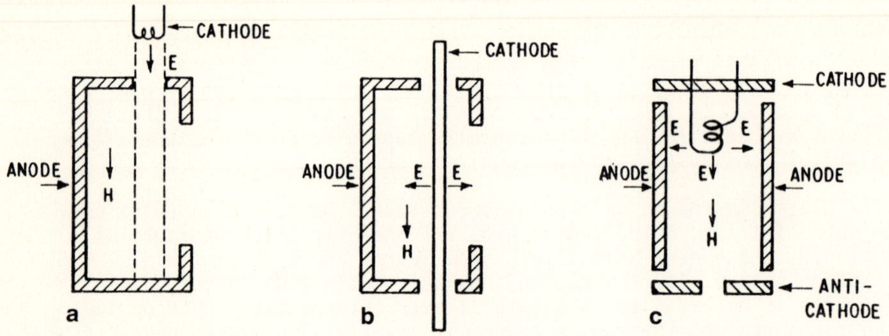


Fig. 1. The control of electron trajectories using a magnetic field

Thus for electrons with an energy component of 50 eV at right angles to a magnetic field of 200 gauss the spiral has a radius of 1.2mm. The electrons scattered by collisions within the plasma are still trapped in the column by the magnetic field.

In Fig. 1b we see a magnetron-type geometry with a rod cathode parallel to the axis of a cylindrical arc chamber. Here the electric field E and the magnetic field H are at right angles and the primary function of the magnetic field is to prevent the electrons travelling directly to the anode in a direction determined by the electric field. The result is a greatly increased path length for the electrons and therefore a reduction in the gas pressure necessary to give a sufficient ionisation probability to maintain the discharge.

In Fig. 1c we have the hot cathode Penning geometry where both the collimating and path-lengthening functions for the magnetic field are present. The collimating effect of the magnetic field allows the electrons to oscillate between the filament and the anti-cathode without being rapidly lost to the anode. As the velocity component towards the anode increases the electrons are forced to oscillate in a spiral path, thus producing efficient ionisation.

A shaped magnetic field can be used to focus the electrons from the cathode in order to create a high electron density in the region of the extraction aperture. This is often used in axial sources; the duoplasmatron is a good example.

## 3.5 The Plasma

If the cathode is heated to thermionic emission temperature, a voltage of 50 - 100V applied between the cathode and anode and a gas or vapour introduced into the arc chamber at a pressure in the range of $10^{-2}$ - $10^{-4}$ torr then an arc discharge can be initiated. The magnetic field necessary is dependent on the type of source, but is generally about 1 kG for collimation and 100 G for path lengthening. Cold cathode sources require voltages in the 1 - 2 kV range in order to maintain a stable discharge, and as the electron flow from the cathode is modest, efficient electron collimation is required to minimise electron loss to the anode, thus requiring high magnetic fields (up to 2 kG).

The discharge created in this way produces a medium with plasma-like properties. A plasma is defined as a neutral isotropic distribution of positive and negative particles (ions and electrons) at constant potential. Therefore the number of positive charges per unit volume is approximately equal to the number of negative charges. In a simple model therefore the number of positive ions per unit volume is approximately equal to the electron density. An interesting feature of a plasma is that it attempts to preserve its internal field-free state by establishing a charged sheath between itself and any other electrode or surface placed in close contact which is not at plasma potential. The neutrality of charge within the plasma is self-regulating since the positive ions and electrons are able to distribute themselves rapidly to compensate any localised field gradients.

At the plasma-sheath boundary the electric field must be nearly zero, as a potential gradient cannot be maintained in the plasma. Therefore where the cathode filament comes into contact with the plasma a double-layer sheath is formed as in Fig.2. The plasma is now protected from the electric field caused by the presence of the cathode by the positive space charge on the plasma side of the sheath. The double-layer sheath is formed because the positive ion concentration is higher on the plasma side of the sheath as they have a low energy (velocity) whereas on the cathode side they have been accelerated and therefore have a higher energy and consequently lower concentration. Similarly the electron density is higher on the cathode side of the sheath.

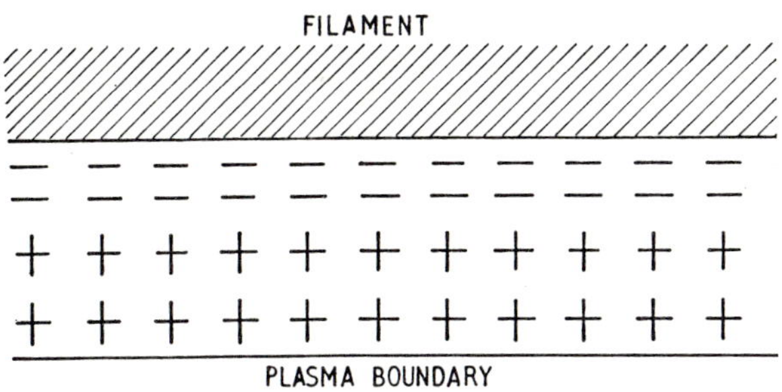


Fig.2. The double-layer sheath at the filament/plasma interface

The high mobility of the low-mass electrons compared with the high-mass positive ions means that the electron current required to maintain space-charge neutrality is much higher. Therefore the majority of the current passing across the sheath is carried by the electrons. The potential of the plasma normally adjusts itself to a value several volts above that of the anode in order to compensate for the charge imbalance that would be caused by the loss of the fast-moving electrons to the anode. The full arc voltage is therefore maintained across the filament sheath, creating a high extraction field where the negative space charge of the electrons is compensated by the positive space charge of the ions. Therefore under suitable conditions the electron current can be the saturation level determined by equation (1). The resulting potential distribution is therefore as in Fig.3. Filament/plasma sheath thicknesses are typically less than 1mm, but the situation is complicated by the presence of a transition region in which the field drops off exponentially into the plasma.

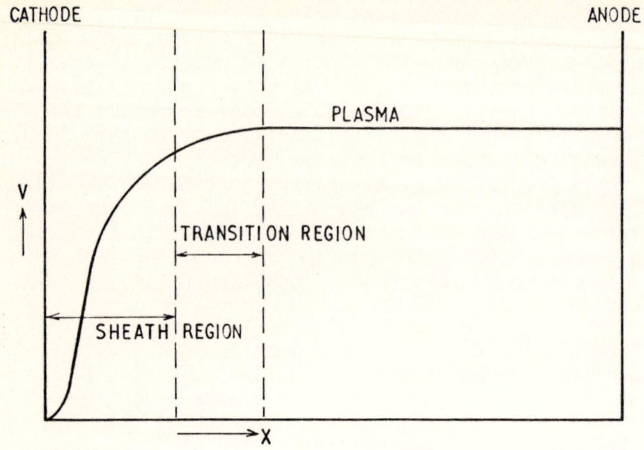


Fig. 3. The potential distribution between the cathode and the anode

## 3.6 The Source Feed System

The pressure in the arc chamber has generally to be maintained in the range $10^{-4}$ to $10^{-2}$ torr. The feed material can be either a gas which is introduced via a needle value or a vapour supplied from a liquid or solid heated in an oven which is directly connected to the source arc chamber by a tube of suitable conductance and temperature. The temperature of the source must be high enough to prevent feed material condensation.

## 4. Beam Extraction

The ions in the plasma generated in the ion source now have to be extracted through an appropriate geometry aperture in order to form a beam. The behaviour of the plasma surface from which the ions are extracted is extremely important in determining the optical quality and the geometric parameters of the beam. It is therefore important to understand the behaviour of the plasma surface in the extraction region. The important factors governing the ion extraction from the plasma surface are:

i) the shape of the plasma surface

ii) the ion current available from the plasma surface

iii) the space-charge limitation to the current that can be extracted.

Figure 4 illustrates the effect of the extraction field on the plasma meniscus (negative voltage on extraction electrode). In Fig.4a the extraction field is so weak that the plasma expands through the aperture because its sheath has no problems coping with the weak external field. It can readily be appreciated that such a geometry for the plasma outside the arc chamber is inherently unstable and easily disturbed by small changes in plasma conditions. If the extraction field is increased (or the plasma density decreased) then the plasma retreats, losing the 'neck' geometry, resulting in a form similar to that in Fig.4b. The surface shape

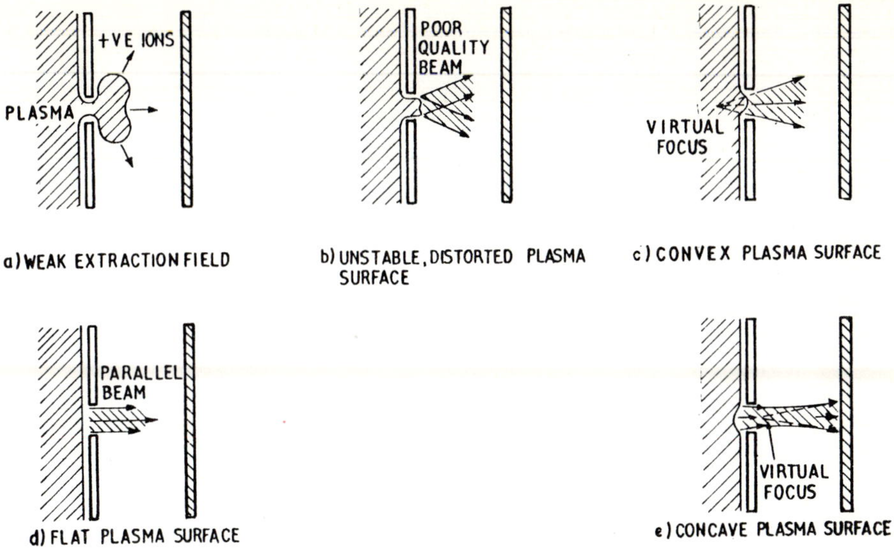


Fig. 4. The effect of extraction field on the plasma meniscus geometry

is complex, unstable and unpredictable and any attempt to produce a beam from such a plasma surface would result in a beam of dubious quality. It will be stressed in the next section that the plasma surface is probably the most important 'lens' in the ion optical system. A further increase in extraction field would produce the form shown in Fig.4c. This simple convex geometry can be stable and gives a virtual focus situated somewhere behind the plasma surface. The apparent size of this virtual object is a major factor determining the quality of the resulting ion beam. Further increases in extraction field give a flat meniscus (Fig. 4d) and eventually a concave meniscus (Fig.4e) which gives a virtual focus in the extraction region. The virtual focus effect in Figs.4c and 4e is useful in achieving high resolving power, as the virtual object can be significantly smaller than the aperture size with a consequent reduction in image size at the resolving slit of the accelerator system.

The amount of ion current available from the plasma surface is approximately the product of positive charge density and their mean velocity component normal to the plasma surface and is given by [1]:

$$\begin{aligned} j &\simeq n\,(kT_e/M)^{1/2} \quad (\text{ions/cm}^2\text{s}) \\ &\simeq 3.5\times10^{-13} n\,(T_e/M)^{1/2}\ \text{mA/cm}^2 \end{aligned} \qquad (5)$$

where $n$ is the positive ion density, $T_e$ is the electron temperature and $M$ the average mass of the positive ions.

The above equation gives an estimate of the ion current available, but the space-charge limitation of equation (2) can now be applied to the ions extracted from the plasma surface giving for the simple parallel meniscus/ extraction electrode geometry, separation d cms :

$$j = 5.5\mathrm{x}10^{-8} \frac{V^{3/2}}{d^2 M^{1/2}} \quad \mathrm{A/cm^2} \ . \qquad (6)$$

5. Beam Formation

As we have seen from Fig.4, a beam can be extracted from an ion source by applying an electric field between the ion source and an electrode which is biased negatively with respect to the ion source. If this electrode has an aperture of the appropriate shape a beam can then pass through this electrode into field-free space. In Fig.5a we have an idealised extraction situation with a flat plasma surface and plane electrodes with a fine mesh grid in the extraction electrode giving a uniform parallel electric field.

In reality things are rather different:

i) the meniscus is not usually flat (see Fig.4) and is not particularly a requirement

ii) the space charge in the beam causes the beam to blow up as in Fig.5b

iii) a grided extraction electrode is generally not practical and an aperture is used - this introduces a divergent lens into the system (see Fig.5c).

The problem of the meniscus shape can be controlled to a certain extent by adjusting the extraction field either by varying the extraction voltage V or the extraction distance d. As both the meniscus stability and the space-charge-limited current improve with increased extraction field this would lead to the conclusion that V should be as large as possible and d as small as possible. In practice considerations of sparking and the required aperture size in the extraction electrode limit how far we can go with this argument.

We can rewrite equation (6)

$$J = 5.5 \times 10^{-8} \left(\frac{V}{d}\right)^{3/2} (Md)^{-1/2} . \qquad (7)$$

If spark initiation is primarily a function of voltage gradient [V/d] then the above equation would suggest that for a given maximum acceptable gradient d should be as small as possible. In reality the minimum value of d is determined by the need to have a sufficiently large aperture in the electrode for efficient beam extraction and an acceptable electrostatic field shape and field strength in the extraction region. The value for the extraction voltage is then determined by the requirements of a stable plasma surface and sufficient space-charge-limited current. A typical high current extraction system uses voltages in the range 30 - 50kV and an extraction gap of 1 - 2 cm.

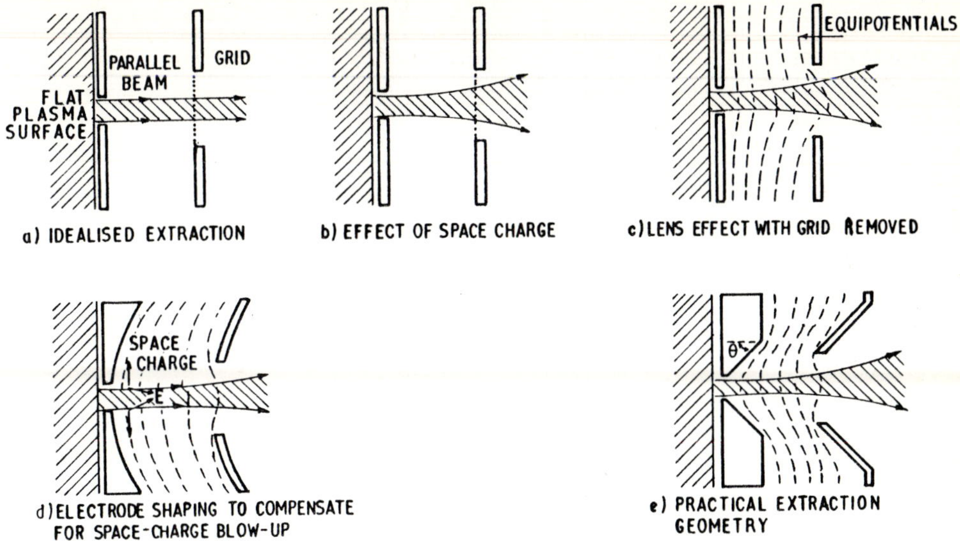


Fig. 5. Beam formation

The space-charge blow-up of the beam as shown in Fig.5b can be compensated by shaping the electrodes as shown in Fig.5d. This produces an accelerating field which has a component which counteracts the outward diverging effect of space charge. In order to maintain this type of convergent field it is important that the ratio of aperture size to extraction distance does not become too large as this would introduce a significant divergent lens into the system. This problem is, however, not too severe as the beam has a low energy during the initial convergent acceleration and a relatively high energy when it reaches the divergent lens, so the effect is relatively minor.

The space-charge situation within the beam is described by a quantity called the beam perveance P.

$$P = I\, V^{-3/2}\, (M/z)^{1/2} \tag{8}$$

where M is the atomic weight of the ion and z is the charge state.

In real situations the shape of the extraction electrode is dominated more by the requirement to minimise sparking rather than any optical considerations. The compensation of space blow-up is achieved by shaping the source as shown in Fig.5e but of course the angle θ can only be optimum for a particular beam perveance.

As we have seen from equation (6) the current density that can be extracted from the plasma surface is limited by space charge. Having selected practical values of V and d, the only option open for increasing the current capability of the system is to increase the area of plasma surface from which the beam is extracted. Unfortunately as the extraction aperture size is increased, the stability of the plasma surface tends to decrease. This effect is generally found to be minimised if one of the dimensions of the aperture is kept small (approximately 2mm for example) and the increase in area is achieved by increasing the length of the aperture. This approach has the advantage of keeping the geometry compatible with the optical requirements of mass analysis and is the approach used in isotope separators and high-current ion implanters.

In systems which do not require mass analysis the extraction area can be increased simply by using a close-packed array of circular apertures in both the source and the extraction electrode.

## 6. Beam Quality

It has been mentioned earlier that a virtual focus can be produced that gives an apparent object size somewhat smaller than the aperture size. This is important in that it potentially improves the beam quality. In the ideal case the apparent object should be a single point (for circular beams) or a line (for ribbon beams). In both cases the object is effectively infinitely small and the beam is therefore capable of being precisely focussed to form for example either an accurately parallel beam or to give a perfect cross-over (as required by mass analysing systems). Such a beam would be said to have zero emittance. At a particular point in a zero emittance beam the ions always pass through in the same direction. If, for example, in the case of a symmetrically circular beam the radial component of momentum is measured as a function of radial position then a momentum/position plot (called a phase-space diagram) would be a straight line for a zero emittance beam. For an imperfect beam the result would be a distribution having a finite area, and this area is defined as the emittance. The current density within this area is referred to as the 'brightness' of the source.

The emittance achievable for a beam extracted from a plasma ion source is primarily going to be determined by the shape of the plasma surface. Other considerations such as space charge and lens abberrations in the extraction region will generally tend to be secondary in importance. The plasma surface is effectively the most important variable in the implanter ion optical system.

The above considerations are mainly concerned with the direction of the beam leaving the extraction area. There are two other aspects of the quality of the beam leaving the source which are of great importance in determining the quality of the beam arriving at the target or resolving slit of an accelerator system. There are:

i) energy spread

ii) current modulation.

Energy spread is important because most accelerator systems require the beam to be mass analysed so that the transmitted beam is of the desired species only. Magnetic analysis is used to achieve this, generally by bending the beam through 60°, 90° or 180°. The analysing magnet is a momentum analyser and therefore only truly analyses mass if the energy of the ions has a single fixed value. Therefore the ability to achieve a sharp cross-over will be adversely affected if there is a significant spread in the energy of the ions leaving the plasma. For a quiescent plasma the ion energy is thermalised and the energy spread is only of the order of a few electron volts. If however the source plasma is unstable then the energy spread can be significantly higher. Sources which use strong magnetic fields and/or use reflecting cathodes which cause electrons to oscillate between the cathode and anti-cathode are particularly subject to high-frequency (10 - 500kHz) plasma oscillations, often referred to as 'hash'. This causes a large energy spread and also causes fluctuations in the geometry of the plasma surface which can lead to catastropic degradation of beam quality, particularly in high-current systems.

In addition these instabilities lead to a modulation of the output current. The consequent changes in perveance not only affect the emittance of the extracted beam but also can have serious consequences with regard to the ability to focus high-current beams through the resolving slit of the analysis system. The successful focussing of high-current beams to the narrow cross-over required by the analysis system relies on space-charge neutralisation of the beam by thermal electrons trapped in the potential well of the beam. The excess electrons are lost when the ion current falls as a result of this beam modulation and are not sufficiently rapidly recovered by ion collision with neutral gas atoms and apertures to neutralise the space charge when the ion current increases. The consequent loss of space-charge neutralisation causes beam blow-up and the degradation in resolving power can be considerable. The need to keep these thermal electrons in the beam is one of the reasons for the use of double electrode extraction systems which enable positive ions to be extracted from the source without accelerating these thermal electrons from the beam towards the source.

## 7. Beam Content

We have mentioned the need for magnetic analysis in order to separate the required ion from the rest of the output of the ion source. Unfortunately, the required ion can often be a small fraction of the total output spectrum. One example of this, commonly found in implanters applied to silicon doping, is that of boron. The element has an extremely low vapour pressure and would require the source to operate at temperatures close to 2000°C. Consequently a boron compound feed material is required. The common choice is boron trifluoride ($BF_3$). This gives an output spectrum as shown in Fig.6a. The required $^{11}B$ beam is diluted by the other boron isotope $^{10}B$, the fluorine ions $^{19}F$, $^{19}F_2$, the molecular ions $^{10}B^{19}F$, $^{11}B^{19}F$, $^{10}B^{19}F_2$, $^{11}B^{19}F_2$.

The advantage of using the element as the feed material is clearly seen from Fig.6b, showing the relatively clean spectrum of phosphorus. The impurity content of the spectrum, the multiply charged and molecular ions of phosphorus, is a significantly smaller fraction of the total output than the case of boron from boron trifluoride. The total extractor-current-to-beam-current (after analysis) ratio for boron from $BF_3$ is typically 12:1 whereas the element feed for phosphorus and arsenic gives a ratio of approximately 2:1.

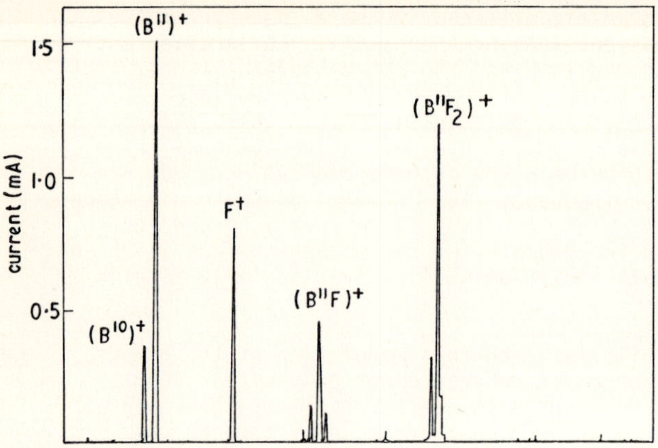


Fig. 6a. The boron trifluoride spectrum

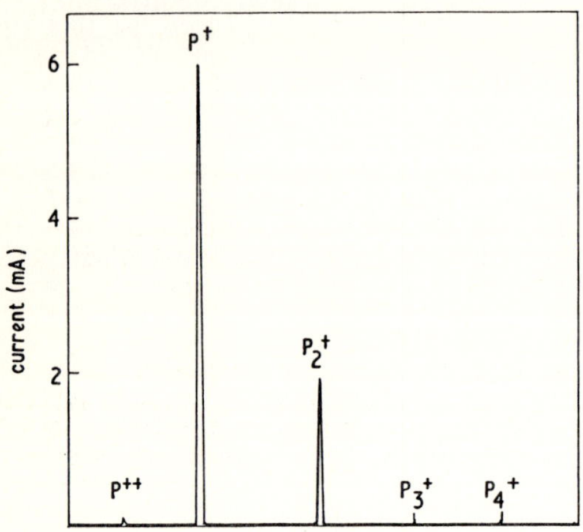


Fig. 6b. The phosphorus spectrum

The relative heights of the peaks in the output spectrum are a function of the ionisation potentials and ionisation cross-sections for the particular ionisation events and are modified by changing the arc conditions in the plasma. For example the multiply charged ions are favoured by a high arc voltage and low operating pressure and the molecular ions by low arc voltage and high pressure.

Elements which have a number of isotopes pose a problem. For example, most commercial ion implanters have a resolving power capability which prevents the transmission of both isotopes of antimony at masses 121 and 123. Therefore implanting the 121 isotope results in a 43% loss in beam current due to the inability to transmit the 123 isotope. This decreases the attraction of antimony as an alternative implanted dopant to arsenic.

It should be appreciated that the use of compounds as feed materials causes a reduction in two ways. There is the obvious dilution effect already mentioned but there is also the space-charge consideration. The space-charge-limited current which can be extracted from the source is proportional to $M^{-\frac{1}{2}}$ for the simple case of singly charged positive ion production of mass M. When a compound is used the heavy-mass species in the spectrum can make the effective value of M greater than that of the ion of interest.

## 8. Ion Source Selection

In this section we will consider how the particular type of implanter determines the nature of the ion source to be used.

Implanters will be divided into the following categories:

i) low-current machines

ii) medium-current machines

iii) high-current machines

iv) surface treatment machines.

### 8.1 Low-Current Machines

We will define a 'low-current machine' as one in which the available beam current at the target is limited by the output of the ion source and where the scanning of the beam over the target is to be achieved by scanning the ion beam.

The interest in this category of machine is largely historical as very few commercial machines of this type are manufactured today. It is clearly not desirable that the performance of a machine should be so severely limited simply because the ion source chosen produces less current than the rest of the machine can handle. Nevertheless early machines of this type were produced and they use cold cathode sources.

As the final beam requirement is a circular spot which can be electrostatically scanned over the area of the target, there is therefore the natural selection of an optical system which handles beams of circular cross-section and therefore a source which produces a beam from a circular aperture. These early machines invariably used a gaseous feed material containing the element of the required ion species and therefore there was no requirement for the source to run at elevated temperatures, and being a cold cathode source there was no need for filament supplies. Consequently the source is simple, requiring only two electrical supplies, one for the arc and the other for the magnetic field coil.

### 8.2 Medium-Current Machines

Medium-current machines will be defined as machines which scan the beam over the target in at least one direction electrostatically. Most machines in this group are X - Y electrostatic but there is the alternative hybrid scanning using electrostatic scanning in one direction and mechanical movement of the target in the other direction.

In these machines the current capability is limited by space-charge problems caused by the electrostatic scanners and are not in general limited by the ion source, although boron ions are a notable exception to this rule.

Consequently the sources tend to be minature high-current sources using hot cathodes, invariably tungsten filaments. The requirement at target is for a circular beam,so a circular aperture at the source might be expected, but as most high-current sources tend to be of the ribbon beam type there is a widespread use of small rectangular apertures with subsequent quadrupole focussing to give a near-circular beam spot at the target.

The use of a rectangular aperture comes from the requirement for increased plasma surface area from which to extract the beam. It has generally been found best to achieve the required plasma surface area by increasing the length of extraction slot rather than by increasing the diameter of a circular aperture because the stability of the plasma surface tends to be controlled by the smaller dimension. Extraction apertures for ribbon beams are rarely much wider than 2mm for this reason.

## 8.3 High-Current Machines

The major requirement for the ion source of a high-current machine other than the obvious current capability is the need to produce a beam which can be space-charge neutralised. For this reason the beam must be 'hash' free so that thermal electrons can be maintained within the beam in the region between the extraction electrode and the resolving slit. The rectangular extraction aperture is almost universally used as this poses few problems in the relatively simple optical systems used in these machines. The space-charge problems mitigate against electrostatic beam scanning and in general the target is scanned through a stationary beam.

## 8.4 Surface Treatment Machines

The ion source requirement for surface treatment machines is very different to the previous three groups. There is generally no need for mass analysis and this removes a number of constraints:

i) energy spread is unimportant

ii) 'hash' has no serious consequences

iii) beam quality is relatively unimportant.

The major requirements are:

i) high current

ii) simple, long-life source

iii) ability to operate with elemental feed materials.

The latter is potentially the most restrictive requirement,particularly if the ion required is that of a very low vapour pressure element. The use of compound feeds without mass analysis may be acceptable in certain circumstances. The major requirement at present appears to be for nitrogen [2] and this clearly poses no serious problems.

## 9. Types of Arc Discharge Source

### 9.1 Calutron Sources

A family of ion sources have developed from the Calutron source geometry which is characterised by a filament at one end of the arc chamber, electron acceleration into the plasma along the lines of force of a powerful collimating magnet and lateral extraction of the beam from the plasma through a long rectangular aperture situated close to the magnetically constrained plasma column.

The Calutron source is a large source originally developed for the separation of uranium isotopes [3]. Resolved currents of the order of 100 mA are achievable with these sources extracting from large rectangular apertures (127 x 4.8 $mm^2$). The layout of the source is shown in Fig.7a. The filament is outside the arc chamber and the electron current is guided through a collimating slit into the arc chamber by the strong magnetic field. These sources were designed to be in the field of the separator magnet and the high degree of parallelism of this strong (>1kG) homogeneous field is important in maintaining the plasma close to and parallel to the extraction aperture. This type of source can have a reflector cathode at the other end of the arc chamber but this is not generally found beneficial.

Attempts to produce smaller ion sources using weaker magnetic fields supplied by a small electromagnet independent of the main analysing magnet have not been particularly successful. One problem is the distortion of the collimating magnetic field by the magnetic field of the filament which can lead to distortions and instabilities in the plasma column. It appears that very strong parallel collimating fields are necessary to minimise these problems, but that this leads to sources which are intrinsically rather prone to plasma instabilities that require tuning out by skillful operation of the source.

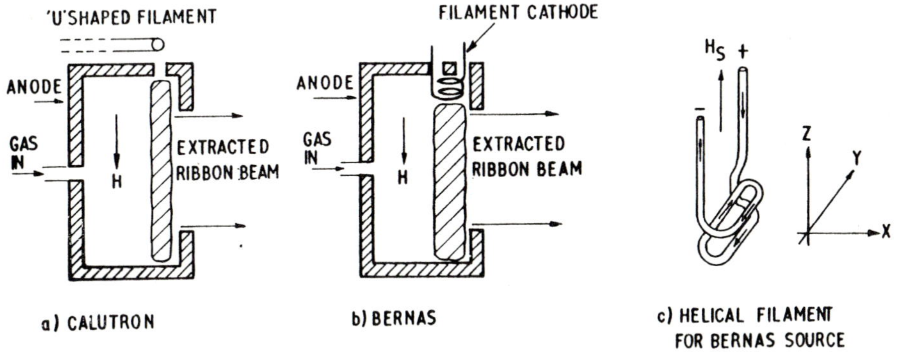


Fig. 7. Caultron-type sources

A more successful exploitation of this arc discharge geometry to a smaller source [4] is the Bernas source shown in Fig.7b. Here the filament is smaller, the collimating slot is omitted and the filament is situated inside the arc chamber. The filament design has been carefully optimised [5] to prevent disturbance of the plasma column either by the magnetic field or the voltage gradient of the directly heated filament. The resulting helical design shown in Fig.7c focusses the plasma column through the helix axis and minimises the electrical bias along the conventional 'U'-shaped cathode design. This allows the source to be run with modest magnetic fields, thus reducing the susceptibility to 'hash', and also reduces the optimum arc voltage for singly charged ion generation [5] with a consequent desirable improvement in the filament life.

Another source which utilises the Calutron geometry, avoids filament interaction and aims to achieve a considerably longer operational life is the indirectly heated cathode source. Figs.8a and 8b show an indirectly heated cathode source developed by Pasztor [6]. The filament is not exposed to the plasma; the electron-emitting component is relatively bulky, allowing for some considerable erosion before the need for replacement. As the heating current does not pass through the cathode the runaway erosion at hot spots (due to reduced cross-section) does not occur.

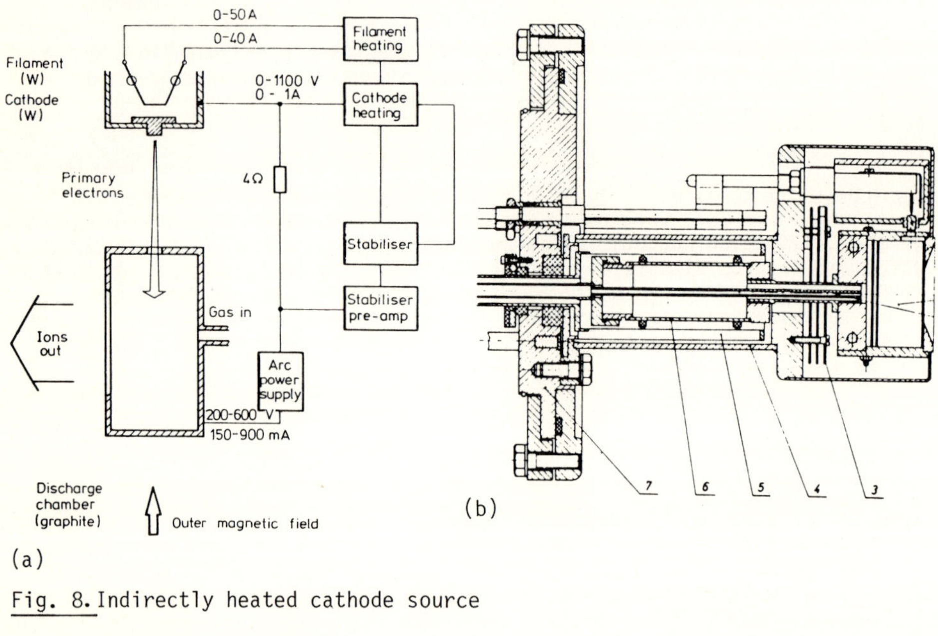


Fig. 8. Indirectly heated cathode source

## 9.2 Magnetron Ion Sources

These sources use a long straight filament parallel to the axis of a cylindrical arc chamber with a relatively weak magnetic field parallel to the filament. The arc voltage is applied between the filament and the arc chamber and therefore the electric field and the magnetic field are at right angles. The normal magnetron geometry with the filament down the centre of the cylindrical arc chamber exhibits a sharp cut-off magnetic field above approximately a few hundred gauss.

The most successful exploitation of this geometry is the Freeman source [7]. This source had its origins in the difficulties found in achieving reproducible behaviour from a scaled down Calutron type of source. One of the most difficult problems with lateral extraction sources is the stable location of the intense plasma column close to the long ($\simeq$ 40mm) extraction aperture. In the Freeman source the magnetron geometry is modified by moving the filament from the axis position to a position close to the extraction slot (see Fig.9). This precisely locates the intense plasma in the required position without the need for strong 'hash'-producing magnetic fields. In this case the compound magnetic fields of the filament and source magnet ($\simeq$ 100G) produces a desirable field geometry for efficient ionisation. This source is notable for the ease with which stable, 'hash'-free beams can be obtained from most elements in the periodic table. The large rod filament (usually about 2mm diameter) heats the arc chamber which with appropriate heat shielding can easily be operated in the temperature range up to 1100°C. This source is extensively used in commercial high-current ion implanters.

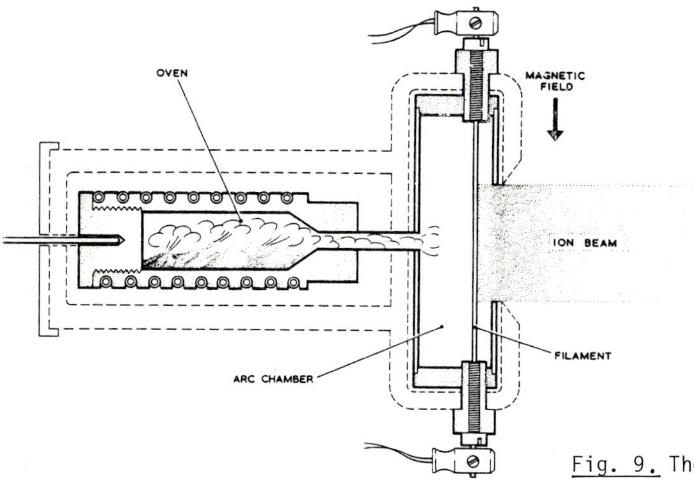


Fig. 9. The Freeman source

## 9.3 Penning Ion Source

It has been mentioned that the Calutron-or Bernas-type source can be equipped with an anti-cathode, either floating or held at cathode potential, which causes the electrons to oscillate between these two electrodes, their loss to the anode being restrained by the strong ($\simeq$ 1kG) magnetic field. The combination of oscillation and the spiral path which occurs as the electron path approaches the anode gives an extremely efficient ionising system. This geometry is so efficient that a low-pressure plasma can be maintained without a hot cathode, electron loss to the anode being compensated by secondary electron generation processes. This type of source is rarely used in high-current systems as the output tends to be 'hashy'.

The main types are:

i) Cold cathode axial extraction sources

ii) Hot cathode axial extraction sources

iii) Cold cathode lateral extraction sources

iv) Hot cathode lateral extraction sources.

9.3.1 Axial Cold Cathode Penning

Used extensively in the early commercial low-current ion implanters for $10^{11}$ ions $cm^{-2}$ type implants into silicon. These are simple sources (see Fig.10a) but their operation can pose problems. They often operate in a variety of discharge modes and the value of the magnetic field can be critical. Sometimes discharge initiation can be a problem. The advantages of this type of source are long life, low power consumption and good corrosion resistance due to the low operating temperature.

9.3.2 Axial Hot Cathode Penning

An efficient source capable of simple reliable operation (see Fig.10b). Often found on medium-current machines and research machines. Capable of producing good ion yields but the 'hashy' output limits the useful output in systems requiring good beam quality.

9.3.3 Cold Cathode Lateral Extraction Source

This type of source (see Fig.10c) is sometimes used in cyclotrons. It is often operated at relatively high arc voltages ($\simeq$1 kV) when it can be an

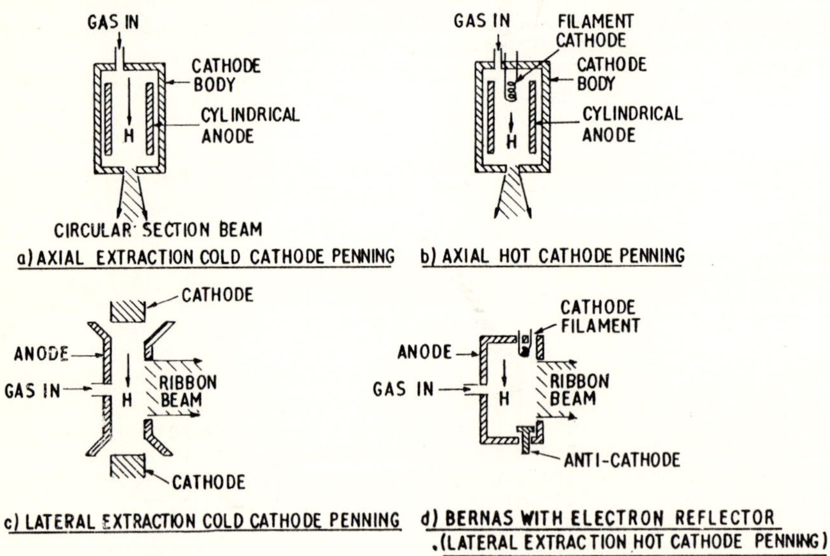


Fig. 10. Penning ion sources

efficient generator of ions with high charge states. For example such a source operating at an arc voltage of 660V can produce a range of xenon ions 70% of which have charge states in the range 4+ to 8+ [7].

## 9.3.4 Hot Cathode Lateral Extraction Source

Calutron, Bernas and indirectly heated cathode sources can be equipped with an electron reflector (see Fig.10d) which effectively converts them to a lateral extraction hot cathode Penning source. These sources become more efficient as a result but there is generally a loss of beam quality due to the susceptibility to 'hash'.

## 9.4 Plasmatron Ion Sources

As the efficiency and output of an ion source is a function of plasma density, any attempt to produce significantly higher performance (in mA $cm^{-2}$ of ions from the plasma surface) requires that the electrons accelerated into the plasma from the cathode are not only collimated but also actively focussed. This can be done by either appropriate shaping of the cathode geometry, the shaping of electric fields external to the plasma or by magnetic field shaping.

This is attempted in the unoplasmatron source [8] by introducing a conically shaped intermediate electrode (see Fig.11a) between the filament cathode and the anode. No magnetic field is used.

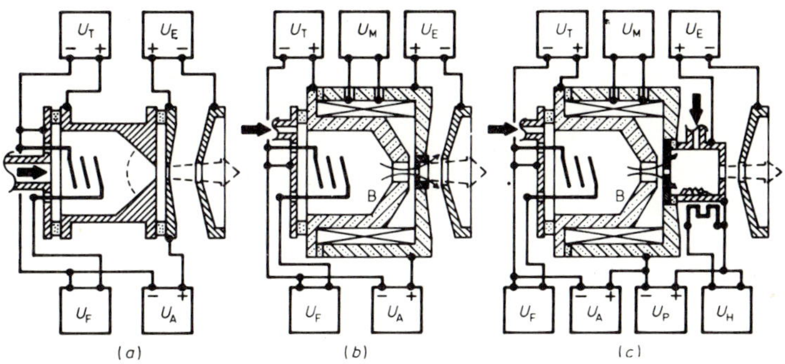


Fig.11. The plasmatron sources (from Sidenius [9])

In the duoplasmatron source (see Fig.11b) magnetic compression is added by making the intermediate electrode from a ferromagnetic material so that a strong focussing field (up to 10kG)is produced between this electrode and the anode [10]. The plasma density can become so high that the normal extraction process is not possible and the plasma is allowed to pass through the anode aperture into an expansion cup where extraction can then take place from a surface an order of magnitude larger in area.
This source has been very successful for producing intense beams from non-corrosive gas feeds but has not been found to be a particularly versatile heavy-ion source as the magnetic components have to be kept cool, thus limiting the use of low vapour pressure materials. Attempts have been made to overcome this by replacing the expansion cup by a high-temperature secondary ionisation chamber (see Fig.11c) where ions of low

vapour pressure materials can be formed by charge exchange reactions [11]. In general the performance of such a source does not justify its complexity. In common with most sources that use high magnetic fields the quality of the beam from plasmatron sources is rather poor and limits its usefulness for low-energy ion implantation.

A variant of the duoplasmatron called the duopigatron is shown in Fig. 12.

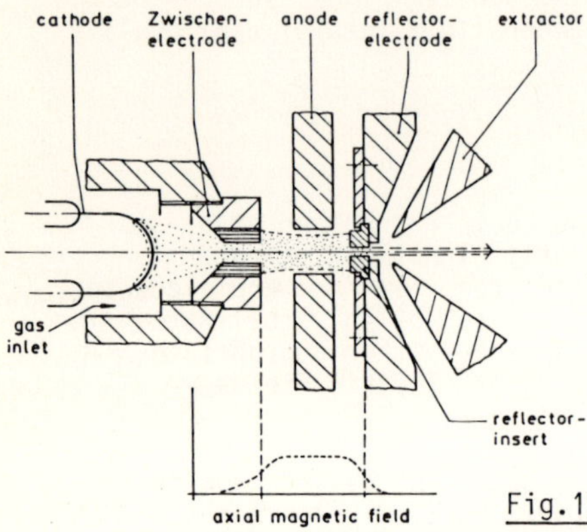


Fig.12. The duopigatron source (from Wolf [12])

It is essentially a duoplasmatron that incorporates an electron reflector electrode thus causing electrons to oscillate. This improves the ionising efficiency and allows the source to operate at a lower pressure. It is an efficient source of multiply charged ions. The reflector electrode can be used as a sputtering electrode by using an extraction aperture insert of the appropriate material.

## 9.5 Hollow Cathode Sources

Hollow cathode sources are a popular choice for research machines where their small size, high efficiency and ability to operate at extremely high temperatures (T >2000°C) makes them very versatile.

There are two versions of this type of source, one with anode extraction (see Fig.13a) and the other with cathode extraction (see Fig.13b). These sources are based on the principle that a hollow cylindrical cathode filled

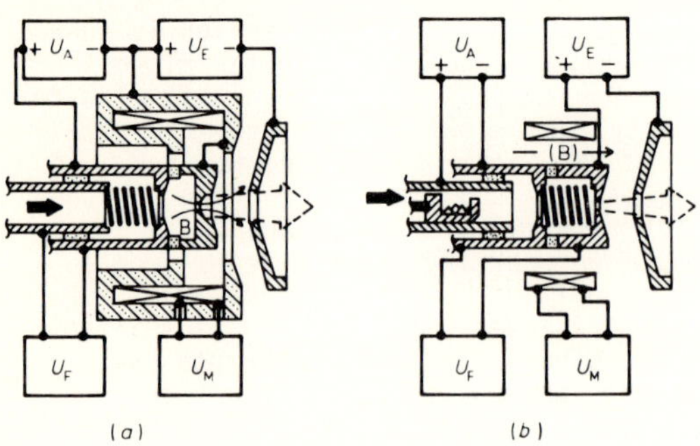


Fig.13. Hollow cathode sources (from Sidenius [9])

with a plasma which is approximately at anode potential focusses the electrons accelerated into the plasma across the cathode sheath to produce an intense plasma along the cylinder axis.

In the anode extraction form [13] a strong convergent magnetic field controls the electron distribution in the plasma and the resulting plasma densities are so high that an expansion cup has to be used. These sources are capable of very efficient operation but have been found to have operational difficulties which have limited their use.

The less efficient cathode extraction form [14] has found a wider applicability. This uses the pure hollow cathode effect and is not complicated by the presence of magnetic fields. Indeed the low-power coil that is used with this source merely has the function of compensating for the field produced by the helical filament coil so that the plasma volume is field free.

It is important to appreciate the small size of these sources - the plasma volume is less than $0.5 cm^3$.

## 9.6 Sputtering Sources

Virtually any plasma discharge source can be adapted to produce metallic ions by placing an electrode of the appropriate material in the plasma region at a negative potential of up to 2 kV. The arc is usually maintained by an inert gas with the optimum sputtering properties and is usually argon, krypton or xenon. A source specifically designed for this mode of operation [15] is shown in Fig.14. These sources are particularly useful for the platinum-group metals.

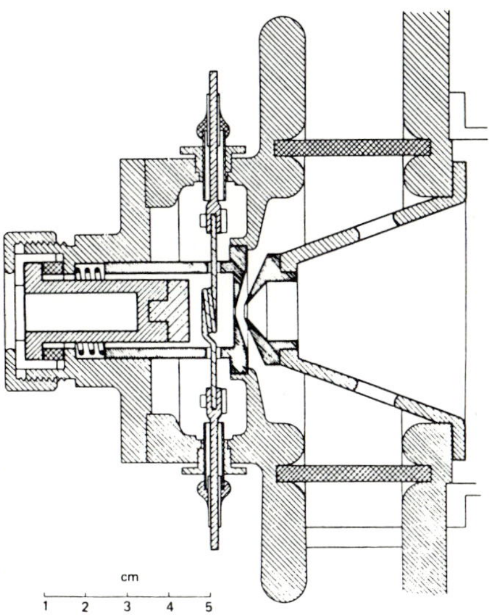


Fig.14. Sputtering source (from [15])

## 9.7 Sources for the Surface Treatment of Metals

The requirements here are simplicity and long operational lifetime. Beam quality is unimportant and mass analysis is not required. One source that seems particularly suitable is the saddle field source [16] shown in Fig.15. This source has two anodes and does not require a magnetic field. Electrons travel in an extended path before being collected on one of the anodes. This mode of operation only strictly applies to the low-pressure or electron-oscillating mode. Under these conditions the source is efficient but the output is limited. As the pressure in the source is increased above $2 \times 10^{-4}$ torr the discharge changes to a transition mode and above $3 \times 10^{-3}$ torr becomes a glow discharge. As the pressure is increased the maximum output capability increases but the efficiency ( = ion current/arc current) decreases. The source is likely to be most useful in the transition mode where there is a reasonable output current and efficiency and the beam profile is more uniform than in the low-pressure mode.

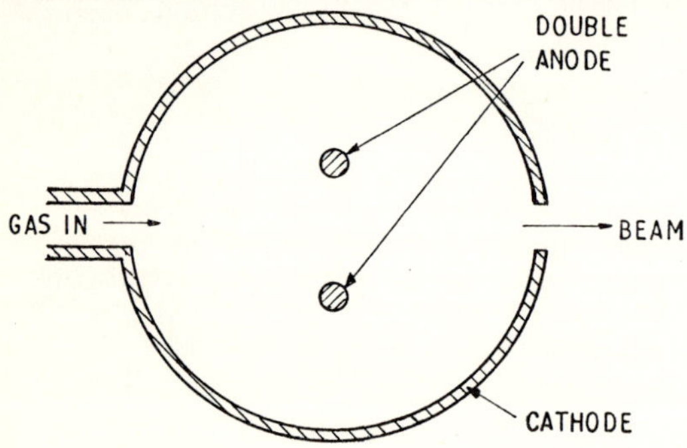


Fig.15. The saddle field source

An unusual feature of this source is that it can operate without an ion extraction system. If the source is run at an anode voltage of 10kV with 70mA discharge current then 2mA of $H_2$ can be extracted from a 10mm diameter aperture with an average energy of 7.2 keV and an energy spread of approximately 1keV [17].

A variety of other sources can be used for surface treatment applications. For example the cold cathode Penning with multiple apertures would be suitable.

## 9.8 Other Sources

It can be seen from the source types described so far that the scope for ingenuity in the development of alternative types of source is almost limitless. A complete account of the entire range of sources reported in the literature would be a major exercise. In this section we will just briefly look at the more important alternatives to the mainstream of plasma source development.

### 9.8.1 Radiofrequency Sources

A plasma can be produced by creating a radiofrequency discharge in the frequency range 10 - 100MHz. The coupling can be either capacitive or inductive. The former creates a discharge between a pair of parallel metal plates. The inductively coupled source is the most popular version of this type of source (see Fig.16) and is unusual in that the plasma formation is electrodeless. The source consists of a glass envelope surrounded by a coil and the RF electric field is created by the alternating magnetic field in the discharge region.

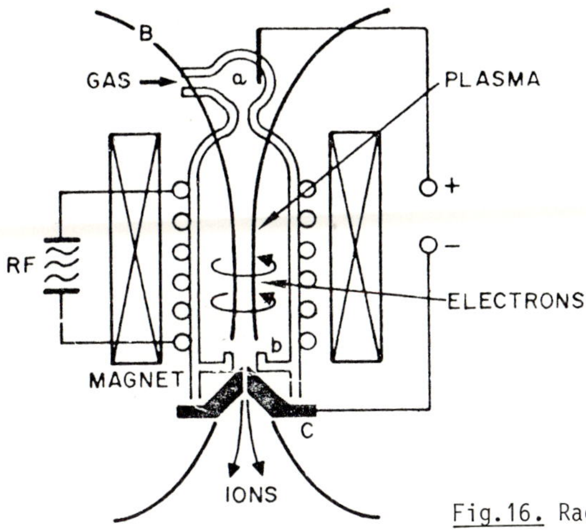


Fig.16. Radiofrequency source (from [18])

In spite of their low power requirements, simplicity and reliability, RF sources have had only a limited application to ion implantation. The main limitations are:

1. they are only suitable for non-metallic ions, as coating of the glass envelope with a conducting layer degrades the performance

2. they have a high energy spread giving rise to resolution problems in mass analysing systems

3. they run a higher pressure than most arc discharge sources.

9.8.2 Microwave Ion Source

A plasma discharge can be initiated and maintained between a pair of electrodes [19] excited by microwave power at a frequency of 2.45GHz (see Fig.17). Wave guide techniques are used to transmit several hundred watts of microwave power to a discharge volume which is defined by a cavity in a dielectric medium. 10mA of $P^+$ can be extracted from a 40 x 2mm² exit aperture using this type of source. The lack of a filament is a desirable feature of this source but the operational lifetime may be somewhat limited by the degradation of the exposed dielectric surfaces.

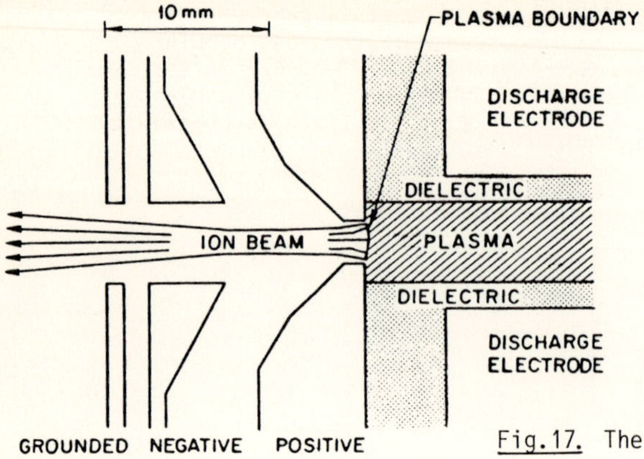


Fig.17. The microwave source (from [19])

### 9.8.3 Laser Sources

Laser sources are potentially an attractive technique but lasers with sufficient power in the required wavelength region for efficient ionisation are not yet available.

### 9.8.4 Spark (Vibration Contact) Sources

A spark source produces a plasma between the tips of a pair of electrodes, one of which is vibrated so that intermittent contact is achieved. A range of voltages and frequencies can be used [20]. These sources could be useful in certain research applications but are unlikely to find any widespread applicability to ion implantation due to their inconsistent performance. An interesting feature of these sources is their ability to operate under high vacuum conditions.

### 9.8.5 Charge Exchange Sources

We have already mentioned a version of the duoplasmatron which uses charge exchange to avoid the use of corrosive materials in the main arc chamber of the source. This principle can be applied to any source but is likely to be restricted to research applications.

### 9.8.6 Negative Ion Sources

Negative ions are of no particular importance in ion implantation although it has been suggested [21] that it could be a convenient way of avoiding surface charging when implanting insulators. The surface charging would be prevented by secondary electron emission.

Negative ions can be extracted directly from discharge sources [22,23] but the most efficient technique is charge exchange using, for example, lithium vapour [24].

The main demand for negative ions is for injection into high-energy tandem accelerators.

### 9.8.7 Very High Charge State Sources

This group of sources includes electron cyclotron resonance plasma sources [25], electron impact ionisation sources [26], and 'Hipac' toroidal sources [27]. They are used for the production of very high energy ions for nuclear physics research and have yet to find application in ion implantation.

## 10. Sources Not Utilising an Arc Discharge

There are a limited number of ion sources which have a totally different ionisation mechanism. Two types will be briefly discussed here.

1. surface ionisation sources
2. molten metal field ion sources.

### 10.1 Surface Ionisation Sources

This is an efficient technique for the production of ions of certain low-ionisation-potential atomic species. It is particularly suitable for the production of positive ions from alkali, alkaline-earth and rare-earth metals. The technique depends on the ionisation of a low-ionisation-potential species when placed in contact with a high work function material at a high temperature.

This technique is very efficient for Cs, Rb and K and is useful with progressively decreasing efficiency for Na, Ba, Li, Sr, In, Al, Ga, Tl and Ca. The ionising surface is a high work function refractory material such as tungsten, iridium, rhenium or osmium.

The most common surface ionisation source is of the type shown in Fig.18a. The metal vapour to be ionised is placed in an oven and the atomic vapour is diffused through a porous tungsten ioniser [28]. The ion beam is emitted from the front surface and extracted in the normal way. This technique is particularly suitable for the high vapour pressure alkali metals with their relatively low ioniser temperature requirements.

There is a diffusion limitation to the temperature at which a porous tungsten ioniser remains porous. For metals which require ioniser temperatures in excess of this value ( $\geq$ 1000°C) a source of the type shown in Fig.18b can be used. This source uses a front feed and the vapour is directed onto the front surface of the ioniser (usually iridium or osmium) from a surrounding cylindrical opening [29]. This produces the required ion beam but for materials such as Al, Ga and In the low efficiency of the ionisation process results in a large neutral beam and for many applications a neutral trap may be an essential requirement. The impurity ion content of the beam can be very low and for many applications magnetic analysis may not be necessary.

In most ion implantation applications the inefficiently ionised metals such as Al, Ga and In are of more interest than the alkali metals. The efficiency can be increased if the work function of the ionising surface can be increased. Figure 18c shows a source where the work function of the surface of a porous tungsten ioniser is increased by spraying the

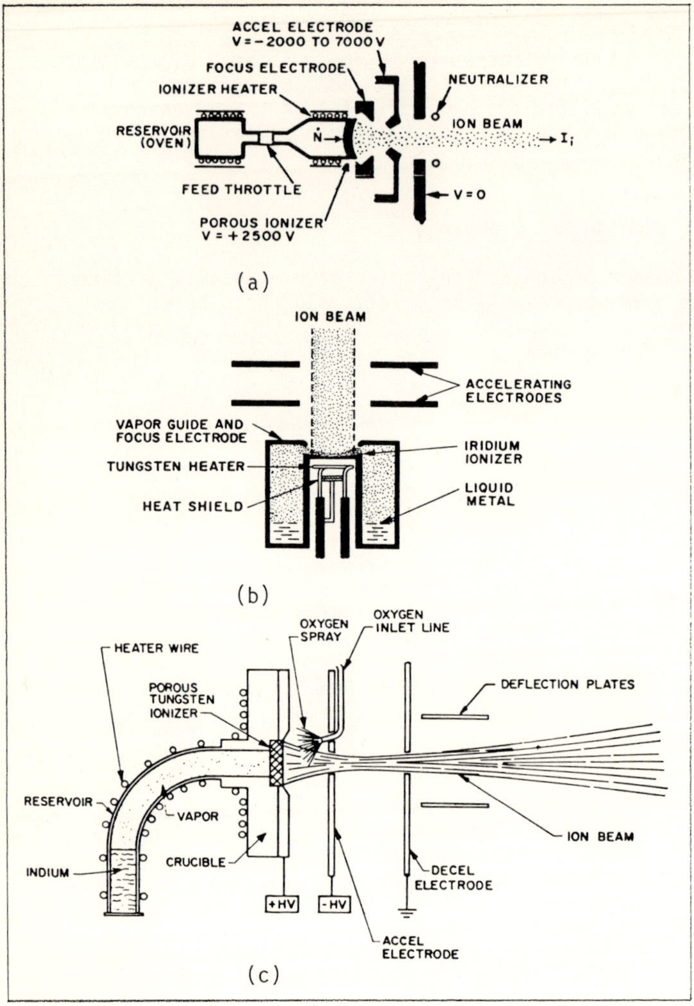


Fig. 18. Surface ionisation sources (from [28,29,30])

surface with oxygen [30]. This increases the work function from 4.6eV to a value in excess of 6eV, thus enabling reasonably efficient ionisation of materials such as indium.

The performance of surface ionisation sources varies from the 10mA $cm^{-2}$ obtainable from a rear - fed porous tungsten caesium source to the range 10 - 100$\mu$A $cm^{-2}$ for In, Ga and Al from surface sources.

## 10.2 Molten Metal Field Emission Ion Sources

Field emission ion sources can be constructed for either gas or liquid metal feeds as shown in Fig.19. The liquid metal sources have received some attention as possible high-brightness sources for ion beam lithography.

Fig. 19. Field emission ion sources

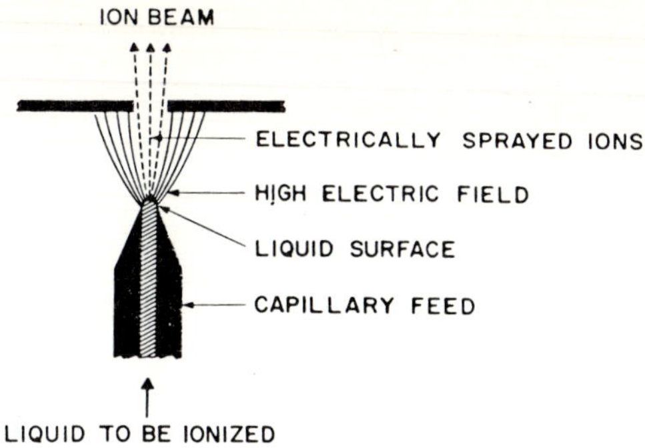


(a)

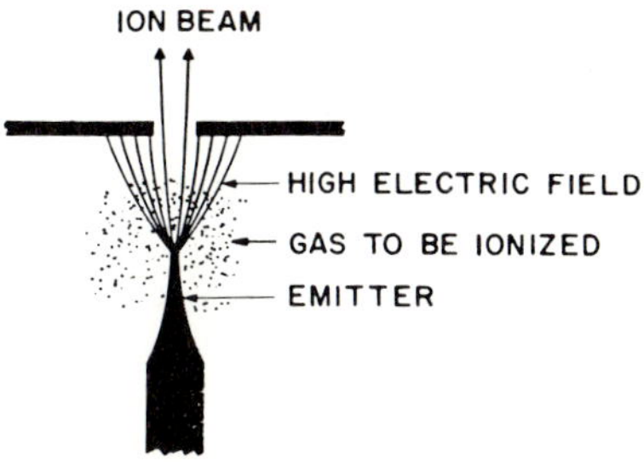


(b)

The liquid metal sources rely on the distortion of the liquid metal surface due to the applied electrostatic field to form a Taylor cone [31]. For example a gallium source with a 10μm radius tip geometry and a needle/electrode distance of 12mm will start to emit at a voltage of just below 7kV. The current capability is in the range 1 - 100μA. Two versions of this type of source have been developed; one uses a wetted needle as the source, the other a capillary as in Fig. 19a.

The ions that can be produced by this technique are limited to elements which have a reasonably low vapour pressure at their melting point and which have acceptable corrosion properties for the range of needle and capillary materials available. Ions of Cs, Cu, Au, In, Hg, Si and Ag have been produced as well as the particularly convenient gallium.

## 11. Operational Characteristics of Arc Discharge Sources

The important operational parameters of an arc discharge ion source are:

1. arc voltage
2. arc current
3. magnetic field strength
4. arc chamber pressure
5. feed material
6. temperature
7. source condition
8. extraction voltage.

The significance of each of these parameters will now be discussed but it must be remembered that the characteristics will vary enormously from one type of source to another.

### 11.1 Arc Voltage

The important aspects determining the selection of arc voltage are:

1. arc initiation
2. arc stability
3. optimisation of the required ion current
4. cathode life.

#### 11.1.1. Arc Initiation

The voltage necessary between the anode and the cathode to initiate the discharge is higher than that necessary to maintain it. This applies to both hot and cold cathode sources. It is therefore desirable to design the ion source power supplies so that the zero current arc voltage is approximately 100 - 150V for a hot cathode source. The cold cathode source starting requirement is very dependent on the detail design and the condition of the source but is typically 1.5 - 2.0kV.

It is sometimes necessary with cold cathode sources to increase the arc chamber pressure and/or the magnetic field strength in order to persuade a reluctant source to fire.

Once the arc has initiated then the arc voltage can be reduced to a value determined by the particular requirements.

#### 11.1.2 Arc Stability

The value to which the arc voltage can be reduced after arc initiation will be determined by the requirement of stable arc conditions and, if maximum output is required, the voltage necessary for optimum output of the required ion. It must be stressed that the total extracted current is not the important parameter to be optimised in a mass-analysed ion implanter. The quantity to be optimised is the beam of the required ion transmitted through the mass resolving slit. This will depend upon beam divergence and beam quality. This optimisation is conveniently achieved with implanters which incorporate a beam sweep facility either by modulating the pre-analysis acceleration voltage or the analysing magnet field, so that a picture of the beam profile at the resolving slit can be displayed on an oscilloscope.

#### 11.1.3 Optimisation of the Required Ion Current

It is important to realise that the optimum arc voltage is different for the various ions in the output spectrum. For example the optimum voltage for the doubly charged ion is significantly higher than for the singly charged ion. It is also different for the various elemental and molecular ions. Figure 20 shows a plot of ionisation efficiency for a range of gases as a function of electron energy. This only gives a rough guide as to the optimum voltage in an ion source because of the complex distribution of electron energy in the source plasma, ranging from the primary electron energy from the cathode to the thermalised electron energy in the plasma. In general, sources show a definite optimum voltage condition for a given ion species.

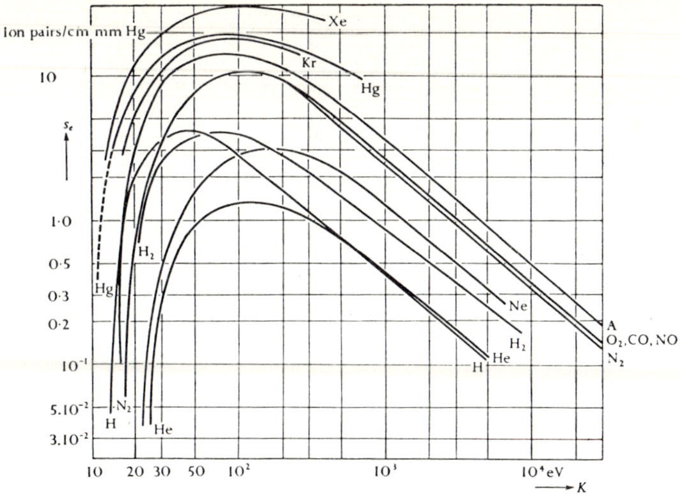


Fig. 20. Ionisation efficiency/electron energy

The following characteristics are usually found:

1. multiply charged ions have a progressively higher optimum voltage with increasing charge stage

2. molecular ions have a progressively lower optimum voltage with increasing molecule size

3. the optimum arc voltage is not a simple function of the appropriate ionisation potential but is dependent upon a complex variety of factors including ionisation cross-section, the particular feed material and virtually all the other ion source parameters.

### 11.1.4 Cathode Life

In the arc voltage range of most ion sources the sputtering erosion of the cathode surface increases with increasing voltage. There is therefore an incentive, particularly in a hot cathode source, to not use an arc voltage any higher than that necessary to achieve the required output. It is also important to remember that sputtering is a more important erosion mechanism than evaporation and therefore arc voltage is more important than cathode temperature in determining the cathode life. This is not necessarily true for feed materials that attack the cathode by a chemical mechanism.

## 11.2 Arc Current

In a cold cathode source the arc current is determined by the arc voltage, the magnetic field strength and the arc chamber pressure. Stability is achieved by arranging the arc voltage to decrease with increasing current.

In hot cathode sources the primary control variable is the electron emission from the cathode, which is a function of cathode temperature.

Therefore in a directly heated cathode source the primary function of the ion source supplies is to control the filament current in order to achieve the required stable arc current. This can sometimes be deceptively difficult to achieve. There are many causes of instability. Plasmas are generally prone to high-frequency instabilities (hash), the various supplies are likely to be modulated at twice the mains frequency and the cathode has a significant thermal inertia. Add to this a plasma electrical impedance which can vary significantly with quite small changes in other parameters and we have a situation which has led to the "black art" of ion source control. For the above reasons the development of ion source supplies is a largely empirical exercise. The ion source is one component in the power supply circuit and it is certainly the most unpredictable.

In ion sources which use large or indirectly heated cathodes, the thermal inertia of the cathode can lead to serious control difficulties.

## 11.3 Magnetic Field Strength

In a cold cathode source the operational characteristics are very sensitive to the magnetic field strength, as efficient electron confinement is required to compensate for the relatively low primary electron current. This usually results in there being a cut-off field below which the arc extinguishes. The value of this field is a function of virtually every other parameter available. If the source is run at a sufficiently high gas pressure this cut-off can be eliminated and the ionisation efficiency becomes insensitive to magnetic field strength.

In hot cathode sources a cut-off field is not generally found. These sources will usually run with no field at all but the output will be low due to the inefficient electron trajectories, and arc initiation may be impossible under these conditions.

In both hot and cold cathode sources the output current will optimise at a particular magnetic field strength. There can be a number of reasons for this. In sources where the electric field and the magnetic field are at right angles (or at least not parallel) there is an optimum electron trajectory for maximum ionising efficiency. In all sources there is an optimum in output current (after mass analysis!) determined by the quality of the extracted beam. This is caused by the onset of serious 'hash' and is a sensitive function of magnetic field strength. The Freeman source for example gives a 'hash-free' output up to a certain magnetic field strength. Above this value a serious degradation in beam quality occurs which will cause an increased extracted beam emittance and defocussing of beams which rely on space-charge neutralisation for efficient transmission through a resolving slit.

## 11.4 Arc Chamber Pressure

The mean free path for a primary electron in a plasma is a function of pressure. The dynamic balance of ionisation and recombination events in the plasma is clearly going to be pressure dependent. The optimum pressure will be different for various charge states, the high charge state being favoured by low pressure (to minimise charge exchange reactions), and molecular ions generally optimise at relatively high pressures.

For normal implanter operation it is usually best to run at the lowest pressure compatible with stable operation and adequate output. There are a number of reasons for this:

1. feed material economy
2. low gas pressure means low gas flow through the extraction aperture and therefore decreased pumping speed requirement
3. low pressure for condensable feed materials reduces the frequency of cleaning operations in the ion source chamber
4. sources generally operate more predictably when the electron current from the filament is space-charge limited. This cannot occur if the pressure is too high.

For a given set of source parameters there is a pressure below which the arc will extinguish. Care must be taken not to operate too close to this limit as this can result in 'hashy', unstable and unpredictable performance.

## 11.5 Feed Material

The feed material has an obvious effect on the composition of the output spectrum from the source. It can also have a significant effect on the operating characteristics of the source. Some feed materials can have a dramatic effect on the cathode performance. This can take the form of activation (increased electron current at a given temperature) or deactivation. The dynamic balance between feed material contamination of the cathode surface and removal by sputtering can result in erratic electron emission characteristics. This is one of the reasons for running sources in the space-charge-limited cathode mode. Chemical reaction with the filament can cause a progressive deterioration in cathode performance.

A good example of the above phenomena is found when boron ions are to be produced from a hot cathode source using boron trichloride as the feed material. Tantalum filaments degrade rapidly due to swelling and loss of emission. Tungsten filaments are much better but the operating pressure is critical. If the operating pressure is above a certain critical value then the contaminating reaction is not compensated by the cleaning effect of sputtering and the filament degrades and soon becomes irreversibly deactivated. For this reason boron trichloride has generally been abandoned as a source of boron ions from hot cathode sources in favour of boron trifluoride.

## 11.6 Temperature

Source temperature is a very important consideration when condensable feed vapours are used. Condensation of the feed material can cause breakdown on insulators and lead to flake formation on the walls of the arc chamber which in turn can lead to erratic behaviour. Flake formation and surface modification can also be caused by chemical corrosion and is probably one of the major causes of the inconsistency in behaviour so characteristic of ion sources. The rate of these chemical changes on the surfaces exposed to the plasma tends to increase with increasing temperature.

## 11.7 Source Condition

The previous section has introduced the importance of the state of the surfaces exposed to the plasma. This leads us to the pheonomenon of "conditioning".

Consider a source which has been thoroughly cleaned and a new filament installed. When the performance of the source is investigated for a particular ion species it is invariably found that the performance improves with time. This is particularly noticeable with the ions of elements with high ionisation potentials and/or low ionisation cross-sections. The reasons are not well understood. There are a number of possible mechanisms:

1. contamination of the feed material by components either outgassed, sputtered or formed by chemical reaction from the structural materials of the ion source. These dilute the required ion by adding extra components to the output spectrum. If they have low ionisation potentials and/or high ionisation cross-sections then they are preferentially ionised

2. the development of a protective, chemically compatible layer over source components which keeps contaminants out and only introduces into the plasma compounds related to the source feed material, which are likely to contain the atoms of the required ion

3. cleaning, by sputtering, of the cathode surface.

For ions such as boron the above conditioning process can lead to significant performance improvement with time. A 50% increase in the boron singly charged ion output after one hour of operation is often found.

After a period of operation the reliability of the source will tend to degrade due to metallisation of insulators (if they are exposed to the plasma) and due to flake formation in the arc chamber. The first problem can be minimised by designing the source in a way which gives the maximum protection possible for these insulators. The formation of flakes of electrically insulating material due to corrosion and sputtering effects can result in charging problems which lead to plasma instabilities.

## 11.8 Extraction Voltage

This is strictly speaking a beam formation rather than an ion source variable. Nevertheless we should always judge ion source performance in terms of the ability to produce a beam which can be transmitted through the resolving slit. There is little point in extracting an intense beam from the source if it cannot be efficiently transmitted through the implanter system.

The most important factor governing the extraction voltage used is the compromise between maximum space-charge-limited current capability (favouring high extraction voltage) and the acceptable sparking behaviour across the extraction gap. Another factor is the need for the extraction field to be strong enough to limit the convex curvature of the plasma meniscus so that beams with acceptable divergence can be produced. At low plasma densities an excessive extraction voltage can produce excessive concave curvature which can also lead to transmission losses.

## 12. Ion Source Feed Materials

The feed materials and the appropriate delivery techniques can be divided into four groups:

1. high vapour pressure materials i.e. gases
2. low vapour pressure materials i.e. materials with a vapour pressure of > $10^{-3}$ torr at 1100°C
3. very low vapour pressure materials that can conveniently be introduced into an arc discharge by sputtering
4. chemical synthesis.

### 12.1 Gaseous Feed Materials

Approximately 10% of the elements are gaseous at room temperature and can readily be introduced into the source through a needle valve or similar metering device. Some of these elements (e.g. the halogens) are corrosive and the gas delivery system must be manufactured from appropriate corrosion-resistant materials.

For ions which cannot be obtained from the gaseous element the alternative is to use a gaseous molecular compound. The choice of compound is determined by a number of considerations:

1. a compound with a high content of the required species is clearly desirable if good yields $\left(\text{i.e. } \frac{\text{current of required ion}}{\text{total current from source}}\right)$ are to be obtained i.e. simple compounds
2. a compound with the lowest possible molecular weight is desirable in order to maximise the space-charge-limited current. (See equation (6).)
3. the other elements in the compound should preferably have high ionisation potentials and low ionisation cross-sections
4. the chemical properties of the compound are important in determining the susceptibility to corrosion of source compounds and activation and deactivation of the cathode.

### 12.2 Low Vapour Pressure Materials

The highest temperature that can conveniently be achieved in a simple ion source design is approximately 1100°C. As arc discharges generally require an arc chamber pressure in the range $10^{-4}$ to $10^{-2}$ torr, this group of feed materials will arbitrarily be defined as those solids and liquids with a vapour pressure > $10^{-3}$ torr at 1100°C. About 23% of elements fall into this group and a large number of compounds, particularly the halides.

The considerations governing the choice of feed material are similar to the gaseous feed materials, but the corrosion problem is likely to be more severe at the higher operating temperatures. Care must be taken to ensure that all parts of the arc chamber operate at a sufficiently high temperature to prevent coating of insulators and the formation of flake on the arc chamber surfaces. Materials at the high vapour pressure end of this range also require particular attention. Thermal radiation from the hot plasma can cause temperature control problems if the required evaporation temperature is low, say $<300°C$. This can be achieved by radiation shields, heat sinks or low conductance tubing between the vaporiser and the arc chamber.

The chlorides are a particularly important group of feed materials. It is essential that the metallic chlorides are completely anhydrous and as many chlorides are very hygroscopic care should be taken to ensure that the compound is stored in suitable manner for the efficient exclusion of water vapour. The dehydration of hydrated chlorides is not always a simple process and conversion to oxychlorides and oxides often occurs when hydrated chlorides are heated in vacuum.

## 12.3 Very Low Vapour Pressure Materials

There are a group of materials, notably the precious metals, which have a very low vapour pressure and do not have convenient stable compounds which can be used as feed materials. A suitable technique for such materials when they are electrical conductors is to use the material as a high negative voltage (1 - 2kV) electrode in the arc chamber, thus introducing the material into the plasma by sputtering.

## 12.4 Chemical Synthesis

The problems already mentioned with the use of anhydrous chlorides have led to the development of in-source chemical synthesis. In this technique a suitable reagent (usually a halide) is passed over the heated metal or oxide and the reaction products fed into the ion source arc chamber.

A useful example of this technique is the chlorination of heated oxides using carbon tetrachloride vapour [32]. To ensure efficient reaction it is normally essential to provide an impedance between the reaction oven and the discharge to maintain an adequate pressure of vapour over the oxide surface. Care must also be taken to prevent the resulting metal chloride from backstreaming along the vapour feed line and condensing, causing a blockage. Temperature control is not critical, provided it is sufficiently high to give the required reaction rate, which can be controlled by the rate of flow of the reactive gas.

There are problems with this technique. The carbon tetrachloride can seriously deactivate and erode the filament of a hot cathode source. The output spectrum is complex, which limits the yield of the required ion and may require a high resolving power in the accelerator mass analysis system in order to ensure a pure beam.

Another similar technique has been developed for the halogenation of the platinum-group metals [33]. This technique places the metal in the arc chamber of the source and the arc is fed with $ClF_3$ vapour. Nickel is a convenient structural material for such a system as it readily passivates after exposure to fluorine.

## 12.5 Feed Materials

A list of feed materials is given in Table 1. The principal sources of information have been the Oak Ridge and A.E.R.E. Harwell separation groups [34, 35]. The temperature for $10^{-3}$ torr vapour pressure is included as a guide to the necessary minimum operating temperature of the ion source arc chamber. A more complete list of feed materials can be found elsewhere [21, 36].

## 13. Materials of Construction

Probably the most important aspect of the practical engineering of an ion source concept is the choice of the appropriate materials. The physical and chemical reactions that take place on surfaces exposed to the plasma can be complex and will have an important effect on the lifetime, reliability and consistency of the source.

The factors governing the choice of material are:

1. strength
2. useful temperature range
3. corrosion resistance
4. sputtering coefficient
5. machinability
6. thermal conductivity
7. thermal expansion
8. cost
9. availability.

The materials problems can be divided into the following categories:

1. arc chamber materials
2. cathode materials
3. insulator materials
4. magnetic materials
5. electrical conductors
6. coating materials.

The following sections will outline the most suitable materials for the various components but it must be stressed that the choice is heavily dominated by the actual ion required and feed material to be used. In some cases the corrosion problems may be so severe that a short useful life for the source components is inevitable.

## 13.1 Arc Chamber Materials

### 13.1.1 High-Density Carbon

This is one of the favoured materials for arc chambers. It is suitable for operation at temperatures up to 3000°C, is easy to machine and is surprisingly tough and durable. It has excellent sputtering properties and good corrosion properties. The only drawback is the ease with which many materials diffuse into carbon and this can give outgassing problems and sometimes causes troublesome "memory" effects when using a source with a succession of different feed materials.

Table 1. Feed materials for arc discharge sources

| Element | At. Wt. | Feed Mat. | Form | Temp. (°C) for $10^{-3}$ torr | Maximum Ionisation cross-sec. ($10^{-16}cm^2$) | Ionisation potential (eV) | Comments |
|---|---|---|---|---|---|---|---|
| Aluminium | 26.98 | Element | Solid | 1082 | 6.2 | 6.0 | Liquid at evaporation temp. |
| | | Al $Cl_3$ | Solid | | | | Volatile solid-sublimes at 180°C |
| | | Al $F_3$ | Solid | | | | - |
| Antimony | 121.75 | Element | Solid | 475 | 7.2 | 8.5 | Toxic |
| | | $Sb_2O_3$ | Solid | 399 | | | Toxic |
| Argon | 39.95 | Element | Gas | | 2.8 | 15.7 | - |
| Arsenic | 74.92 | Element | Solid | 237 | 5.0 | 10.5 | Toxic |
| | | $AsH_3$ | Gas | | | | Very toxic - extreme care |
| | | $AsF_3$ | Gas | | | | Very toxic - extreme care |
| | | GaAs | Solid | | | | - |
| Beryllium | 9.01 | Element | Solid | 1395 | 3.2 | 9.3 | Toxic (BeO) |
| | | $BeCl_2$ | Solid | 176 | | | |
| Boron | 10.81 | Element | Solid | 1867 | 2.6 | 8.3 | - |
| | | $BF_3$ | Gas | | | | Preferred |
| | | $BCl_3$ | Gas/liquid | | | | Fil. poisoning problems |
| | | $B_2H_6$ | Gas | | | | Unstable |
| Carbon | 12.01 | Element | Solid | | 2.0 | 11.2 | - |
| | | $CO_2$ | Gas | | | | - |
| | | CO | Gas | | | | Best yield, toxic |

Table 1. Feed materials for arc discharge sources (cont.)

| Element | At. Wt. | Feed Mat. | Form | Temp. (°C) for $10^{-3}$ torr | Maximum Ionisation cross-sec. ($10^{-16}cm^2$) | Ionisation potential (eV) | Comments |
|---|---|---|---|---|---|---|---|
| Chlorine | 35.45 | Element<br>NaCl | Gas<br>Solid | 850 | 3.4 | 13.0 | -<br>- |
| Chromium | 52.00 | Element<br>$CrCl_3$<br>$Cr_2O_3+CCl_4$ | Solid<br>Solid<br>- | 1267 | 5.1 | 6.7 | -<br>Very hygroscopic<br>Use freshly prepared low-temp. oxide |
| Copper | 63.55 | Element<br>$CuCl_2$<br>$CuO+CCl_4$ | Solid<br>Solid<br>- | 1130 | 3.8 | 7.7 | -<br>Decomposes controllably to $Cu_2Cl_2$<br>- |
| Gallium | 69.72 | Element<br>$Ga_2O_3+CCl_4$ | Solid/Liquid<br>- | 1007 | 5.9 | 6.0 | Liquid at evaporation temp.<br>- |
| Germanium | 72.59 | Element<br>$GeCl_4$ | Solid<br>liquid | 1257 | 5.7 | 8.1 | -<br>- |
| Gold | 196.97 | Element | Solid | 1252 | 5.9 | 9.2 | Sputter source |
| Helium | 4.00 | Element | Gas | | 0.21 | 24.5 | - |
| Hydrogen | 1.01 | Element<br>$H_2O$ | Gas<br>liquid | | 0.22 | 13.5 | Explosive mixture with air<br>- |
| Indium | 114.82 | Element<br>$InCl_3$ | Solid<br>Solid | 837<br>224 | 7.7 | 5.8 | Liquid at evaporation temp.<br>- |

Table 1. Feed materials for arc discharge sources (cont.)

| Element | At. Wt. | Feed Mat. | Form | Temp. (°C) for $10^{-3}$ torr | Maximum Ionisation cross-sec. ($10^{-16}cm^2$) | Ionisation potential (eV) | Comments |
|---|---|---|---|---|---|---|---|
| Iron | 55.85 | Element<br>$FeCl_2$<br>$Fe_2O_3+CCl_4$ | Solid<br>Solid<br>- | 1342<br>428 | 6.3 | 7.8 | -<br>-<br>- |
| Krypton | 83.80 | Element | Gas | | 4.1 | 13.9 | - |
| Molybdenum | 95.94 | Element<br>$MoO_3+CCl_4$ | Solid<br>- | 2307 | 6.9 | 7.4 | Small beams from Mo fil.<br>Careful temp. control required |
| Neon | 20.18 | Element | Gas | | 0.82 | 21.5 | - |
| Nickel | 58.71 | Element<br>$NiCl_2$ | Solid<br>Solid | 1382<br>491 | 5.5 | 7.6 | -<br>Anhydrous $NiCl_2$ |
| Nitrogen | 14.01 | Element | Gas | | 1.5 | 14.5 | - |
| Oxygen | 16.00 | Element<br>$CO_2$ | Gas<br>Gas | | 1.3 | 13.6 | Oxidation of source and fil.<br>Convenient |
| Platinum | 195.09 | Element<br>$Pt + ClF_3$ | Solid | 1907 | 6.6 | 8.9 | Sputter source |
| Phosphorus | 30.97 | Element (red)<br>$PH_3$<br>$PCl_3$<br>$PF_3$ | Solid<br>Gas<br>Gas<br>Gas | | 4.5 | 10.9 | Convenient<br>Very toxic<br>Filament poisoning problem<br>- |
| Rhenium | 186.2 | Element<br>$Re + O_2$ | Solid<br>- | 2807 | 9.1 | 7.9 | Sputter source or from fil.<br>- |

Table 1. Feed materials for arc discharge sources (cont.)

| Element | At. Wt. | Feed Mat. | Form | Temp. (°C) for $10^{-3}$ torr | Maximum Ionisation cross-sec. ($10^{-16}cm^2$) | Ionisation potential (eV) | Comments |
|---|---|---|---|---|---|---|---|
| Selenium | 78.96 | Element | Solid | 199 | 5.0 | 9.7 | Toxic |
| | | CdSe | Solid | 588 | | | Convenient |
| | | $SeO_2$ | Solid | | | | - |
| Silicon | 28.09 | Element | Solid | 1472 | 5.4 | 8.1 | - |
| | | $SiH_4$ | Gas | | | | Spontaneously inflammable with air |
| | | $SiCl_4$ | Gas | | | | - |
| | | $SiF_4$ | Gas | | | | Convenient |
| Sulphur | 32.06 | Element | Solid | 80 | 3.9 | 10.3 | - |
| | | $SO_2$ | Gas | | | | - |
| | | Cds | Solid | 610 | | | - |
| Tantalum | 180.95 | Element | Solid | 2807 | 9.6 | 7.9 | Small beams from fil. |
| | | $Ta_2O_5 + CCl_4$ | - | | | | Preferred |
| Tellurium | 127.60 | Element | Solid | 323 | 7.1 | 9.0 | - |
| | | CdTe | Solid | | | | $Cd^+$, $Te^+$ resolution problem |
| Tungsten | 183.85 | Element | Solid | 2977 | 9.2 | 8.1 | Small currents from fil. |
| | | $WO_3+CCl_4$ | - | | | | - |
| Xenon | 131.30 | Element | Gas | | 6.2 | 12.1 | - |

### 13.1.2 Tantalum

Suitable for applications up to 2600°C. It is ductile, easily machinable, weldable and readily available in sheet and wire form and is therefore ideal for the manufacture of complex components. It does, however, readily embrittle under most operational conditions and is particularly susceptible to reaction with oxygen and nitrogen at elevated temperatures.

### 13.1.3 Molybdenum

For T<2000°C. Readily machined but cannot easily be welded. Its corrosion and embrittlement properties are generally superior to tantalum and it is significantly lower in cost.

### 13.1.4 Stainless Steel

Commonly used as a general vacuum component material. It has good outgassing properties and is generally suitable for components up to 800°C. It is really only suitable as an arc chamber material for non-corrosive gases.

### 13.1.5 Tungsten

Limited in usefulness by its poor machining characteristics. It recrystallises above 1000°C and becomes very brittle. It does have reasonable resistance to halogens at elevated temperatures but reacts readily with oxygen to form a volatile oxide.

## 13.2 Cathode Materials

### 13.2.1 Tungsten

Tungsten has the best combination of mechanical strength and low vapour pressure at electron-emitting temperatures. Coiled filaments are easily manufactured from tungsten wire up to about 1mm diameter. Tungsten rod is readily available for thicker filaments but fabrication of complex shapes is a problem. Tungsten generally shows a good resistance to filament poisoning.

### 13.2.2 Tantalum

Tantalum has the advantage of easier fabrication. It is readily available in sheet and wire form and it is ductile, weldable and machinable. It is generally less resistant to poisoning by feed materials such as boron trichloride.

### 13.2.3 Rhenium

Extremely expensive. Its superior resistance to halogens and to the formation of low melting point alloys can be an advantage. Rhenium has a low coefficient of resistance compared with tungsten and this helps to minimise excessive temperature and consequent increased erosion rate at reduced cross-section parts of a filament structure.

### 13.2.4 Other Materials

1. Molybdenum
2. Carbon - too fragile for most directly heated applications
3. Coated or activated cathodes - too easily poisoned to be of any general applicability.

## 13.3 Insulator Materials

1. Alumina - an excellent dielectric material
2. Boron nitride - has the advantage of being a machinable insulator suitable for operation up to approximately 2000°C. It decomposes slowly at this temperature. Its main disadvantages are its poor strength, a tendency to react with corrosive feed materials and an ability to absorb large quantities of water vapour. This latter property causes outgassing and if this material is heated too rapidly in vacuum it can lead to catastrophic blistering. A unique feature of this material is its high thermal conductivity which is useful for the heat sinking of electrically isolated components.
3. Silicon nitride - this has superior mechanical properties to boron nitride but it has to be machined in the green state before firing.
4. Machinable glasses - brittle, suitable for $T<1000°C$ in unstressed use.

## 13.4 Magnetic Materials

Iron is almost universally used for magnetic circuits in ion sources. If the poles are exposed in the arc chamber then it may be necessary to have a protective coating. Permanent magnets are rarely used in ion sources, as the magnetic field is usually a critical variable necessary for source optimisation.

## 13.5 Electrical Conductors

Copper is usually used for high-current conductors. A protective coating will often be found necessary as it sputters easily and has a poor resistance to chemical corrosion.

## 13.6 Coating Materials

1. Nickel - gives good protection against halogens.
2. Gold - useful for $T<1000°C$. Good chemical corrosion resistance but easily sputtered.
3. Platinum - similar to gold but higher temperature range ($T<1600°C$).

## 14. Ion Sources in Commercial Ion Implanters

The dominant commercial application for ion implanters has been for the doping of silicon.

The major requirements are:

1. Ions of B,P,As,Sb - dopants
   Argon - gettering
2. Dose range $10^{11}$ - $10^{16}$ ions $cm^{-2}$
3. Simple, reliable ion sources requiring minimum operator skill.

It has been mentioned that implanters can be divided into three groups:

1. low-current machines - beam current limited by the ion source (up to 300μA).

2. medium-current machines - beam current at the target limited by space-charge problems in the electrostatic scanning system (300μA - 3mA).
3. high-current machines - beam current limited by the source (>3mA).

## 14.1 Sources for Low-Current Machines

This group of machines almost universally use cold cathode Penning sources. One of the most commonly found examples of this type of source is the PD-50 source produced by Extrion [37] and has been used in their 200kV and 400kV low-current machines. A schematic of this source is shown in Fig.21a and a photograph of the internal electrode structure in Fig.21b. The extraction cathode and hollow anti-cathode are at body potential and the electrically isolated anode typically runs at <2kV.

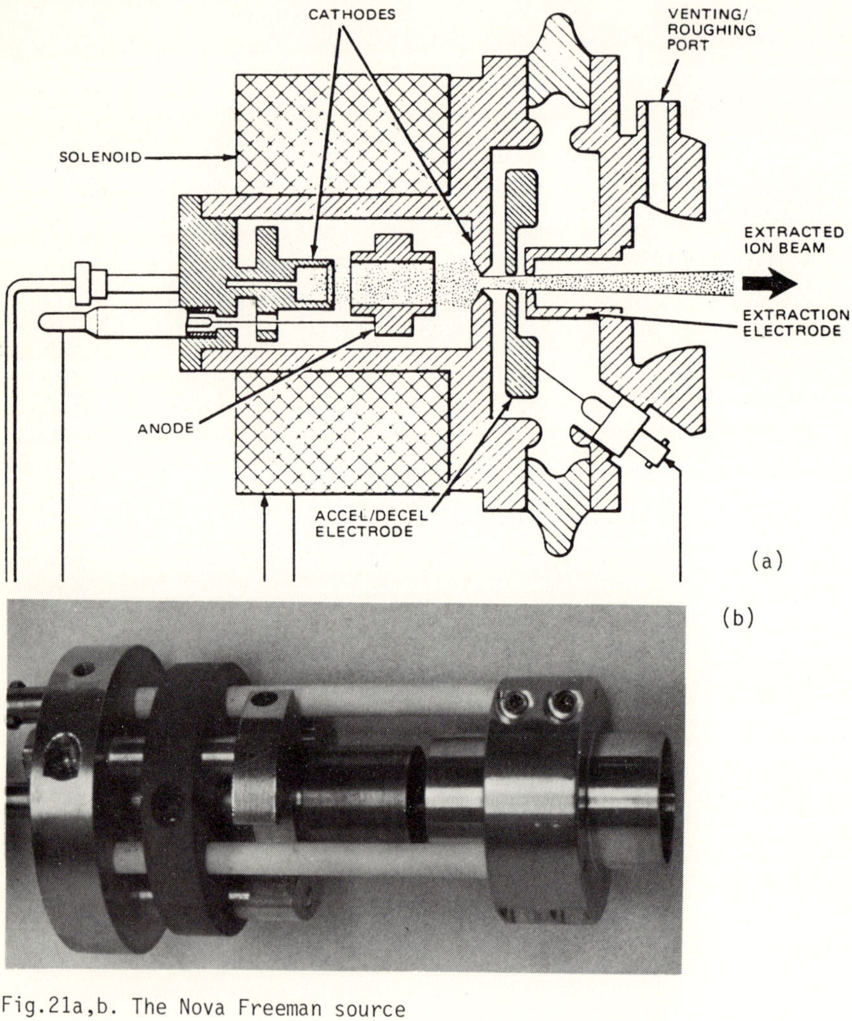


Fig.21a,b. The Nova Freeman source

## 14.2 Medium-Current Machines

These machines use hot filament sources. They generally divide into two groups:

1. axial extraction hot cathode e.g. Accelerators Inc, Balzers [38,39]
2. lateral extraction miniaturised high-current sources e.g. Extrion, Kasper (Nova) [40].

## 14.3 High-Current Machines

The requirement for simple, easily operated, 'hash'-free sources has eliminated most of the exotic sources (such as duoplasmatrons) in favour of the Freeman and Bernas types. Extrion, Nova and A.I.T.(Lintott) [41] use the Freeman source in their high-current machines and Balzers use a Bernas source.

The Freeman source used in the Nova NV-10 machine is shown in Figs.22a and 22b. In Fig.22a the front plate of the source has been removed and

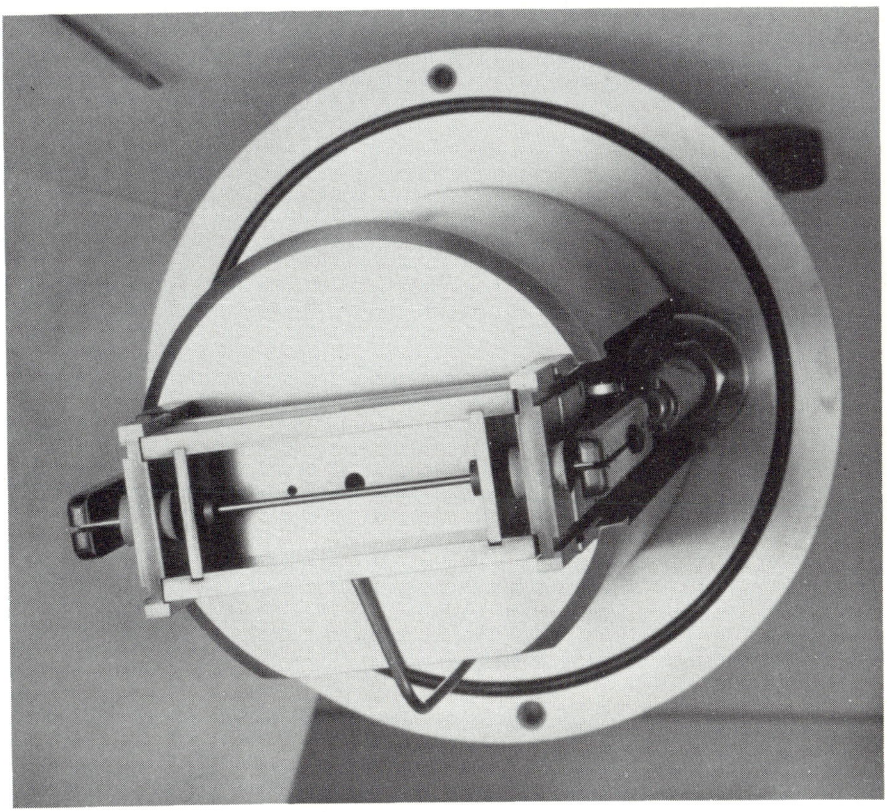

Fig. 22a.The Nova Freeman source

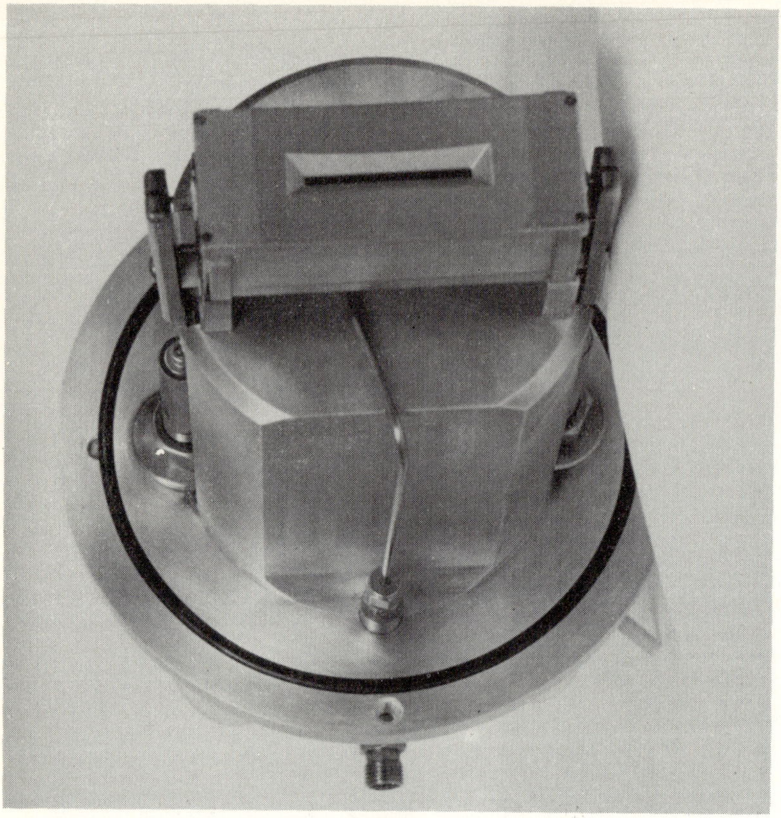

Fig. 22b. The Nova Freeman source

the filament assembly can be seen. The source manufactured by Applied Implant Technology (Lintott) is shown in Figs. 23a and 23b. This source is used in the IIIA machine with a 60 x 2mm extraction slot and in the IIIX with a 90 x 2mm slot. In both the Nova and the A.I.T. sources the front plate of the arc chamber is a removable component with a curved extraction slot geometry (in order to reduce the height of the beam at the analysing magnet). These sources are capable of producing $P^+$ and $As^+$ beams in excess of 10mA. Vaporisers are standard for the Nova and A.I.T. sources (the latter in fact has twin vaporisers) and the Extrion source has the vaporiser as a standard accessory. The vaporisers are primarily required for $P^+$ and $As^+$ from the element and $Sb^+$ from the metal or the oxide. This requires a vaporiser capability of up to 700°C (for antimony) and the arc chamber needs to operate at this temperature in order to prevent feed condensation. $B^+$ and $A^+$ are obtained from $BF_3$ and argon gas feeds.

Balzers use a Bernas source in the Scanibal SCI 218 machine. There are four versions of the source, two are gas feed sources and two vaporiser sources. Fig. 24 shows one of the vaporiser versions.

(a)

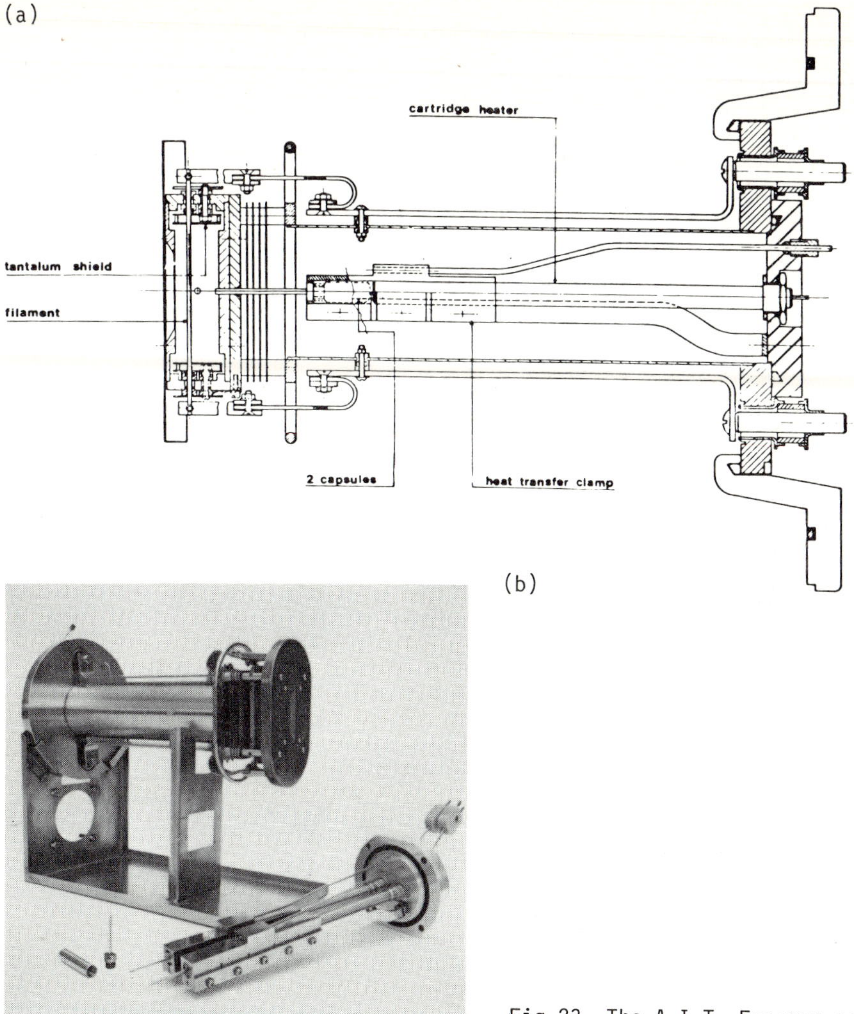


(b)

Fig. 23. The A.I.T. Freeman source

Fig. 24. Balzers Bernas source

## 15. Conclusions

The intention has been to give an introduction to the physics and technology of ion sources. The references given do not necessarily refer to the originators of the ideas and designs under discussion but are chosen for their suitability for further reading. For further more detailed information a number of books and review articles are recommended. These include Freeman [21], Wilson and Brewer [29], Alton [36] and Sidenius [9].

## References


1. D. Bohm: The Characteristics of Electrical Discharges in Magnetic Fields N.N.E.S. 1 - 5, (McGraw - Hill, New York 1949)

2. G. Dearnaley, P.D. Goode, Inst. Phys. Conf. Ser. No. 54, 26 (1980)

3. Separation of Isotopes in Calutron Units, ed., H.W Savage, Nuclear Energy Series (1951)

4. I. Chavet, R. Bernas, Nucl. Instr. & Meth. 51,77 (1967)

5. I. Chavet, M. Kanter, M. Menat, Nucl. Instr. & Meth. 139, 47 (1976)

6. E. Pasztor, Inst. Phys. Conf. Ser. No. 54, 345 (1980)

7. J.R.J. Bennett, Proc. Int. Conf. on Ion Sources, 571 (I.N.S.T.N - Saclay, France)

8. N.J. Freeman, W.A.P. Young, R.W.D. Hardy, H.W. Wilson, E.M. Separation of Radioactive Isotopes, 83 (Springer - Verlag, Vienna 1961)

9. G. Sidenius, Inst. Phys. Conf. Ser. No. 38, 1 (1978)

10. P.H. Rose and A. Galejs, Prog. In Nucl. Techniques and Instrumentation 1965

11. R. Masic, R.J. Warnecke, J.M. Sautter, Proc. Int. Conf. on Ion Sources, 387 (I.N.S.T.N - Saclay, France 1969)

12. B.H. Wolf, Nucl. Instr. & Methods 139, 13 (1976)

13. G. Sidenius, Proc. Int. Conf. on Ion Sources,401 (I.N.S.T.N - Saclay, France 1969)

14. G. Sidenius, Proc. Int. Conf. on Electromagnetic Isotope Separators and Techs. of Apps (1970), 423, Forschungsbericht K70 - 28

15. K.J. Hill, R.S. Nelson, Nucl. Instr & Meth. 38,15 (1965)

16. G.J. Rushton, K.R. o'Shea, R.K. Fitch, J. Phys. D: Appl. Phys. 6 , 1167 (1973)

17. L. Romathiod, D. Henry, Y. Arnal, R. Boswell, Inst. Phys Conf. Ser. No. 54, 309 (1980)

18. C.J. Cook et al., Rev. Sci. Instr. 33, 649 (1962)

19. N. Sakudo, K. Tokiguchi, H. Hoike, I. Kanomata, Rev. Sci. Instr., 49, 940 (1978)

20. V.S. Venkatasurbramian, H.E. Duckworth, Can.J. Phys. 41, 234 (1963)

21. J.H. Freeman, Ion Implantation, 366, North Holland Pub. Co. (1973)

22. C.D. Moak, H.E. Banta, J.N. Thurston, J.W. Johnson, R.F. King, Rev. Sci. Instr. 30,694 (1959)

23. A.B. Wittkower, R.P. Bastide,N.B. Brooks, P.H. Rose, Phys. Letters 3, 336 (1963)

24. J.H. Freeman, W. Temple, D. Chivers, A.E.R.E. Harwell Report No. M2491 (1971)

25. R. Geller, IEEE Trans. Nucl. Sci NS-23, 2, 904 (1976)

26. R.D. Donets, IEEE Trans. Nucl. Sci NS-23, 2 , 879 (1976)

27. J.D. Dougherty, J.E. Eniger, G.S. Janes, R.H. Levy, Proc. Int. Conf Ion Sources, 643 (I.N.S.T.N - Saclay, France 1969)

28. G.R. Brewer, Ion Propulsion (Gordon and Breach, New York, 1970)

29. R.G. Wilson, G.R. Brewer, Ion Beams with Application to Ion Implantation, 73 (Robert E. Krieger, New York, 1979)

30. H.L. Daley, J. Percl, R.H. Vernon, Rev. Sci. Instr. 37, 473 (1966)

31. A. Von Engel, Ionised Gases, 63 (Clarendon Press, Oxford 1965)

32. G. Sidenius, O. Skilbreid, Electromagnetic Separation of Radioactive Isotopes, 243 (Springer-Verlag, Vienna 1961)

33. E.W. Newman, et. al, Nucl. Instr & Meths. 139, 87 (1976)

34. Electromagnetic Isotope Separations (ORNL Manual)

35. J.H. Freeman, G.A. Gard, W. Temple, A.E.R.E. Harwell Report No. R6758

36. G.D. Alton, Nucl. Instr. & Meths. 189, 15 (1981)

37. Extrion Division, Blackburn Industrial Park, Gloucester, MA 01930, USA

38. Accelerators Inc, Austin, Texas, USA

39. Balzers, Aktiengesellschaft, Fürstentum, Liechtenstein

40. Eaton Semiconductor Equipment, Ion Implantation Division (Nova), Beverley, MA01915, USA

41. Applied Implant Technology Inc., Foundry Lane, Horsham, West Sussex, England: Kifer Road, Santa Clara, California, USA

# Faraday Cup Designs for Ion Implantation

Charles M. McKenna

Hughes Research Laboratories
Malibu, CA 90265, USA

## 1. Introduction

As a general materials processing approach, ion implantation possesses tremendous leverage due to its insensitivity to chemical reactions exhibited by bulk material diffusions, its extendability to large target dimensions, and its ability to be mask-defined with small lateral spreading and controlled depths [1-3]. In its practical execution, ion implantation employs the electromagnetic properties of the ions to regulate the energy, species, total flux and areal distribution of the dopant material. A given target response to the incident ions can be used as a continuous monitor of the implant status and, under proper conditions, can provide a means for closed-loop process control.

While several target responses have been investigated and accurate measurement techniques demonstrated [4-7], by far the most commonly used ion implantation monitor is the ion-beam-induced current flow between the target and ground. This approach is preferred because of its ease of implementation, its compatibility with a variety of target materials, and its applicability over a wide range of beam energies and currents.

In this chapter we will review the basic assumptions implicit in an electrical measurement approach, some of the conditions which can invalidate these assumptions, and the general principles involved in Faraday cup design for current measurement. We will then look at the application of these principles for accurate dose measurement on the two standard types of ion implantation systems; i.e., where the dopant material is distributed (a) by scanning the ion beam across the target (low/medium current) or (b) by scanning the target through the beam (high current). Since most processes are dependent not only on total dose, but also on dose uniformity, we will also look at techniques for obtaining in situ information and/or controlling the spatial variation of the dose on the above systems. In general we will assume that measurement of total dose implies a Faraday cup which integrates the current over the entire implant area of interest. Sampling techniques, in which total dose is inferred by area normalization, are usually designed for control of dose uniformity, and will be covered in the discussion of that topic.

## 2. Dose Control by Current Measurement

### 2.1 Assumptions

In its most simplistic form, electrical measurement of ion implantation dose involves placing a current integrator between the target and ground, and counting the total charge collected from the incident ions. This collected charge is then related to the total dose by means of the charge on each incoming ion. That is

$$\text{Dose} = \frac{Q_{TOTAL}}{q_{ion}} = \frac{1}{q_{ion}} \int_0^T I_B \, dt \ . \qquad (1)$$

Since most implanted samples are characterized over only a small portion of their surface, area-normalized dose values are generally used. Thus

$$\text{Dose/cm}^2 = \frac{1}{A(\text{cm}^2)} \int_0^T \frac{I_B}{q_{ion}} \, dt \ . \qquad (2)$$

There are several assumptions implicit in the relationship expressed in Eq.(2). Most obviously, this relationship is based on the contention that the current $I_B$ measured on the meter or integrator accurately reflects the dopant-material arrival rate. That is

$$I_B \equiv q_{ion} \frac{d\,N(\text{dopant})}{dt} \ . \qquad (3)$$

Thus, all arriving atoms are ions of charge $q_{ion}$ and are ions of the correct dopant species. Less obviously, this identity also assumes that all the incident ions are retained in the implanted sample and that all the incoming charge flow, and only that charge flow, constitutes the measured current $I_B$. Finally, Eq.(2) implies that the electronics correctly integrate the current, responding to all temporal variations in intensity, and that the implant area, A, is well determined.

## 2.2 Discrepancies

It is apparent to anyone who has worked with ion implanters that the above assumptions are not always (and, in some cases, never) achievable in actual system performance. Therefore, the accuracy of such measurements should be treated with a healthy degree of skepticism based on an understanding of the various conditions which can invalidate the relationship expressed in Eq.(2). These conditions can be separated into three categories: (a) inadequacies in the beam transport system resulting in an incoming beam consisting of a mixture of ion species, charge states and/or energies, or in an uneven distribution of the beam over the target; (b) physical interactions between the ions and the target; (c) improper design or operation of the Faraday cup and electronics used to integrate the ion beam current. We will not attempt a thorough discussion of the first two categories which have been reviewed in detail elsewhere [1,2][8-10]. However, we will briefly review some examples of which operators and users of ion implantation equipment should be aware.

Within the generation and selection process, the beam can become "contaminated" with ions of similar charge/mass ratios and of identical energy, and also by ions of different energy created by charge exchange or dissociation of molecular ions (e.g., Aston bands) [1,2,8]. Proper setup and operation of the optics of most ion implanters generally provide adequate mass resolution (typically ≈ 1 part in 100). However, certain situations may require special precautions. For example, most mass analyzers will not resolve a $^{28}Si^+$ beam from $N_2^+$ or $CO^+$, so the operator must be careful to flush the system prior to start-up to reduce the level of background atmospheric gases. Similarly, when implanting with multiply charged ions, the operator should maintain the lowest possible pressure in the pre-analysis region to minimize the probability of charge exchange. Charge exchange in the post-analysis region of the beamline can produce a neutral beam component [11,12]. These neutrals not only destroy the correlation between integrated current and total dose, but for those systems which distribute the dopant over the target by electrostatic scanning, they can introduce severe dose non-uniformity [4,8,10]. For this reason, such systems incorporate a final steering plate which diverts the ions away from the pre-scan beam axis. This final deflection "traps" out the unscanned neutral beam component (Fig.1). Additional scan-related sources of dose non-uniformity, such as non-linearites in the scan waveform, target tilt, incident-beam-angle changes during scan, pattern overlap, etc. have been addressed in a number of papers [13-17]. The reader is directed to these references for details.

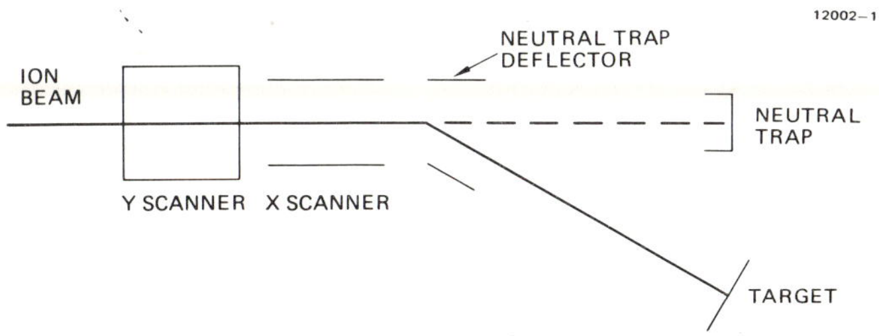


Fig.1. Neutral beam trap for electrostatically scanned beam

In the second category of effects mentioned above, dose discrepancies can arise due to the reflection of the incident ions [9,18] and sputtering of previously implanted dopant atoms by subsequent ion impact [8,18,19]. Thus, the assumption that all incident ions are retained in the sample is never truly achieved. The former effect generally dominates for light ions while the second becomes significant for heavy ions implanted at high doses. Generally these effects are small (< 1-5%) and do not affect uniformity. Desired retained dose levels can be adjusted empirically.

The final category, consisting of those effects involving the Faraday cup design and operation, will be covered in detail in the following sections. First, we will itemize the most common sources of error, and where applicable, provide some general description of the mechanisms involved. Based on this information we will establish some guidelines for Faraday cup design. Then we will look at the implementation of these guidelines for systems having either scanned beams or scanned targets. Finally, we will consider Faraday cup designs which can provide direct information on dose uniformity and their potential use for real-time implant monitoring or closed-loop feedback control.

## 3. Limitations in Dose Measurements with Faraday Cups

The principal factors which can adversely affect the performance of a Faraday cup are: (1) space charge of the ion beam; (2) secondary and tertiary particle emissions from the target and surrounding beamline elements; (3) inaccurate computation of the implant area; (4) improper electrical isolation of the Faraday cup; and (5) variations in the continuity or response of the current integration circuit.

## 3.1 Ion Beam Space-Charge Effects

The charge carried by the ions sets up a potential difference between the center and edge of the beam. This potential difference can be calculated [20] for a long cylindrical beam of radius R according to the formula:

$$\Delta\phi_{beam} = \int_0^R E_r \, dr \qquad (4)$$

where $E_r$ is the radial electric field within the beam. For a beam with uniform current density j (= $I/\pi R^2$), $E_r$ can be determined from Gauss's law:

$$\int \vec{E} \cdot d\vec{S} = \int 4\pi \frac{j}{v} \, dV \qquad (5)$$

where v is the ion velocity. As a result:

$$E_r = 2\pi r \frac{j}{v} \qquad (6)$$

and

$$\Delta\phi = \frac{I}{v} \quad . \qquad (7)$$

Thus, for a given ion current, this potential will be largest for low-energy, heavy ions and smallest for high-energy, light ions. For the milliampere-current heavy-ion beams used in ion implantation, potentials of several hundred volts can be readily attained.

In a perfect vacuum, the radial electric field would produce a transverse acceleration of the ions, observable as a "blow-up" of the ion beam. The resultant reduction ion beam density would, in turn, reduce $E_r$. In practice, the beam, passing through the residual gas in the beamline (molecular density ≈ $10^{11}/cm^3$ at $10^{-6}$ Torr), ionizes a certain fraction of the molecules. The "space-charge" potential expels the ionized molecules while trapping the electrons. Also, secondary electrons created by the beam striking collimating apertures, mass-analysis slits, etc. also are captured in the beam. By this process, the beam achieves a certain level of "space-charge" neutralization [21,22]. That is, the ions have not captured electrons and become fast neutrals (such as occurs in charge exchange) but rather the fast-moving ions are shielded from the electric field due to neighboring ions by a background density of relatively slow-moving electrons. In this fashion, the potential $\Delta\phi$ is reduced:

$$\Delta\phi = \eta \frac{I}{v} \qquad (8)$$

where $\eta$ represents the fraction of the beam whose space charge is not neutralized [22]. Thus, some level of space-charge neutralization is beneficial for the efficient transport of the high-current ion beams now prevalent in ion implantation.

However, in order to achieve an accurate assessment of the dopant ion flow into the Faraday cup, the trapped electrons must be removed from the beam prior to its entrance into the cup. This function is usually performed by a negatively biased electrode immediately beyond the aperture which defines the Faraday cup opening, or by a transverse magnetic field applied in the same region. These two arrangements are shown in Fig.2(a) and (b).

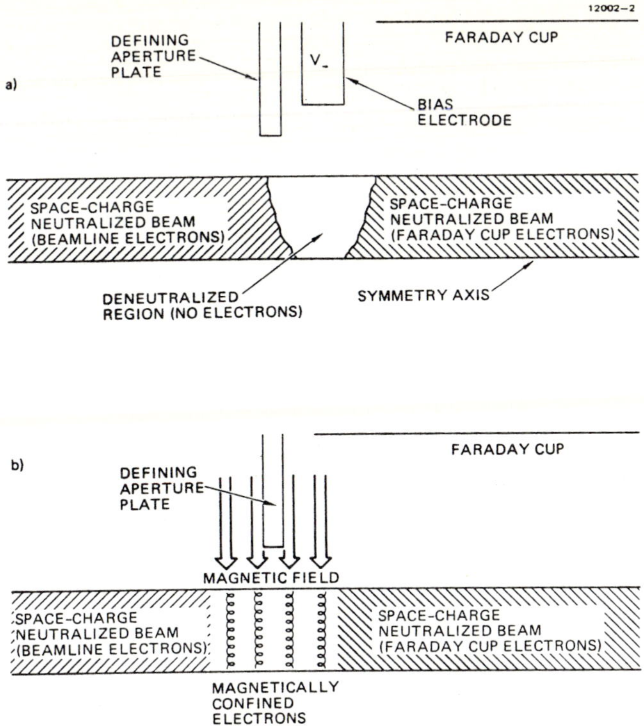


Fig.2. Beam deneutralization (a) by electrostatic potential (b) by magnetic field

With the biased electrode approach, two points should be noted. First, the bias on the electrode must overcome both the potential within the ion beam and the potential between the ion beam and the electrode surface. The latter term is:

$$\Delta\phi = 2\,\frac{I}{v}\,\ln\left(\frac{A}{R}\right) \tag{9}$$

for a cylindrical beam of radius R, where A is the radius of the electrode opening. This term increases linearly with beam current, and can easily be larger than the potential drop in the beam (Eq.(7)) particularly for electrodes on systems using scanned beams to implant large-area targets, 5" to 6" in diameter. The current dependence of this term and its impact on the effectiveness of a deneutralizing electrode potential are illustrated in Fig.3(a), (b), (c) [23]. A 1 cm radius, 100 keV $As^+$ ion beam is shown passing through a series of apertures and an electrode biased at -500 V at the entrance of a Faraday cup. (As shown, the horizontal scale is ≃ 1/4 the vertical scale). For a current of 10 μA (Fig.3(a)), all electrons with energies <100 eV are effectively gated out of the beam. As the current increases (Fig.3(b)), the beam space-charge potential begins to perturb the electrode potential until, at a current of 1 mA (Fig.3(c)), essentially thermal electrons (E < 5 eV) are free to move throughout the length of the beam. Thus, care must be taken that the electrode

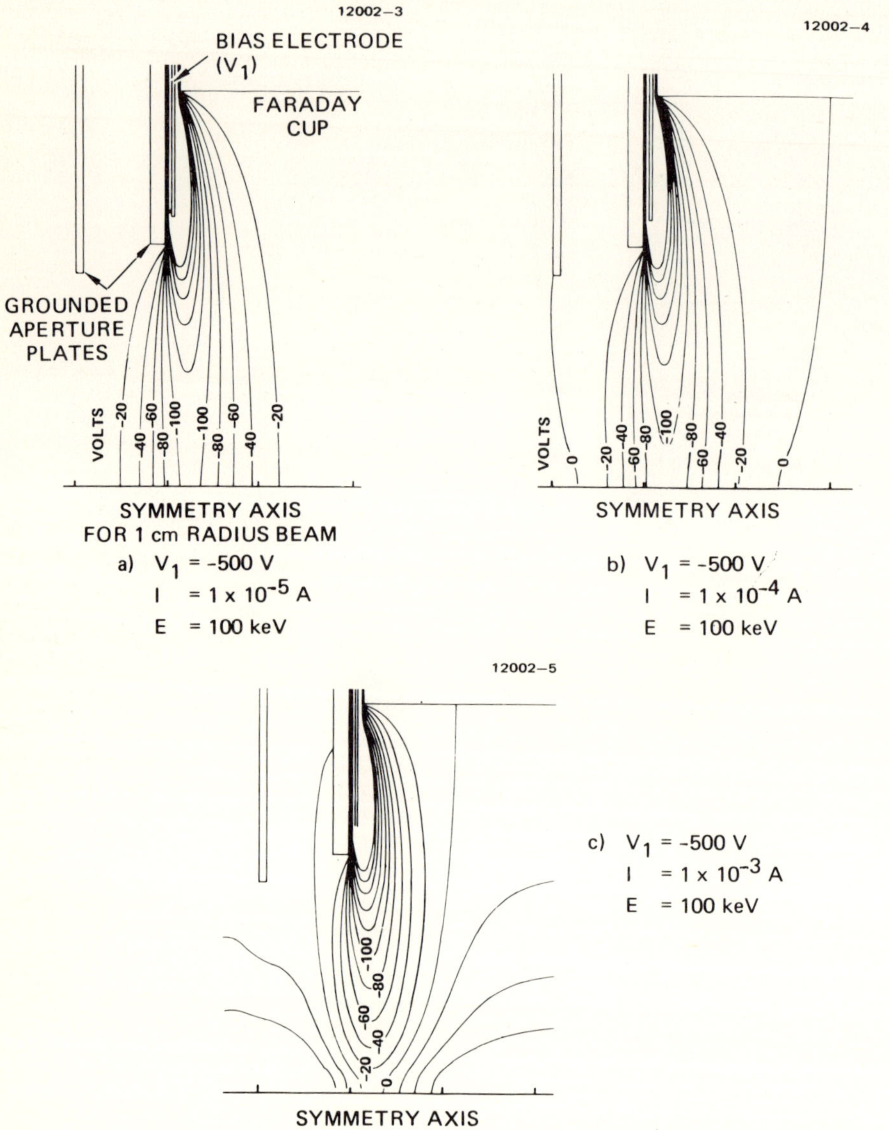


Fig.3. Effect on potential distribution of biased electrode from space charge of 1 cm radius ion beam of current: (a) 10 μm; (b) 100 μA; (c) 1 mA

configuration and bias will function properly with beams of the maximum machine-rated current density.

A second point is that the region over which the beam is deneutralized should be kept to a minimum in order to limit increases in beam size due to "space-charge blow-up". Significant changes in beam size require larger scan fields, thus reducing the effective beam current, and can lead to uncertainty in the implanted target area. Therefore, the biased electrode must be shielded from the beam upstream of the cup entrance (as in Fig.3) to avoid deneutralizing the beam over an extended region. Also, if this electrode is looped in the charge collection circuit (as described later) this shielding will prevent it from collecting slow ions created in the ionization of the residual gas by the beam in this upstream region. Such ions would contribute an erroneous component to the current measurement.

For a transverse magnetic field, the field strength must be sufficient to bend electrons with the maximum kinetic energy in a radius ($\rho$) smaller than the separation between the defining aperture and the Faraday cup, or smaller than the distance between the beam and the Faraday cup wall, whichever is less. For $\rho$ = 1 cm and $E_{electron}$ = 100 eV, B = 34 G in the beamline. The magnetic field approach has the advantage that the beam can maintain some degree of space-charge neutralilzation throughout the beam path, if the fields are not too strong [22]. The principal difficulty with this technique, particularly for scanned beams requiring large beamline apertures, is confinement of the magnetic field to a reasonably small region. Fringing fields from magnetic poles separated by a gap d tend to extend distances as large as 2-3 d on either side of the uniform field region [24]. Conditions of reduced beam neutralization due to the impeded mobility of the trapped electrons can still permit significant space-charge blow-up when they exist over such large distances. It is possible for some of the electrons to cross the magnetic field lines by collision-induced diffusion, but there will be no preferred direction to this flow (time-averaged value equal zero) if the level of beam neutralization is the same on both sides of the field region. Thus, both the magnetic and electric field techniques, properly applied, also serve to prevent electrons trapped in the beam, on the cup side of the field region, from leaving the cup.

## 3.2 Secondary and Tertiary Particle Emissions in the Faraday Cup

When the ion beam impinges upon the target, several effects can be observed. Secondary electrons, sputtered neutral and ionized (both positively and negatively) atoms, photons, and adsorbed gas molecules are emitted due to energy transfer from the incident ions to the target surface (Fig.4). In addition, under conditions where the bulk target temperature is significantly increased by the absorbed ion beam power, vapors evolved in the heating process are also released. The latter situation is particularly prevalent for targets with polymer coatings such as photoresist (Fig.5).

Secondary electron emission coefficients (number of electrons emitted/incident ion $\equiv \gamma_{SE}$) for materials typically used in targets and target holders are of the order of 2-20 at ion energies common for ion implantation [25,26]. The secondary electrons are dominantly of low energy (most probable energy < 5 eV) but the energy distribution can extend out to energies of several hundred eV [26]. For all practical cases, less than 0.1% of all the electrons have energies greater than 100 eV. The secondary electrons generally will be captured by the space-charge potential of the ion beam, if it is not already neutralized. Otherwise, in the absence of

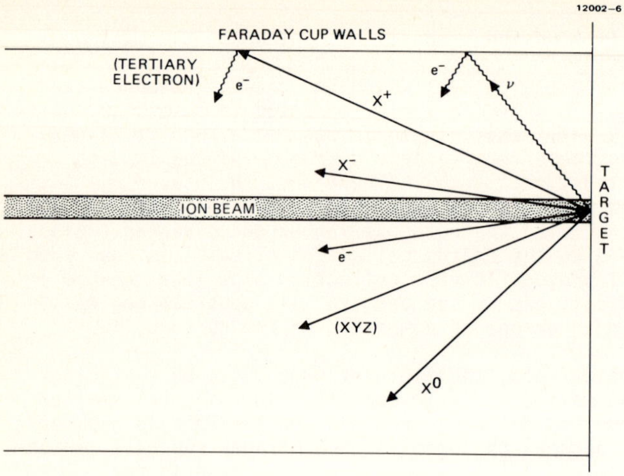


Fig.4. Secondary and tertiary emissions during ion bombardment

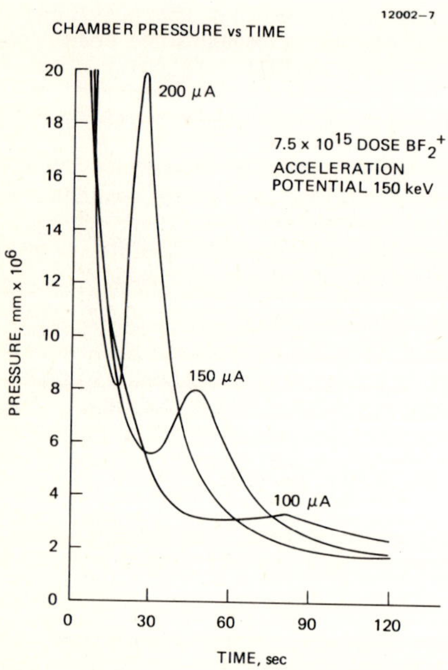


Fig.5. Pressure changes due to outgassing of photoresist during ion implantation at various ion beam current levels

any applied electric or magnetic fields, they will travel in straight lines away from the target back down the beamline on into the surrounding walls. Obviously, if these electrons are not collected, the measured target current will be grossly in error, i.e.,

$$I_{meas} = (1 + \gamma_{SE})\, I_{beam} \quad . \tag{10}$$

The loss of secondary electrons not only destroys the accuracy of the current measurement process but, if the target is non-conducting, it results in a rapid charge buildup on the target surface. Theoretically an insulating surface can charge up until it reaches a potential equal to the beam-accelerating voltage. Practically, however, once the target surface potential exceeds the beam space-charge potential, it will deneutralize the beam by drawing off the trapped electrons. As a result, the beam will then expand due to its own space charge. This expansion can result in the beam striking a conducting surface which then acts as a source of secondary electrons for neutralization. If not, the target will continue to charge until it arc-discharges to a conducting surface (Fig.6) or until further increases in the beam space charge, due to the decelerating effect of the target potential, produce enough beam divergence for it to reach a conducting surface. As a rule, such charging is not acceptable for ion implantation processes, and can be reduced to acceptable levels by returning the secondary electrons to the target surface. For certain combinations of high-current beams and sensitive dielectric layers, the target surface must be forcibly neutralized. At present, these conditions are unique to systems with mechanically scanned targets, and the problems associated with surface charge neutralization (or electron flooding) will be covered in the section on these systems.

Uncollected sputtered ions also will result in erroneous current measurements. While these ions constitute only a few percent of the sputtered atoms, the sputtering ratio (S = atoms/ion) can easily be ~ 20 for ion-target combinations and beam energies used in ion implantation [27], as illustrated in Fig.7. Thus, errors due to uncollected sputtered ions can run as high as 20% [12,28]. These errors are compounded

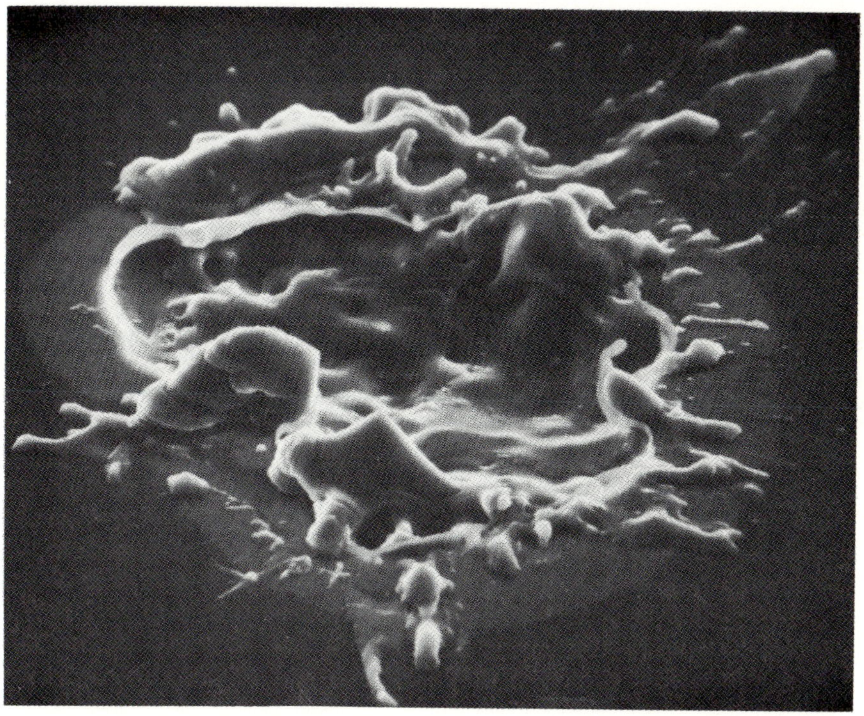

Fig.6. Insulated layer damage caused by arc discharge of surface charge produced during ion implantation to underlying substrate

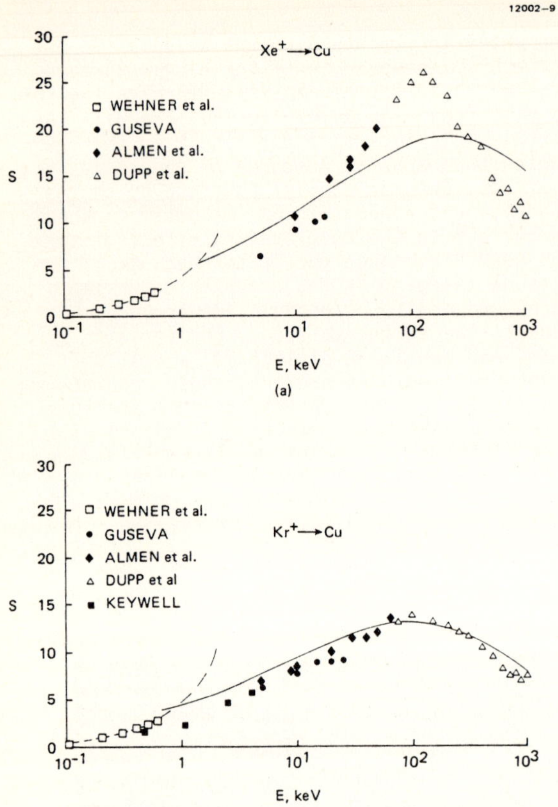


Fig.7. Sputtering yield, s(atoms/ion), for copper substrate as a function of incident energy of Xe and Kr ions (Reference 27 )

by tertiary processes, particularly electron emission, which occur when the secondary particles strike surrounding beamline components or walls (Fig.4).

Several techniques have been applied to eliminate the loss of these secondary charged particles, involving the mechanical configuration of the cup as well as the application of electrical and/or magnetic fields near the target surface. The traditional structure of a Faraday cup as a long tube of small diameter ($L/D \approx 10$) was developed on accelerators which generated small, tightly collimated, high-energy beams of low-mass particles for use in atomic and nuclear physics. In this structure, the walls are electrically in common with the target and the efficiency of secondary/tertiary particle collection is based on the small solid angle for escape ($\Delta\Omega \approx D^2/4L^2$). In general this structure has not proved suitable for ion implanters since the continual loading/unloading of substrates demands high pumping speeds at the target for high throughput. High pump speed is also desirable for rapidly removing vapors released during the implantation process to maintain the high vacuum and minimize contamination. Since the pumping speed of most target stations is limited by the conductance F of the beamline, which varies as $F \propto D^3/8L$, a high vacuum is more rapidly achieved with short beamlines tubes with large diameters. As a result, use of such beamline tubes requires that an electrical or magnetic field be imposed to eliminate secondary and tertiary particle losses. For the electric field approach, a biased electrode or tube is placed in front of the target. The bias is usually negative so as to reflect the secondary electrons (the largest component in the secondary charged particle flux) back to the target; this approach also reduces the surface-charging rate for insulator targets. To efficiently confine >99.9% of all the secondary electrons, the biased electrode(s) should establish an equipotential of at least -100 V across the entire cup opening.

For accurate current readings, it is important that current flowing to this secondary electron suppressor be included in the integrations circuit for the target. This arrangement will prevent loss of the positively charged sputtered ions which are attracted to this electrode and loop out the current due to tertiary electrons, ejected during secondary ion impact, which flow to the target [12,27,28,30]. On some implanters, this electrode has been positively biased to collect secondary electrons and reflect sputtered ions. The electrode current is included in the integration circuit and this arrangement does provide accurate current readings. It also will reduce tertiary currents and minimize contamination arising from electrode sputtering by ions emitted from the target. However, it is not suitable for implanting insulators since the current flow will severely aggravate surface charging of the target and catastrophic effects may result [29].

Transverse magnetic fields can also be used to suppress secondary electrons [1,2,30]. These fields must be strong enough to bend electrons with the highest energy to be collected into a circle of radius less than the length of the Faraday cup. For this purpose, very weak fields are satisfactory. For example, to confine 100 eV electrons to a 10 cm deep Faraday cup, a transverse field of only a few (<5)G is necessary. If the field is also required to suppress secondary ions of comparable energy, it may have to be 500-1000 G depending on the ion mass. In general, magnetic suppression traps charged particles and restricts them to cycloidal motion along the field lines. Thus, secondary electrons are not necessarily returned to the target surface unless the target is one of the poles such that the field lines terminate there.

### 3.3 Target Area Effects

In general, the implanted area consists of the area of the substrates plus the area necessary for overscan. Usually this latter region is kept to a minimum to reduce the probability of target contamination by material sputtered or evaporated from this area. This problem is a particular concern for machines which perform implants with a variety of ion species. Fig.8 is an illustration of such cross contamination [23]. This figure shows a Rutherford backscattering spectrum [31,32] of a wafer implanted to a dose of $2 \times 10^{16}$ $Ar^+/cm^2$ in a system normally used to perform $As^+$ implants. A target surface layer of $\approx 1 \times 10^{14}$ $As/cm^2$ has resulted from sputtered As from previously implanted regions on the target holder.

Accurate determination of the implanted area is necessary to control the areal density of the implanted dose. The beam must be permitted to strike only a well-defined target area and must be shielded from striking the cup walls or suppression electrodes. Errors can arise from changes in the beam profile (space-charge blowup, electrostatic scan geometry [13,30,33]) or from improper control of the scan area. Sources of these errors are strongly related to the scanning technique involved, and details will be covered in the analysis of the specific configuration.

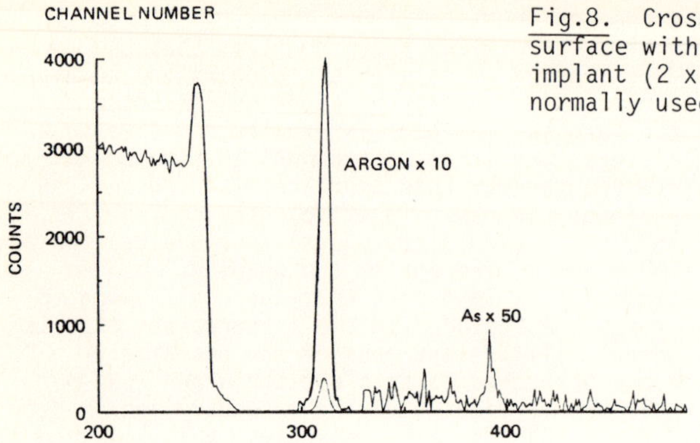


Fig.8. Cross contamination of target surface with As, produced during an Ar implant (2 x $10^{16}/cm^2$) in a chamber normally used for As implants

## 3.4 Electrical Errors

The final sources of potential problems in current measurement with Faraday cups are related to the integrity of the electric circuit. The existence of low-impedance current paths between the Faraday cup and ground, other than through the integrator, will result in the loss of some fraction of the incoming beam current. These paths usually develop along leaky insulators, which may be either standoffs which have acquired conductive coatings of sputtered material, or other normally insulating links, such as target-cooling-fluid lines or couplings in a mechanical drive, whose dielectric properties have deteriorated with time or improper use [34,35,36]. These conditions can lead to severe errors when electrically biased electrodes are looped in the integration circuit, since current drain from the bias supply will appear as beam current. Such effects should be readily observable by running the integrator with all biases and target connections operating, and the beam blanked off.

To reduce the probability of insulating standoffs becoming coated with sputtered or evaporated material, it is a good general practice to shield these insulators to prevent line-of-sight depositions and/or to convolute their surfaces to increase the surface path length to ground (see Fig.9).

More subtle discrepancies occur when the swept beam-current signal exceeds the frequency response of the integrator [36], or when noise either on the beam, or in the bias supplies under load, or picked up from other system supplies distorts normal integrator performance. Use of shielded cables for all connections and thorough characterization of the circuit frequency response with a precision pulse generator are advisable to eliminate these sources of error.

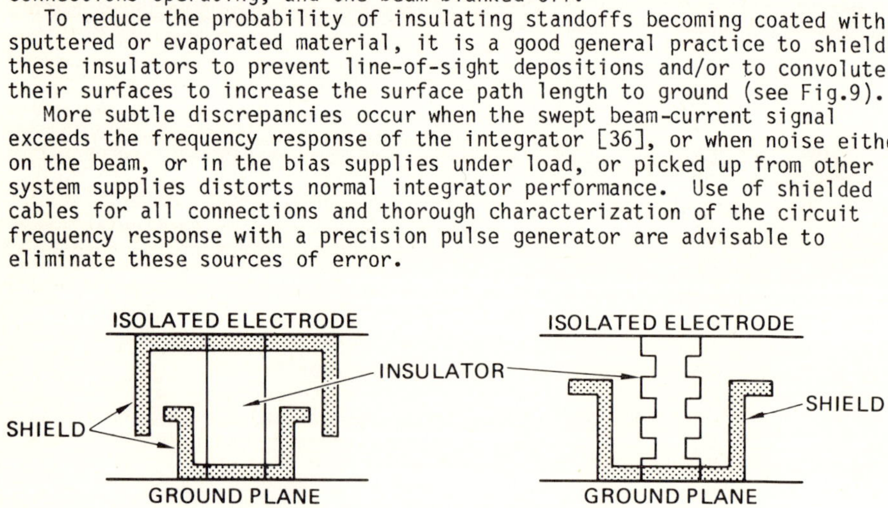


Fig.9. Insulator shielding to reduce leakage caused by sputter-deposited coatings

## 4. Design Principles for Faraday Cups

We can summarize the preceding discussions in the form of general principles for Faraday cup designs:

1. Only the ion beam enters the cup; trapped electrons are removed or blocked over a small region at the cup entrance.

2. All secondary, tertiary charged particles created in the cup are collected, with secondary electrons being returned to the target.

3. Implant area is well defined.

4. Faraday cup is electrically isolated ($\approx 10^{12}\Omega$) from ground except for a low impedance path through the integrator.

5. Integrator and bias electronics are tested for noise and proper response for beam-sweep frequency and current intensity.

These principles must be satisfied within a configuration which also permits target loading at high throughput, and hence high conductance pumping of the target chamber. Furthermore, consideration should also be given in the design for the capability to achieve real-time control of dose and dose uniformity. We shall next evaluate particular structures in light of these principles, and attempt to identify optimum designs for scanned beams and scanned targets.

## 5. Faraday Cup Design for Dose Measurement for Scanned Beams

Systems using electrostatically scanned beams represent approximately 75% of all ion implantation machines sold in recent years [37]. Machine throughput has been enhanced by increased ion beam current and target scan area. As previously noted, both of these increases make it more difficult to achieve accurate dose measurements with Faraday cups. We shall look at some conventional Faraday cup designs as they are affected by these factors and how they perform in terms of the general principles defined in the previous section.

### 5.1 Beam Space Charge and Secondary Particle Collection

In Fig.10, two conventional Faraday cup structures are illustrated. The first (Fig.10(a)) is a traditional long tube, electrically in common with

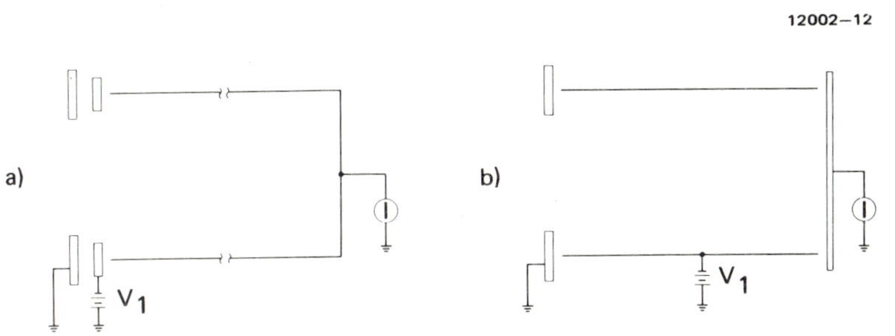


Fig.10. Conventional Faraday cup configuration on implanters with electrostatically scanned beams

the target, with a deneutralizing electrode at the entrance similar to that shown in Fig.3. While this structure will provide accurate dosimetry for low-current beams, we have seen that the biased electrode configuration is ineffective in deneutralizing mA-current ion beams. As a result, electrons trapped in these higher-current beams can travel into and out of the cup freely. Also, the reduced potential near the beam permits some secondary electrons to escape. These effects are illustrated in Fig.11 which shows computer-calculated electron trajectories based on the electrostatic potential distribution in the cup set up by the 1 mA, 100 keV $As^+$ beam and the electrode bias [23]. Proper performance would require increases in electrode bias since the small electrode thickness and its proximity to the grounded aperture reduce the effect of the electrode bias on the potential at the cup center. Furthermore, the biased electrode is not included in the integration circuit, so slow ions from ionized residual gas and sputtered ions from the target, which are both collected by this electrode, are lost. Generally, this discrepancy will be negligible due to the small region over which the potential extends and the small solid angle the electrode subtends for sputtered ions from the target. However, errors of several percent are possible at high cup pressures ($10^{-3}$-$10^{-4}$ Torr). In addition to these shortcomings, the cup design does not return secondary electrons to the target, but collects them on the cup walls. Also, the large length-to-diameter ratio for this cup (horizontal scale in Fig.11 is 1/4th the vertical scale) reduces the pumping speed at the target.

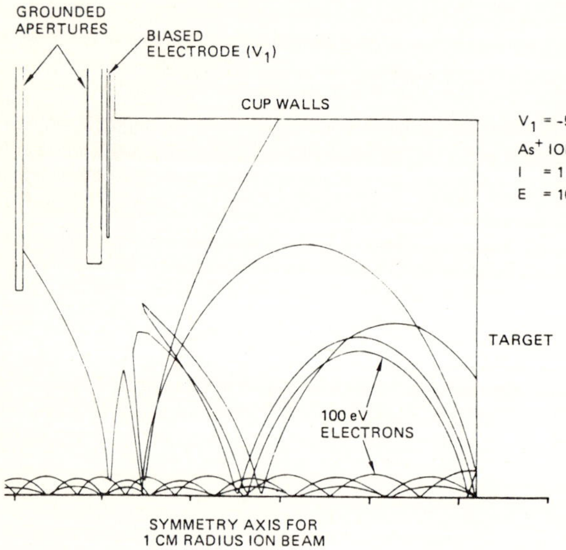


Fig.11. Trajectories of 100 eV secondary electrons in Faraday cup shown in Fig. 10(a) showing electron losses caused by space charge of 1 mA, 100 keV $As^+$ ion beam

The second structure (Fig.10(b)) reduces the length-to-diameter ratio, and effectively returns all secondary electrons to the target by means of a long, negatively based cylindrical electrode. The potential distribution set up by this electrode, biased at -500 V, and by a 1 mA, 1 cm radius $As^+$ beam at 100 keV is shown in Fig.12 [23]. While this arrangement eliminates some of the exceptions taken to the cup in Fig.10(a), it is still subject to several problems. Most significantly, the biased electrode is not looped in the integration circuit, so ion current to this electrode (which can exceed 20% of the beam current in this configuration) [12,28] is lost.

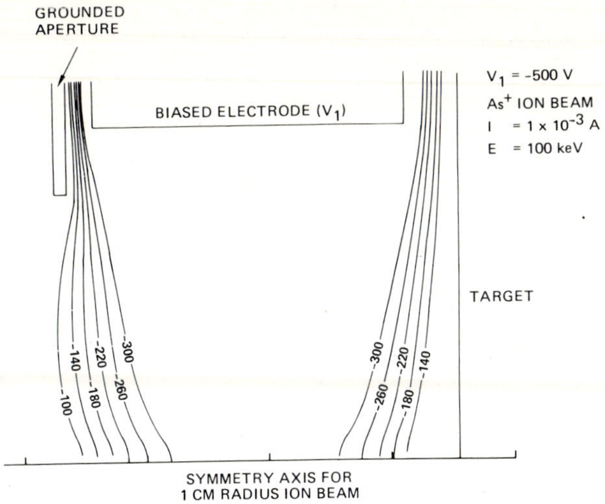


Fig.12. Potential distribution in Faraday cup shown in Fig. 10(b) set up biased electrode and 1 mA, 100 keV $As^+$ ion beam

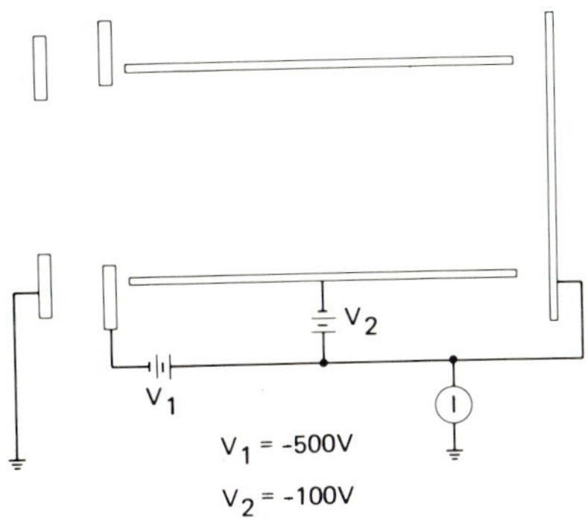


Fig.13. Recommended biasing arrangement for Faraday cup measurements

Correcting this situation, however, does not eliminate certain other difficulties. First, the potential distribution shown in Fig.12 will deneutralize the beam over the entire length of the cup (horizontal scale ≈ 1/2 vertical scale in Fig.12). This condition can lead to significant space-charge blow-up in higher-current beams [30]. Furthermore, with this potential distribution, electrons generated in the ionization of residual gas in the front half of the cup will be lost out the cup entrance. This loss mechanism also holds for tertiary electrons created by ion bombardment of the front half of the biased electrode. Finally, ions collected at the biased electrode are accelerated to energies at which they can sputter the biased electrode and produce target contamination.

A preferred Faraday cup structure for scanned beams is shown in Fig.13. In this design, the beam is deneutralized at the cup entrance by a short electrode biased at -500 V. (Magnetic fields are not recommended for the large openings required for scanned beams because of the difficulty in confining the field to a small section of the beam path.) This electrode is shielded from the beam by a grounded aperture which also defines the target area scanned by the beam. A second, longer electrode, biased at

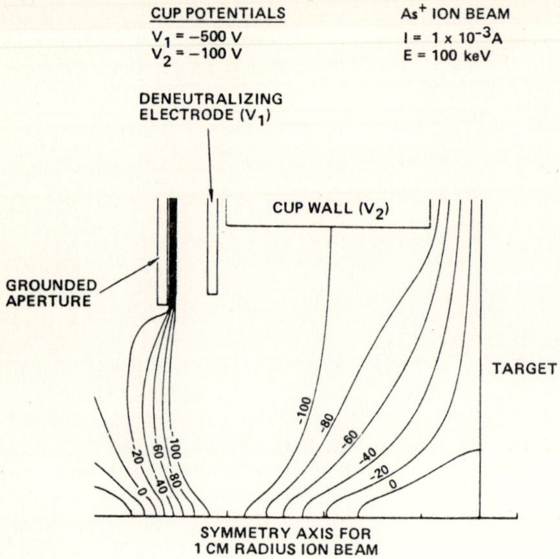


Fig.14. Potential distribution in Faraday cup shown in Fig.13 set up by biased electrodes and 1 mA, 100 keV $As^+$ ion beam

-100V, constitutes the cup walls. Both electrodes are looped in the integration circuit with the target. The potential distribution for this arrangement is shown in Fig.14, again with a 1 mA, 1 cm radius $As^+$ beam at 100 keV. As can be seen, the deneutralization region is reduced, the secondary electrons will be efficiently returned to the target, ions will be collected on the longer electrode at low energies, and ion-electron pairs created in the ionization of residual gas anywhere in the cup are accounted for. The inclusion of the deneutralizing electrode in the integrator circuit could cause some errors due to the collection of slow ions from a small region just before the cup entrance. This contribution, however, should be much smaller than that from ions created within the cup. This latter current would be lost if the deneutralizing electrode were not included. In either case, current to this electrode becomes significant (> 1%) only for high background pressures and very high current beams.

## 5.2 Implant Area

The area defined by the Faraday cup aperture must be corrected for the beam-scan angle to determine the implant area at the target [10,36]. The maximum beam deflection off-center, in a given direction, at the target is:

$$r_{T,\ max} = \ell_T \frac{r_a}{\ell_a} \tag{11}$$

where $r_a$ is the aperture radius, $\ell_a$ is the distance from the center of the appropriate deflection plates to the aperture, and $\ell_T$ is the distance from the deflection plates to the target, as shown in Fig.15. Since the beam is deflected in both x and y by scanners which, on most systems, do not occupy the same location in the beamline, the ratio $\ell_T/\ell_a$ is different for the x and y axes. As a result, the target area corresponding to a circular defining aperture is slightly elliptical. A good approximation (< 2%) of this area, however, is:

$$A_T = \pi \langle r_{T,max} \rangle^2 = \pi\, r_a^2 \left[\frac{\langle \ell_T \rangle}{\langle \ell_a \rangle}\right]^2 \qquad (12)$$

where, for example

$$\langle \ell_T \rangle = \frac{\ell_{T,x} + \ell_{T,y}}{2} \quad . \qquad (13)$$

The target area should be somewhat larger than the substrate to be implanted for optimum coverage, and electrodes should be located such that there is no possibility that they can be struck by the beam. Redeposition of material sputtered from the area surrounding the substrate could be reduced by recessing the area slightly as shown in Fig.15.

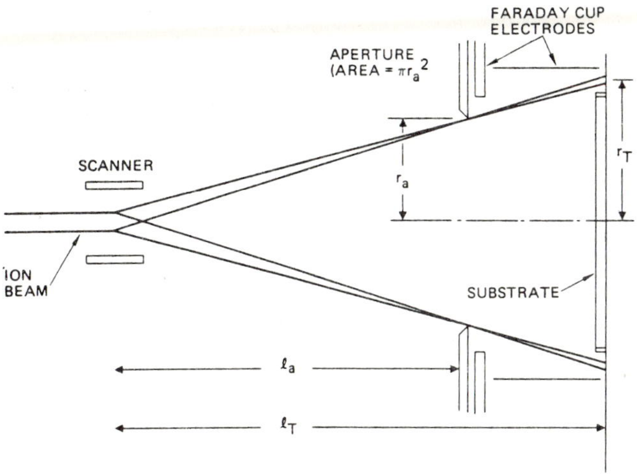


Fig.15. Implanted area geometry for scanned beam systems

## 5.3 Electrical Considerations

For scanned beam systems, leakage caused by coated insulators is generally not a problem, since the cup geometry itself usually provides reasonable shielding of electrode and target standoffs, and the principal pumping path is through (rather than around) the cup. A more common source of difficulty is the target cooling now being required when implanting resist-coated targets with near mA-current ion beams. The coolant lines can provide a leakage path to ground, particularly if a fluid with poor dielectric properties (such as jacket cooling water) is used. A closed-loop freon, deionized water or gas heat exchanger is more satisfactory. Also, films used between the target and cooled holder to improve thermal conductivity may be relatively insulating, electrically. Good electrical contact must be achieved between the clamps used to secure the target and the target holder. Connections should be checked for continuity and isolation from ground with both dc and ac signals.

A final concern in the integration circuit for scanned beams is the ability of the integrator to accurately track the rapid change in current intensity over several orders of magnitude. Jamba [36] has reported that substantial integrator errors are possible for certain ranges of scan frequency and/or beam current. As a rule, the integrator must be set on a scale which is well below saturation (~ 75% scale typically) for the unscanned (dc) beam current intensity. For integrators found on most commercial systems, frequency response is not a concern. However, if operation at frequencies well above or below standard values is desired, integrator response should be verified with a pulse generator. For complete accuracy, the test signal should be fed through the target holder connections to observe the behavior of the entire circuit.

## 6. Faraday Cup Design for Dose Measurement for Scanned Targets

Electrode geometry in Faraday cups for target chambers using mechanical scanning has varied significantly with the configuration of the scan mechanism. Since the ion beam is stationary for fully mechanically scanned targets, the electrodes can be brought radially much closer to the beam axis. This arrangement improves the effectiveness of the electrode biases, particularly the deneutralizing electrode. However, with the reduced dimensions of the electrode apertures, pumping through the cup becomes less efficient. Therefore, vacuum systems for such target chambers are usually designed to pump through the openings between the electrode and the target holder, and/or around a limited Faraday cup region to the remainder of the scanned surfaces.

Historically a number of different scanning techniques have been developed [37], with varying degrees of success. However, in recent years the rotating disc approach [38] has evolved as a near industry-wide standard (e.g., Western Electric, IBM, Eaton/Nova, Varian/Extrion, etc.). This preference is based on the advantages provided by the rotating disc concept: (1) faster scan speeds, which reduce target heating and surface charging; (2) constant angle between beam and target holder during implant; (3) simpler mechanism with fewer moving parts in vacuum (lower contamination, easier electrical isolation and better circuit continuity). Accordingly, we will limit those aspects of the discussion related to the scan mechanism to the disc approach.

Current levels available in systems using mechanical scanning can exceed 10 mA with cross-sectional areas of only 2-5 $cm^2$. The dimensions required for disc size and motion, the electrode spacing necessary to provide vacuum pumping, and the considerations involved in surface charging with the intense beam currents all make accurate Faraday cup dose measurements more difficult.

### 6.1 Beam Space Charge, Surface Neutralization and Secondary Particle Collection

In keeping with the general principles defined in Section 3.4, the initial electrode in the Faraday cup must deneutralize the incoming beam by blocking transmission of electrons trapped in the beam. Also, this electrode must stop forward-scattered secondary electrons from the aperture in the grounded electrode used for shielding and, on these systems, for physically defining the beam dimensions. (These electrons will generally have a larger high-energy population than the backscattered secondaries [39].) Although beam currents are generally 2 to 20 times those in scanned beam systems, the use of smaller biased electrode apertures (typically ≈ 25% larger than the beam dimensions set by the defining aperture in the shielding electrode) permits voltages comparable to those used on scanned beam systems to perform satisfactorily [23]. With these smaller openings, the application of magnetic fields also becomes more attractive [30,40], particularly since the risk of space-charge blow-up of the beam can be reduced within a well-confined magnetic field region [41].

In either event, however, the deneutralizing electrode presents a limitation in dealing with a significant problem for high-current beams in mechanically scanned systems - charging of insulated target surfaces. For these systems, unlike for scanned beam systems, it may not be sufficient to return all secondary electrons to the target to keep surface charging to a reasonable level, particularly for sensitive semiconductor devices [23,29]. As previously described, when the ion beam strikes an insulating surface, the surface charges to a potential equal to the space-charge potential of the beam and then begins to draw electrons out of the beam. However, when the beam is totally confined to the insulator, the number of electrons available (≤ electron density in the neutralized beam x the beam volume in the cup) and the rate of generation of free electrons by beam-induced ionization of the background gas in the cup will not be sufficient to neutralize the amount of charge delivered by the beam. For instance, take a 5 mA $Ar^+$ beam at 50 keV, 2 cm wide x 1 cm high, in a Faraday cup 5 cm long. The particle density in the cup is $\sim 10^9/cm^3$. If the beam were completely neutralized in the cup (which it would normally never be), then $n_e = n_i$ and $N_e = n_e x$ cup volume $\approx 10^{10}$ electrons. Mechanical scan velocities are typically $\approx 10^3$ cm/sec (~ 1/10th those of scanned beam systems). During the time it takes a 1 cm x 1 cm portion of the target to traverse the beam at this velocity, it receives $> 1 \times 10^{14}$ ions. Thus, the cup electrons could not neutralize even 0.1% of the delivered charge. Also, beam-induced residual gas ionization rates at typical cup pressures would be several orders of magnitude too low to provide the needed electron flow. The limits on the beam volume from which electrons can be drawn due to the deneutralizing electrode, coupled with the slower scan speeds, thereby result in increased insulated target charging rates. This condition can lead to discharges through the insulating layer, such as shown in Fig.6, or arcs to open conducting regions for patterned layers, as shown in Fig.16. To eliminate this condition, a means for generating additional electrons and efficiently delivering them to the target is necessary. For accurate dose measurement, all these electrons must be generated in the cup, properly collected and looped out in the integration circuit.

Holmes [22] has analyzed techniques for space-charge neutralizing a steady-state ion beam for efficient beam transport (good collimation). These techniques included: (a) electron injection from an emitter outside the beam; (b) electron injection from an emitter inside the beam; (c) electron injection from a plasma bridge; and (d) ionization of residual gas at high background pressures ($10^{-4}$ to $10^{-3}$ Torr). The degrees of neutralization which could be achieved were reported as: (a) 60%; (b) 91%; (c) 97%; and (d) 99%, respectively. Based on these results, it would appear that high background gas levels in the Faraday cup would provide the best solution to eliminate target charging.

However, several liabilities with this approach can be identified. First, the Faraday cup would have to be extremely long to provide an adequate beam volume and electron generation rate to draw upon. For typical cup geometries, conductance for pumping the high background pressures would be poor and contamination by redeposition of sputtered material could become a problem. Second, stripping the beam of its neutralizing electrons over a sizeable distance could lead to beam blow-up [30] and loss of effective beam current due to increased scan requirements. Third, Holmes has shown that the slow ions, created in the ionization process, limit the ultimate level to which the beam space-charge potential can be reduced. This level can be 10-40 V, being worse for the higher-mass, low-velocity

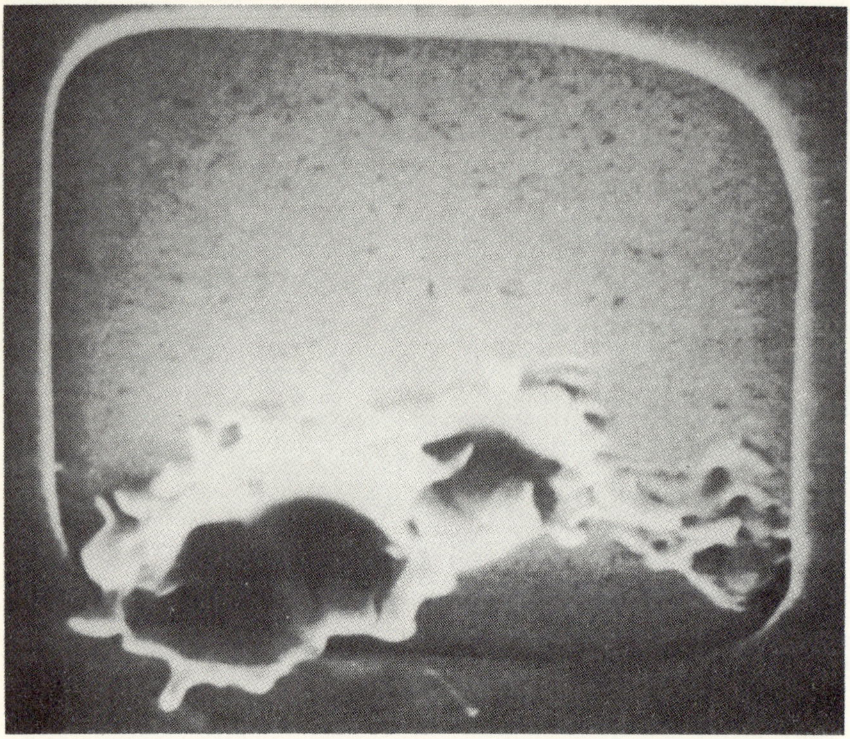

Fig.16. Device damage produced by arc discharge of ion-implantation–induced surface charge on neighboring insulated layer

ions used in ion implantation. These potentials would represent the lowest value to which the target surface would be controlled [42] and they would increase as the electron density is depleted by the target surface. Even these potentials may be too large for certain high-density, sensitive devices. (For example, 400 Å gate oxides for FET devices with dielectric strengths of 5 x $10^6$ V/cm will break down at potentials of 20 V.) Furthermore, in the absence of additional biases, the electron flow will be limited by the potential gradient in the beam and the finite mobility of the electrons. For these low-energy electrons, the presence of even very weak ($\lesssim$ 10 G) stray magnetic fields can impede their mobility leading to reduced neutralization efficiency [43]. Thus, systems relying on this approach must carefully shield the beam from fringe fields due to analyzing magnets or magnetic suppression electrodes. Proper electrode biases can increase the removal rate of slow ions from the beam and, at levels sufficient to overcome the space-charge potential of the beam, improve the electron transport to the target. Nevertheless, the electron generation rate will still limit the ability of this approach to neutralize insulating surfaces.

Ideally we would like to provide an equal amount of ion and electron current at each spot where the beam strikes the target, for a net target current of zero. In practice, this situation is essentially impossible to achieve. Generation of additional electrons within the Faraday cup, using one of the approaches (a, b, or c) described by Holmes, is reasonably straightforward. The difficulty lies in transporting these electrons to the target at the proper rate, energy, and location to accurately control the surface potential. The immersed emitter method (b) can be immediately rejected because the sputtering of the cathode by the ion beam would result in poor reliability and create of source of target contamination. McKenna has described a Faraday cup arrangement [23] using an external emitter which controls insulator-target surface potential to $0 < V_T < -50$ V. This electron-flooded cup design is shown in Fig.17. The concept of this design is to produce a net negative current at the target with a minimum value:

$$I_{net} = -\frac{A_{cup}}{A_{beam}} I_{beam} , \tag{14}$$

to allow for the fact that the electron flow is directed out over the entire cross-sectional area of the cup ($A_{cup}$) rather than just over the beam dimensions ($A_{beam}$). Under these conditions, the target surface will charge negatively until $V_{target}$ reaches a value:

$$V_{target} = V_C = -50 \text{ V} , \tag{15}$$

at which point the electrons reach the target with zero kinetic energy, and the electron flow becomes self-regulating. (This approach has also been used in scanning electron microscopy (SEM) systems, where low-energy ions are provided to neutralize insulator specimens [44].)

The -50 V bias was set up to prevent electron losses along the gap between the target and the cup electrode [23]. It was subsequently found

FARADAY CUP WITH ELECTRON FLOODING.

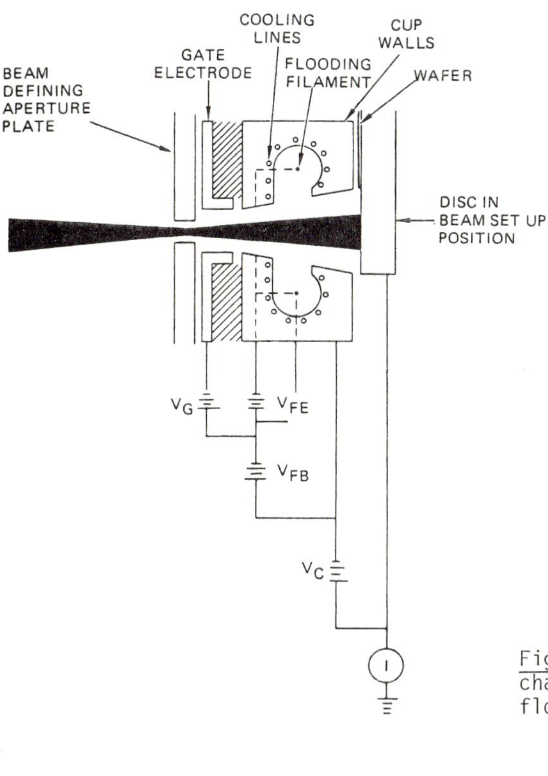


Fig.17. Faraday cup with surface charge neutralization by electron flooding

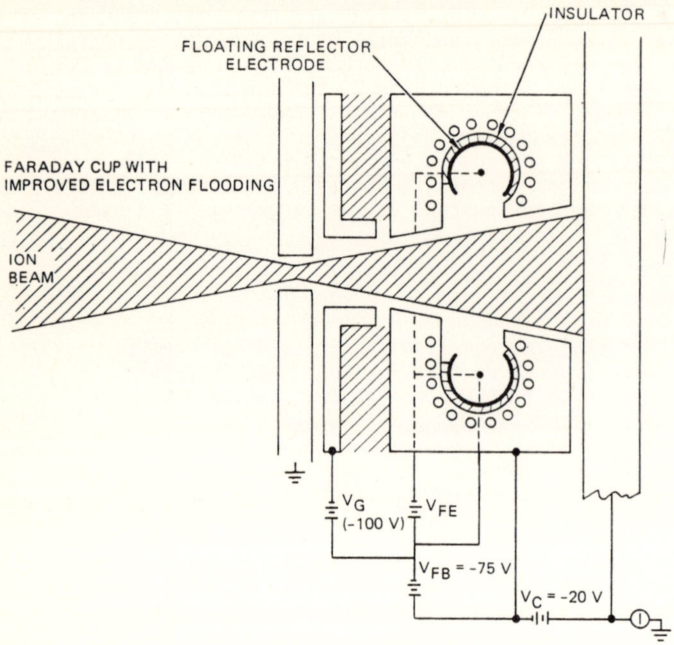


Fig.18. Faraday cup with increased electron emission at low cup wall biases for improved control of target surface potential

that these losses occurred only in the beam-set-up position at the extreme outside edge of the disc, where the disc did not extend over the lower portion of the cup electrode surface (Fig.17). For the disc positions during actual implants, biases as low as -10 V were adequate to prevent electrons from escaping along this gap and thus to provide accurate beam measurement. Using these lower cup wall biases ($-10\ V \leqq V_C \leqq -20\ V$) in an attempt to limit target surface potentials to a similar value, it was found that these potentials did not provide a sufficient potential gradient between the cup walls and the beam to achieve the desired electron flow from the filament region. In a modified design [45], the surface immediately surrounding the filament was floated such that it would charge up to the filament potential and divert the emitted electrons into the beam (Fig.18). This design demonstrated sufficient electron flow to neutralize ion currents > 4 mA and maintain target surface potentials to $|V_T| < 20$ V. As with the original unit, this design also included coolant loops to dissipate the heat from the filament, in order to prevent target contamination by material evaporated or outgassed from the cup walls. Electrode structures and biases for both designs set up Faraday cup potentials similar to those in Fig.14, but in a more compact geometry. Accordingly, these designs both satisfy the general principles of efficient beam deneutralization and secondary particle collection, with the majority of secondary electrons returned to the target.

A plasma bridge neutralizer [46] approach could provide an even higher electron flow capability, particularly if the potential gradient can be determined by the Faraday cup biases rather than by the beam space-charge

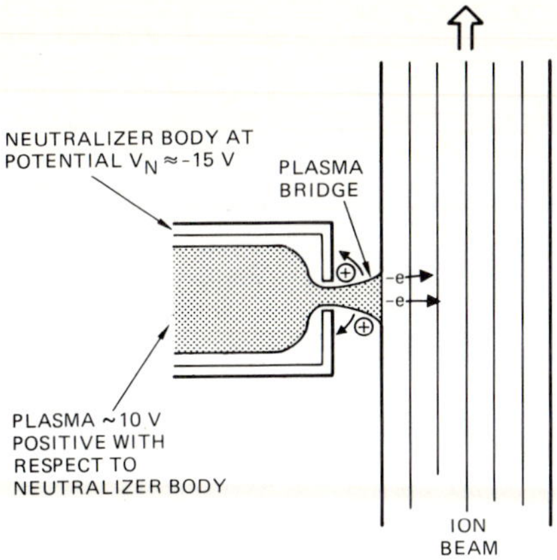


Fig.19. Plasma bridge neutralizer arrangement

potential. The arrangement for beam neutralization using a plasma bridge is shown in Fig.19. On ion thrusters, this technique has been shown to reduce the space-charge potential in 500 mA ion beams to 12 V $\leq$ V $\leq$ 40 V, indicating the capability of providing high electron injection rates at moderate potential differences. The principal liabilities with the plasma bridge concept are the requirement of an appreciable neutral gas flow in the plasma bridge and the potential for target contamination by sputtered material from the neutralizer discharge or from bombardment of the neutralizer by slow ions created in the bridge region. It may be possible to locate the plasma bridge further upstream away from the target such that the target is not directly exposed to the neutralizer. Under these conditions, however, it is not obvious whether the target surface charging will provide a sufficient potential gradient in the beam to produce the necessary electron flow to the target. Ward and King have shown that a plasma bridge neutralizer, at a potential of $\approx$ -15 V, controlled the potential of a floating target, being bombarded by a 500 mA ion beam, to < +15 V with the plasma bridge located 5 meters upstream of the target. (The target potential was lower than the beam space-charge potential measured at the plasma bridge region due to the reduced current density at the target. This reduced current density resulted from the beam divergence acting over the 5 meter path.) As with beam neutralization by residual gas, this technique will be adversely affected by stray magnetic fields, due to the reduced mobility of the low-energy electrons in diffusing across the magnetic field lines [43].

Furthermore, Ward and King obtained their data using Hg vapor in the plasma bridge neutralizer. The low Hg ionization potential (10.43 V) permitted the use of a low plasma potential in the neutralizer. For a gas such as Ar, which is more suitable for use in an ion implantation environment but which has a higher ionization potential (15.76 V), the ionization rate required to sustain the plasma bridge will require a higher neutral flow and a higher plasma potential which could raise the beam space-charge potential by several volts. While some adjustment is available by varying the neutralizer potential, the minimum space-charge potential will be set by the need to confine the electrons accelerated across the plasma bridge and prevent their escape to the beamline walls [22].

## 6.2 Implant Area

The implanted dose distribution on a properly controlled disc scanner (see section on dose uniformity) is shown in Fig.20. The implant area consists of: (1) a central region which is scanned completely through the beam and (2) on either side of this central region, bands whose width is equal to the beam dimension perpendicular to the scan motion. Generally these areas are kept small compared to the central region for optimum scan efficiency, and targets are positioned only within the central region. However, the area in the fringe regions must be properly accounted for in computing the total dose. With a homogeneous beam, the total area can be computed as:

$$A = \pi \left(R_2^2 - R_1^2\right) \tag{16}$$

where $R_2$, $R_1$ are the radial positions of the beam center at the outer and inner edges of the scan, respectively. Thus, if a "beam-burn" pattern on the disc is used to define the implant area, $R_1$ and $R_2$ are located a distance in from the pattern edges equal to half the beam dimension perpendicular to the scan.

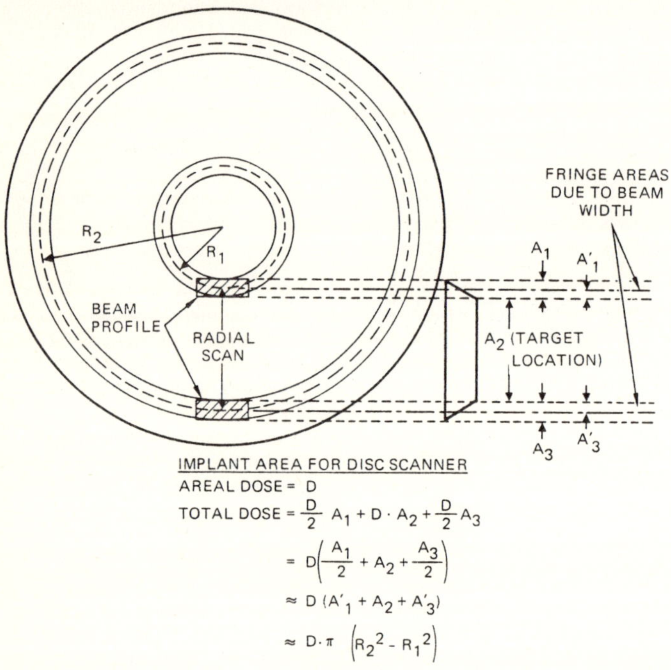


IMPLANT AREA FOR DISC SCANNER

AREAL DOSE = D

$$\text{TOTAL DOSE} = \frac{D}{2} A_1 + D \cdot A_2 + \frac{D}{2} A_3$$

$$= D\left(\frac{A_1}{2} + A_2 + \frac{A_3}{2}\right)$$

$$\approx D\,(A'_1 + A_2 + A'_3)$$

$$\approx D \cdot \pi \left(R_2^2 - R_1^2\right)$$

$$A = \pi \left(R_2^2 - R_1^2\right)$$

Fig.20. Implant area for disc scanners

## 6.3 Electrical Considerations

The high-current beam flux, the close proximity of the electrodes to the beam, and the practice of pumping around the Faraday cup rather than through it all serve to accelerate the coating of electrode-supporting insulators for scanned target systems. Use of shielded insulators, such as shown in Fig.9, is crucial for system reliability. Again, insulator

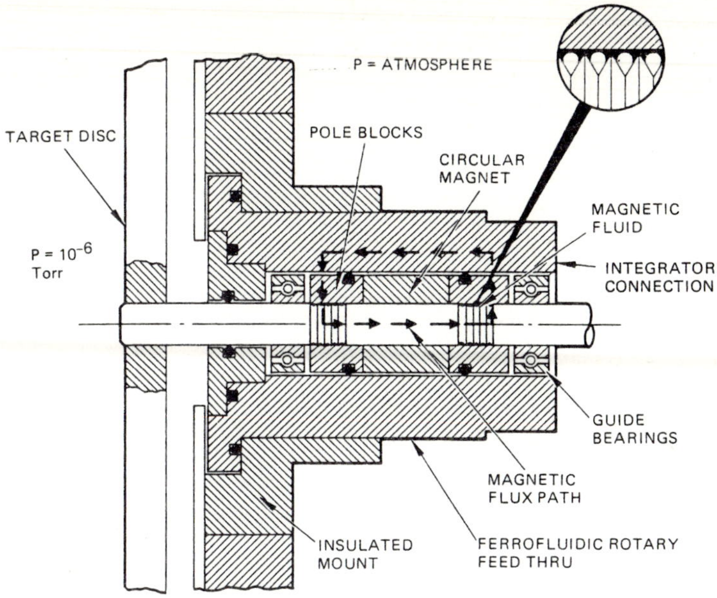


Fig.21. Ferrofluidic seal mounting for isolating rotary feed-through

leakage can be tested by observing current drain from a biasing power supply connected to the appropriate electrode.

A more serious problem for scanned targets is maintaining a low impedance current path between the moving target and the integrator. For disc scanners, this path must include a connection to the rotating drive for the disc. Contacts through conventional bearing surfaces, gears, or commutator brushes tend to be noisy and intermittent. As a result, these types of connectors can compromise the integration accuracy. Ferrofluidic [47] seals, used in several rotary disc scanners [34,37,48], provide continuous conducting contact to the rotating drive shaft, such that, if the seal is insulated from the vacuum housing, as shown in Fig.21, an electrical connection to the stationary portion of the seal can be used to read the target current. Again, the integrity of this connection should be tested using an oscilloscope to observe a dc current source fed to the rotating disc.

## 7. In Situ Monitoring of Dose Uniformity

The ultimate performance of ion implantation in dosimetry control involves the capability of obtaining continuous, real-time information on the uniformity of the doping process. Procedures for obtaining this information consist of sampling the dose over selected regions of the implant area and spatially differentiating the sampled doses.

### 7.1 Uniformity Control for Scanned Beams

Dose sampling on scanned beam systems is typically achieved by small Faraday cups located in the region outside the main target area, as shown

IN-SITU DOSE UNIFORMITY MEASUREMENT FOR SCANNED BEAMS

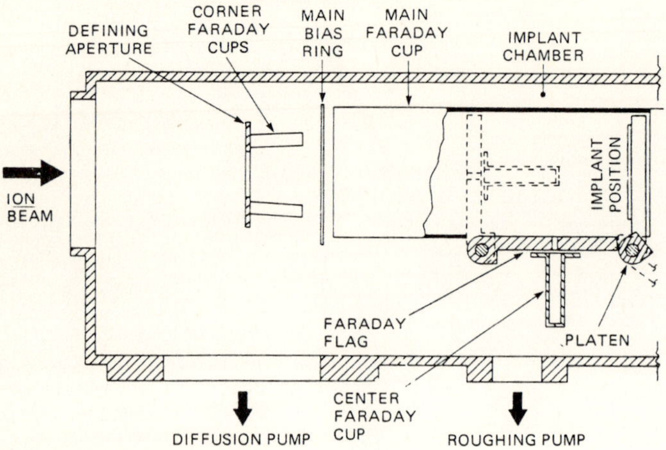


Fig.22. Corner Faraday cup arrangement for dose sampling measurements of implant uniformity

in Fig.22 [49]. In this design, small Faraday cups are placed at each of the four corners of the x-y scan area. Care must be taken to ensure that the cross-sectional area and distance from the unscanned beam axis are identical for each cup/suppressor electrode combination. This will guarantee that the cups present equal acceptance angles for the scanned beam. Also, the cup locations must lie within the deflection range covered by the linear portion of the scan voltage waveform in order to provide an accurate representation of the target scan conditions.

Proper implementation of the design shown in Fig.22 supplies several controls in establishing and maintaining implant uniformity. First, the removable center cup provides the ability to align the unscanned beam with the cup axis. Therefore, the beam scanning will be symmetric about the center of the target which will result in optimum uniformity [13]. Once scanning is initiated, the scan voltage amplitude can be set by requiring equal current readings for all the cups. Then, once the implant has begun, the relative current intensities from opposite pairs of corner cups can be used to monitor beam alignment, and ultimately provide a closed-loop feedback signal to control offset voltages for the scan plates [36]. Also, in this design, the main Faraday cup is used to measure the target dose. However, with proper normalization, the corner cups could provide accurate evaluation of the total implant dose, and should give more reproducible results since they are not subject to the variations introduced by different target material. Using this approach, the main bias ring could then be grounded and the beam could remain space-charge neutralized all the way to the target, which may be advantageous for implantation of insulated targets. This dosimetry control design on a microprocessor- or microcomputer-controlled machine, coupled with improved scanning techniques [15-17], should provide the capability for continual readout of the dose uniformity and for implant-interrupt protection (if not feedback control) for such critical factors as the linearity of the scan waveform and beam offset.

An assessment of the doping uniformity for scanned beams can also be obtained by observing a visual display of the analog signal of the instantaneous main target current from the integrator versus scan voltage [36]. This display aids in establishing proper scan conditions during beam setup and can serve as a qualitative uniformity monitor during implant, assuming proper Faraday cup performance. Quantitative information on dose uniformity using this technique would require rather sophisticated electronics capable of generating a dose matrix array corresponding to the spatial dimensions of the target. Integrated charge, from an integrator with a high-frequency ($\geq$50 kHz) digital output, could be stored incrementally in this array with the matrix-element address being determined by clock signals which are proportional to the respective scan frequencies.

### 7.2 Uniformity Control for Scanned Targets

Dose sampling on scanned target systems is performed directly from the instantaneous current on the moving target or in stationary cups located behind the target using apertures in the target holder [2,35,37]. Hammer and Michel [50] have demonstrated the ability to perform in situ measurements of dose uniformity for a linear-drive x-y scan mechanism. A natural extension of this technique is to feed back the instantaneous measurements to proportionally adjust the drive-motor speed and thereby control the final implant uniformity.

For the practical case of the disc scanner, the scan speeds are not uniform in either axis. The tangental velocity (fast scan) varies directly with R and the radial velocity (slow scan) must therefore be varied inversely with R [34,38,48] to achieve a uniform rate of areal doping. Also this slow scan speed is usually modified to compensate for changes in the instantaneous current intensity. These control techniques have demonstrated the capability of controlling implanted dose non-uniformity to $\sim$ 1% [34,51].

Ryding and Farley [35] have recently described a novel dose-sampling technique which provides feedback control of both slow scan speed adjustments simultaneously. The technique consists of monitoring the intermittent current transmitted to a Faraday cup behind the disc through a precision slot of constant width in the disc (Fig.23). The transmitted current signal will vary inversely with the speed at which the slot passes through the beam ($\propto 1/\omega R$) and directly with the instantaneous current intensity ($\propto I$). By means of an appropriate scaling constant k, the radial scan velocity can then be directly controlled by this transmitted current signal. This technique has the added benefits that the beam remains space-charge neutralized all the way to the target and that high pumping speeds can be attained at the target surface in the absence of conductance-limiting Faraday cup structures. The more extensive beam volume available for supplying electrons to a charging insulated surface should provide reasonable control of target potential, subject to the limitations imposed by the beam space charge as previously discussed.

More recently, Ryding et al. [49] have reported observation of dosimetry errors in this system at the high background pressures suitable for optimum neutralization of the ion beam by residual gas ionization. These discrepancies were interpreted as arising from the appreciable ion and electron densities generated in the residual gas ionization process and their response to the high negative electrode biases needed to deneutralize the beam at the Faradary cup entrance. A biasing arrangement similar to that illustrated in Fig.13 was shown to give improved results, as did use of a magnetic field of $\approx$ 400 G at the Faraday cup entrance. For effective target neutralization by electrons from the beam, this magnetic field must be completely shielded from the target region (to levels $\ll$ 10 G).

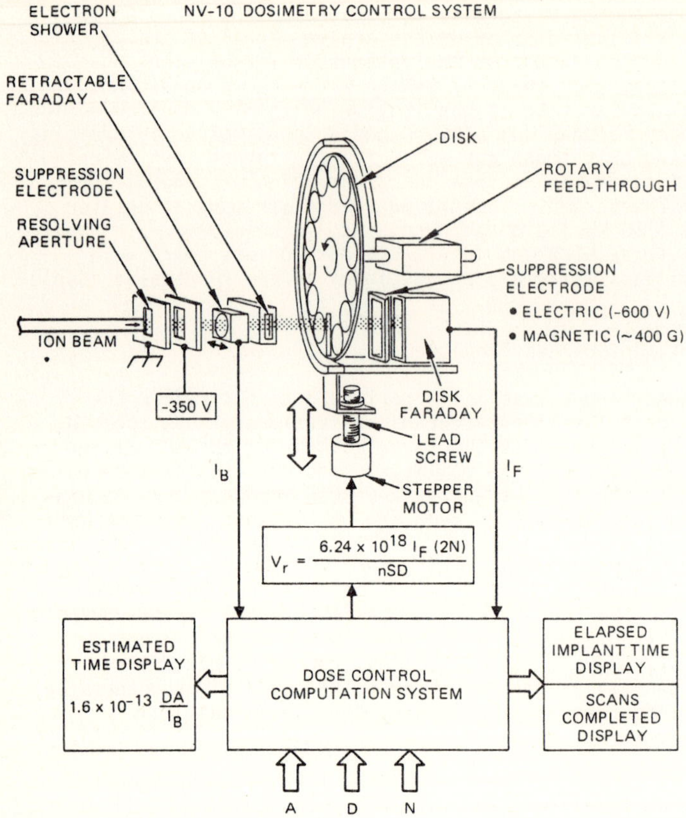


Fig.23. NV-10 dose-sampling Faraday cup used to control dose uniformity on disc scanner

## 8. Hybrid Systems

Hybrid systems [33,48], in which the beam is swept in one scan axis and the target is translated in the other (orthogonal) axis, have not been directly discussed. Certain aspects of the discussions both for scanned beams and for scanned targets will apply to these systems. In general, Faraday cup design for these systems will be more difficult since the cup cross section must be larger to accommodate the scanned beam while the current densities will require target neutralization techniques similar to scanned target systems. These conflicting requirements suggest that the beam sampling approach would be a much more advantageous dosimetry arrangement for these systems. Even so, the electromagnetic scanner will eliminate regions of the beam from which electrons can be drawn to neutralize insulated targets, so some forced neutralization scheme will still be necessary to control the surface potential of such targets to low levels.

## 9. Summary

We have evaluated techniques for obtaining accurate measurement of implanted dose and dose uniformity using Faraday cups, and reviewed the major difficulties encountered in such measurements. Particular attention was paid to neutralization techniques for high beam-current implants of insulated targets.

Operator and process engineers must be familiar with all aspects of machine performance which can affect the implant results. By the same token they should also understand the measurement techniques [32,52] used to evaluate machine performance and the effects that other variables such as crystal axis alignment to the target surface normal, thermal gradients in annealing furnaces, etc. can have on these results. Only then can implanter performance be properly evaluated and controlled.

## Acknowledgement

The author would like to express his thanks for the following technical contributions: G. Ryding for providing a preprint of the paper presented at the 1982 ECS meeting in Montreal, and for information on the NOVA-10 Faraday cup design; J.W. Ward for discussions of plasma bridge neutralizers; and D.M. Jamba for a technical review of the manuscript. Also, the author appreciates the clearance provided by IBM, East Fishkill for publication of some previously unreported results of experiments which he performed there.

## References

1. G. Dearnaley, J.H. Freeman, R.S. Nelson, J. Stephen: Ion Implantation (North Holland Publishing Co., American Elsevier Publishing Co. Inc., New York 1973).

2. R.G. Wilson and G.R. Brewer: Ion Beams (John Wiley and Sons, New York 1973).

3. J.W. Mayer, L. Eriksson, J.A. Davis: Ion Implantation in Semiconductors (Academic Press, New York 1970).

4. P.L. F. Hemment: Low Energy Ion Beams, Salford 1977, Inst. of Phys. Conf. Ser. No. 38, 117 (1978).

5. C. Thomann, J.E. Benn: Nucl. Inst. and Meth. 138, 293 (1971).

6. A. Laurio, J.F. Ziegler: Appl. Phys. Lett. 31, No. 7, 1, 482 (1977).

7. I.V. Mitchell, K. Barfoot, H.L. Eschbach: Fifth Symposium on Activation Analysis, Oxford (1978).

8. J.H. Freeman: Application of Ion Beams to Materials, Warwick, 1975, Inst. of Phys. Conf. Ser. No. 28, 340 (1976).

9. J. Bottiger, J.A. Davies: Rad. Eff. 11, 61 (1971).

10. A.B.Wittkower: Sol. State Technol., Nov. 61 (1978).

11. K. Leyland, D.G. Armour, G. Carter, J.H. Freeman: Low Energy Ion Beams, Salford 1977, Inst. Phys. Conf. Ser. No. 38, 175 (1978).

12. P.L.F. Hemment: Rad. Eff. 44, 31 (1979).

13. J.H. Keller: Rad. Eff. 44, 71 (1979).

14. D.S. Perloff, F.E. Wahl, J.T. Kerr: In Proceedings of 7th Internat. Conf. on Electron and Ion Beam Science and Technology ed. by R. Bakish, 464 (1976).

15. H. Glawischnig, K. Hoerschelmann, W. Holtschmidt, W. Wenzig: Nucl. Instr. and Meth. 189, 291 (1981).

16. E.J. Rogers: Nucl. Inst. and Meth. 189, 305 (1981).

17. N. Turner: Nucl. Instr. and Meth. 189, 311 (1981).

18. W.H. Gries: Internat. Jour. Mass Spectrom. and Ion Phys., 30, 97, 113 (1979).

19. Z.L. Liau, J.W. Mayer: J. Vac. Sci. Technol. 15, 1629 (1978).

20. J.E. Osher: Low Energy Ion Beams, Salford, 1977, Inst. Phys. Conf. Ser. No. 38, 201 (1978).

21. R.G. Wilson: In Applied Charged Particle Optics, ed. by A. Septier (Academic Press, New York 1980).

22. A.J. Holmes: Rad. Eff. 44, 47 (1979).

23. C.M. McKenna: Rad. Eff. 44, 93 (1979).

24. H.A. Enge: In Focusing of Charged Particles Vol. II ed. by A. Septier (Academic Press, New York 1967) p. 203.

25. S.C. Brown: Basic Data of Plasma Physics (MIT Press, Cambridge, Mass. 1966).

26. B. Svensson, G. Holmen: J. Appl. Phys. 52, No. 11, 6928 (1981).

27. H.H. Andersen: In Proc. of Seventh Symposium on Physics of Ionized Gases (Yugoslavia, 1974).

28. D. Jamba: Rev. Sci. Instr. 49, No. 5, 634 (1978).

29. K.H. Nicholas: J. Phys. D, Appl. Phys. 9, 393 (1976).

30. W.C. Ko, E. Sawatzky: In Proc. of 7th Internat. Conf. on Electron and Ion Beam Science and Technology, ed. by R. Bakish (E.C.S., Washington 1976).

31. W.K. Chu, J.W. Mayer, M.A. Nicolet: Backscattering Spectroscopy (Academic Press, New York 1978).

32. P. Eichinger, H. Ryssel: This publication.

33. J. Chaumont: Lecture at Ion Implantation School, Third International Conference on Ion Implantation Equipment and Techniques (Kingston, Canada, 1980).

34. J.R. Kranik: Rad. Eff. 44, 81 (1979).

35. G. Ryding, M. Farley: Nucl. Instr. and Meth. 189, 295 (1981).

36. D.M. Jamba: NBS Special Publication 400-39, 26 (1977).

37. G. Ryding, Nucl. Instr. and Meth. 189, 239 (1981).

38. G.I. Robertson: J. Electrochem. Soc. 122, 796 (1975).

39. N. Oda, F. Nishimura, Y. Yamazaki, S. Tsurubuchi: Nucl. Instr. and Meth. 170, 571 (1980).

40. G. Ryding, M. Farley, M. Mack, K. Steeples and V. Gillis: Eleventh International Conference on Electron and Ion Beam Science and Technology, Electrochem. Soc. Meeting, Montreal 1982.

41. R. Booth, H.W. Lefevre: Nucl. Instr. and Meth. 151, 143 (1978).

42. P.L. F. Hemment: In Low Energy Ion Beams, Bath, 1980, Inst. of Phys. Ser. No. 54, 77 (1981).

43. H.R. Kaufman: NASA CR-159814, January 1980.

44. C.K. Crawford: Scanning Electron Microscopy II, 31 (1979); Scanning Electron Microscopy IV, 11 (1980).

45. C.M. McKenna, G. Popp: unpublished.

46. J. Ward, H. King: AIAA paper No. 67-671 (1967).

47. Ferrofluidics Corporation, Burlington, Massachusetts, USA.

48. P.R. Hanley, Nucl. Instr. and Meth. 189, 227 (1981).

49. Extrion DF4 Manual, Varian Associates, Gloucester, Mass.

50. W.N. Hammer and A.E. Michel, J. Appl. Phys. 47, No. 5, 2161 (1976).

51. J.H. Jackson, H.M.B. Bird, J.P. Fleming, G.J. Hofer, J.C. McCallum, P.J. Mostek, G.I. Robertson, A.F. Rodde, B. Weissman, N. Williams: Rad. Eff. 44, 59 (1979).

52. P.L.F. Hemment, this publication.

# Safety and Ion Implanters

R. Bustin and P.H. Rose

Eaton Corporation, Ion Implantation Division
Beverly, MA 01915, USA

Synopsis

Ion implanters can be dangerous pieces of equipment unless they are designed with attention to the potential hazards and operated in a safe manner. X-radiation, toxic chemicals, high voltages and mechanical forces capable of causing serious harm all exist. Manufacturers build equipment with interlock systems which are capable of protecting the operator from harm. Maintenance procedures are also safe, provided that proper procedures are followed, interlocks are not cheated, and dangerous materials are treated with great care.

## 1. Radiation

Of all the hazards associated with implanters, x-radiation is the most insidious, and the least dangerous in a properly designed implanter. Because the factors leading to exposure to x-rays are often poorly understood, a considerable section is devoted to this problem.

All implanters contain regions of electric fields which accelerate beams of positive ions. X-rays are produced whenever electrons get caught up in these same fields. The trapped electrons are accelerated toward the positive end of the high-voltage gap, and come to rest against the anode. The collision of high-speed electrons with the anode generates x-rays; thus the "hot" region of any implanter is around the extraction aperture of the ion source, or the high-voltage end of the acceleration tube in a two-stage or medium-current machine (see Fig. 1).

### 1.1 Mechanisms of X-Ray Production

The vast majority of electrons come to rest through a series of "glancing" collisions with the atoms of the target material. X-radiation is produced each time an electron is decelerated by a collision, and as a result is called bremsstrahlung or "braking radiation", which is the major safety hazard as far as radiation is concerned. Bremsstrahlung x-rays have a continuous energy spectrum, from infrared up to a maximum value, $E_o$, which is the energy of the incoming electron stream.

Characteristic radiation is generated when the energy of an incoming electron corresponds to one of the transition energies of the electrons in the target atoms, and shows up as sharp peaks in the x-ray energy spectrum. While characteristic peaks may have an intensity far greater than the bremsstrahlung background, their energy spread is quite narrow. Therefore, when considering the total x-ray energy emitted by a source, bremsstrahlung radiation clearly dominates, as Fig.2 shows.

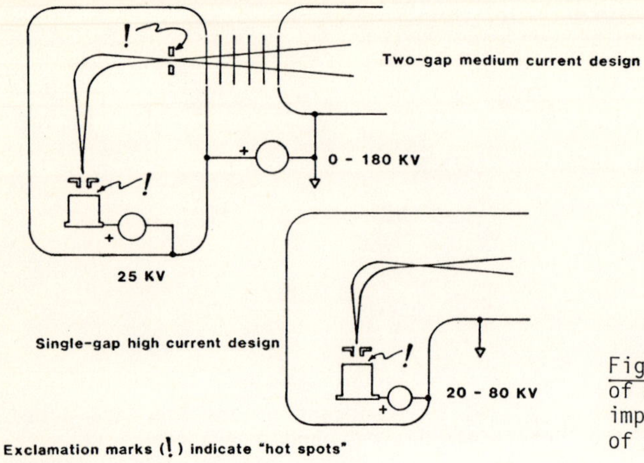


Fig. 1. Schematic diagrams of medium- and high-current implanters, showing regions of x-ray generation

## 1.2 X-Ray Production Efficiency

The total x-ray energy produced will depend on the target material, the total electron current striking the target, and the maximum energy of these electrons.

Target efficiency is directly proportional to the atomic number, Z, of the target material. This is one of the reasons why tungsten (Z=74) makes a good target material for an x-ray tube. This is also why, in a two-gap implanter design, the beamline near the positive end of the acceleration tube is sometimes covered with a low-Z material such as beryllium (Z=4).

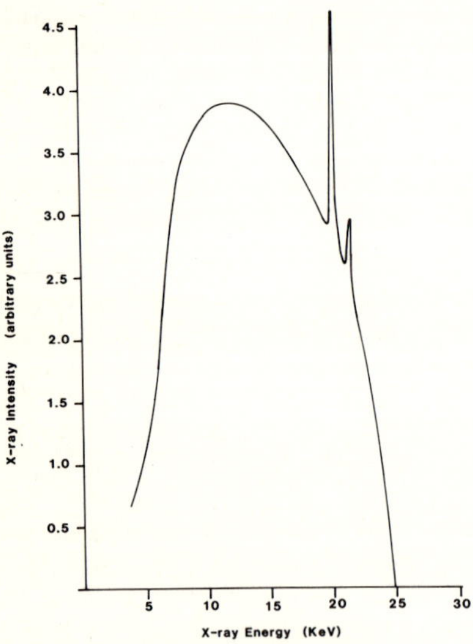


Fig.2. Typical spectrum of x-ray intensity vs energy for a high-Z target

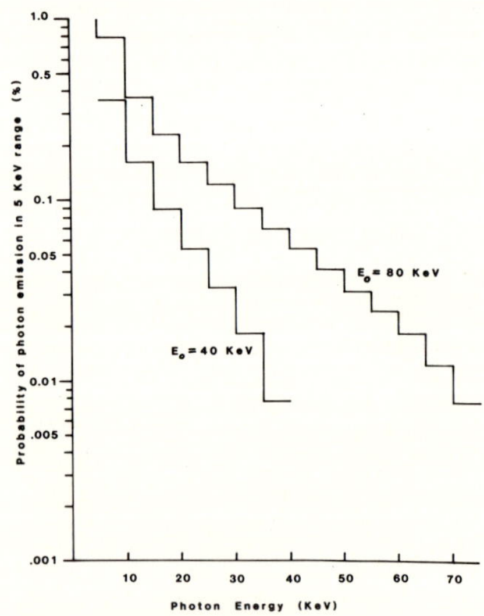


Fig. 3. Probability of photon emission vs energy, calculated from Eq.(1)

The other "hot " surface in an implanter is the front plate or extraction aperture of the ion-source arc chamber. The target material in this case is usually molybdenum (Z=42) or carbon (Z=6). It is unfortunate that the x-ray output from molybdenum - preferred source material - is seven times higher than that from carbon.

The distribution of x-ray intensity as a function of energy is given approximately by the expression [1]:

$$dP_x = 2.5 \times 10^{-9} Z \frac{E_o - E_x}{E_x} dE_x \tag{1}$$

where $dP_x$ = probability of photon emission in the energy range from $E_x$ to $E_x + dE_x$,
$Z$ = atomic number of target, and
$E_o$ = energy of the incoming electron in electron volts.

Equation (1) is plotted in Fig. 3, where $dE_x$ is arbitrarily chosen as 5 keV. The total fraction of energy emitted in x-radiation is obtained by integrating Eq.(1); see for example Beatty [2] and Kramers [3]:

$$W_{tot} = \int_0^{E_o} P_x E_x dE_x = 1.25 \times 10^{-9} ZE_o^2 I \tag{2}$$

where I is the incident beam current in amperes and $W_{tot}$ is the integrated x-ray energy from a target with atomic number Z.

Equation (2) shows that the x-ray energy generated is proportional to $E_o^2$ and only amounts to a few percent of the electron-beam energy in the energy range up to 200 keV.

## 1.3 Generation of the Reverse Electron Flow Responsible for X-Ray Production

A beam of particles, as it is being accelerated and as it moves through the drift space, strikes molecules of residual gas which are thereby ionized, generating free electrons. Some of the ions may be neutralized by capturing an electron during a collision and proceed to strike the walls of the vacuum chamber. These processes produce low-energy electrons which may be trapped in the potential well created by the positive space charge of the ion beam. Unless contained by an orthogonal magnetic field, these electrons are free to move along the beam and can enter a field region, where they are accelerated.

The electrons may be prevented from entering field regions by bias electrodes such as shown in Fig. 4. Without suppression, the electron current drained from the beam plasma destroys the plasma and the space-charge neutralization is lost, the beam blows up and strikes the vacuum chamber or accelerating electrodes, generating even more electrons to contribute to the x-radiation. An interlock on bias supplies to show they are operative is therefore an excellent protective device (Fig. 5). The interlock should preferably measure the voltage on the bias plate as shown, rather than making this measurement in the power supply itself.

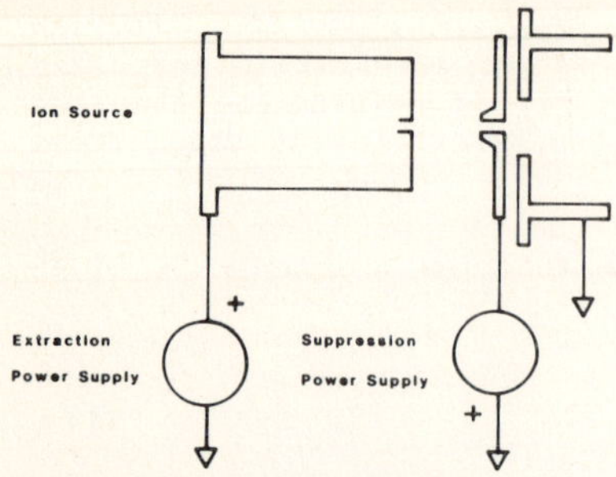


Fig. 4. Typical geometry of extraction system, showing suppression electrode

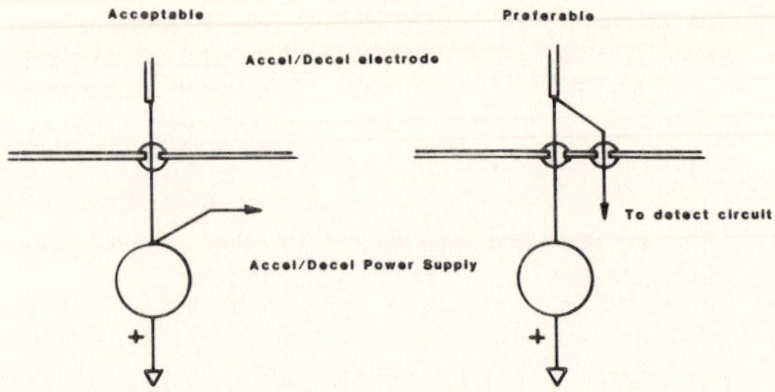


Fig. 5. Preferred method of connection for suppression-detect circuitry

In the extraction gap of a well-operated implanter, the return electron current may be as high as 10 percent of the extraction current. The electron current in the acceleration tube of a two-stage implanter is usually only a few percent of the accelerated beam current. The electron current can of course be made higher if there is no suppression system, or if there is a poor vacuum or an electrode misalignment. The design of a radiation shield should take the worst case into account.

## 1.4 Mechanisms of X-Ray Absorption

Below 1.02 MeV, x-rays may be absorbed either by photoelectric absorption or Compton scattering. Figure 6 shows that the relative importance of

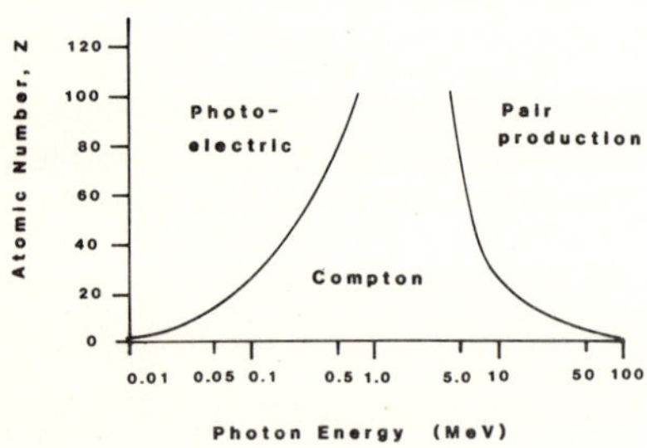


Fig.6. Plot showing regions of dominance of major absorption mechanisms as a function of energy and atomic number, Z, of absorber

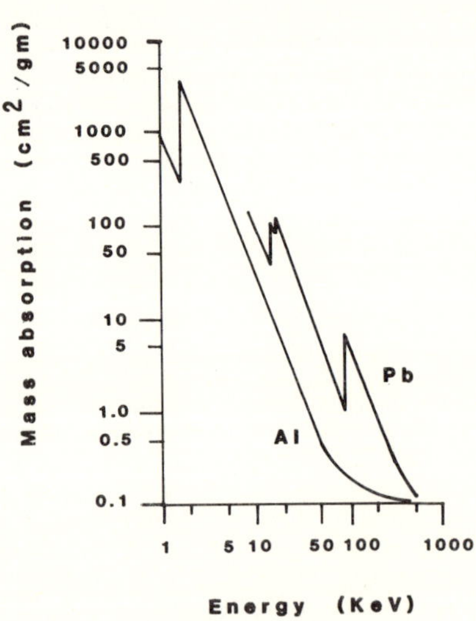


Fig.7. Photoelectric mass absorption coefficients vs energy for Al and Pb

these two effects is a function of the incoming photon energy and the atomic number, Z, of the absorbing material.

In the photoelectric process, an incoming photon transfers its energy directly to an electron, exciting it to a higher energy level or actually ejecting the electron from its bound state in the target atom. Consequently, absorption edges are observed just at those energies at which, in the reverse process of x-ray production, characteristic x-ray peaks are observed. Photoelectric excitation is the dominant process for low-energy photon absorption. In Fig. 7 we have plotted the photoelectric mass absorption coefficient, $\mu_e$, versus energy for the elements Al and Pb. The sharp peaks in both curves are due to the characteristic shell transitions for these elements, as discussed earlier.

As photon energy increases beyond the K-shell for a given absorber, photoelectric absorption loses significance because the binding of the electron is small relative to the incident photon energy, and Compton scattering becomes dominant; see Fig. 6 [4]. In this process, a high-energy photon collides with an electron; some of the energy is imparted to the electron and the remainder appears as a lower-energy photon. The mechanics of a single Compton collision are rather easily explained in terms of conservation of energy and momentum, and the relativistic equivalence of mass and energy.

The linear absorption coefficient, $\mu_C$, for Compton scattering can be calculated using the Klein-Nishima formula, and shows relatively little variation in the energy range of 10 keV to 500 keV, while the photoelectric coefficient declines rapidly with increasing energy in this range. Figures 8a and 8b [5] show the relative importance of the two effects for a light element, Al, and a heavy element, Pb.

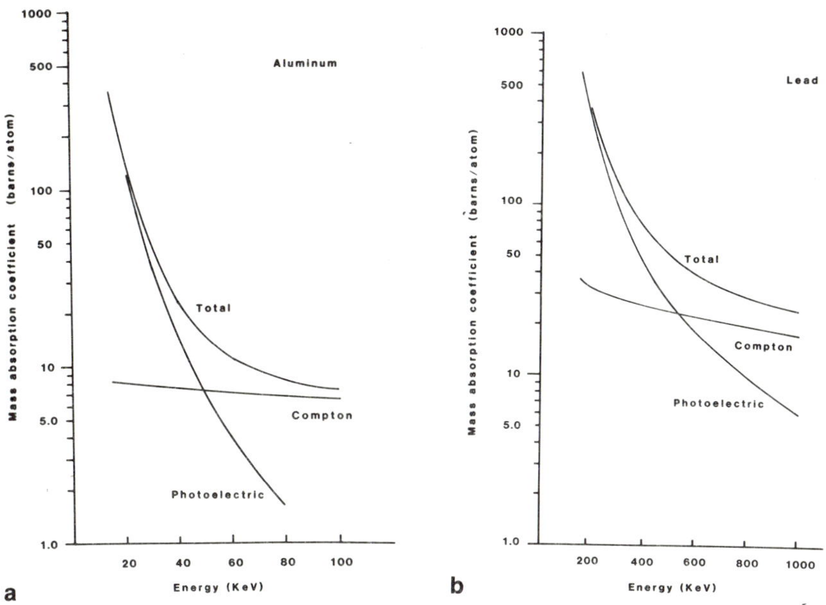


Fig. 8. Mass absorption coefficients for Al (Fig. 8a) and Pb (Fig. 8b), showing relative contributions of photoelectric and Compton scattering

Discontinuities in the photoelectric absorption curve do not appear in either figure, because the energy range plotted in each case is well above the K-shell energy for that element. Notice also that the "crossover" point occurs at 50 keV for Al and about 500 keV for Pb, in accordance with the curve of Figure 6. Materials chosen for shielding are generally heavy elements and in an implanter shielded with lead, the photoelectric absorption still dominates at energies up to 200 keV, even though this energy is well above the K-shell energy for lead.

A monochromatic x-ray beam is attenuated in an absorber according to the relation:

$$I = I_o \exp(-\mu_x x) \tag{3}$$

where $I$ = the intensity of x-rays emerging from an absorber of thickness x,
$I_o$ = the incident intensity, and
$\mu_x$ = the linear absorption coefficient which is the sum of $\mu_e$ and $\mu_c$.

The radiation from an implanter will, however, have a continuous energy distribution extending up to the maximum electron energy. Since the linear absorption coefficient $\mu_x$ diminishes steadily with energy in the region of interest, the selective absorption of the softer x-rays shifts the spectrum towards the high-energy end of the spectrum. Thus, any direct x-ray beam, after sufficient initial absorption, behaves as if it were a monochromatic beam having an energy equal to the highest energy in the continuous spectrum. This approximation is almost exact in shielding situations where the incident radiation is reduced by many orders of magnitude. For reference, Table 1 gives the absorption of aluminum and lead for a range of energies.

Table 1. Linear x-ray absorption coefficients for aluminum and lead as a function of x-ray energy. For example, the attenuation of 80 keV x-rays in 1.5 cm of aluminum is $I/I_o = \exp-(0.53 \times 1.5) = 0.45$. The mass absorption coefficient $\mu_m = \mu_x/\rho$, where $\rho$ is the density of the element in g/cm$^3$ ($\rho$=2.7 for Al and 11.4 for Pb)

| X-Ray Energy (keV) | $\mu_x$ for Al (cm$^{-1}$) | $\mu_x$ for Pb (cm$^{-1}$) |
|---|---|---|
| 10 | 72.4 | 964 |
| 20 | 9.40 | 821 |
| 30 | 3.05 | 268 |
| 40 | 1.51 | 120 |
| 50 | 0.97 | 65.3 |
| 60 | 0.73 | 40.5 |
| 80 | 0.53 | 18.9 |
| 100 | 0.46 | 62.4 |
| 200 | 0.33 | 10.7 |
| 300 | 0.28 | 4.30 |
| 400 | 0.25 | 2.51 |
| 500 | 0.23 | 1.73 |
| 600 | 0.21 | 1.36 |

## 1.5 Radiation Units

The term "radiation" covers electromagnetic radiation from soft x-rays to gamma rays, all types of charged particles and neutrons. Radiation therefore has a wide spectrum of energies and many types of interaction with matter. To measure radiation and radiation effects quantitatively, it is necessary to use several units. These fall into two classes: those which measure quanta or numbers of particles, such as flux and the curie, and a second, more familiar group of units including the Roentgen, rad or rem, which measure ionization or biological damage. In this paper only x-radiation is considered, since this is the only form of radiation that presents any danger in commercially used implanters.

The Roentgen (R) is by far the most commonly used and best defined unit for radiation dose. As defined by the International Commission on Radiological Units and Measurements, one Roentgen is:

> "...that quantity of x- or gamma radiation such that the associated corpuscular emission per 0.001293 gram of dry air produces, in air, ions carrying 1 esu of quantity of electricity of either sign." [6]

It is important to note that radiation can only be measured by its effects on matter. In this case, the effect is the creation by ionization of 1 esu ($3.3 \times 10^{-10}$ coulombs) of positive and negative charges in 1 $cm^3$ of air at standard temperature and pressure (.001293 grams). Alternatively, the Roentgen can be defined in terms of energy deposited per gram of air, assuming a value for the ionization energy of air (34 eV). In this case, one Roentgen can be defined as that amount of radiation that deposits 84 ergs per gram of air traversed.

When biological effects of radiation are considered, the energy-absorbing medium is living matter, usually soft tissue having a slightly higher ionization potential than air. In this case, one Roentgen liberates 93 ergs/gm of tissue.

The rad (r) is a radiation unit which is independent of the absorbing medium, and is defined as that amount of radiation which deposits 100 ergs/gm in a material. In Table 2, radiation safety standards of several countries are given.

Table 2. Radiation safety standards

| | |
|---|---|
| USA | 0.25 mr/hr |
| Japan | 0.06 mr/hr |
| Germany | 0.25 mr/hr |

It is well known that one rad of fast neutrons produces much more biological damage than one rad of x- or gamma rays. For this reason, different types of radiation have been assigned values of relative biological effectiveness (RBE), and another radiation unit, the rem (acronym for Roentgen Equivalent Man) was devised. Rads, rems and RBE's are related as follows:

$$\text{No. of rems} = \text{No. of rads} \times \text{RBE}$$

The RBE for x-rays is unity, by definition. Fast neutrons have an RBE of 20. Thus, a rad of x-rays equals one rem, while a rad of fast neutrons equals 20 rem.

It should be noted that the Roentgen, rad and rem are quite large units. Table 3 summarizes the physiological effects for homo sapiens of whole-body exposure received in a period of a few hours.

Federal and state regulations generally use the rem as a measurement unit. For this reason, it is useful to note that the quantity of radiation defined by the Roentgen in air or in tissue, and by the rad, are all approximately the same to within ten or fifteen percent.

Table 3. Effects of whole-body exposure received in a few hours

| | |
|---|---|
| 1 rem | No detectable change |
| 10 rem | Blood changes detectable |
| 100 rem | Some injury; no disability |
| 200 rem | Injury and some disability |
| 300 rem | Injury and disability |
| 400 rem | 50 % deaths occur within 30 days |
| 600 rem | 100% deaths occur within 30 days |
| 10,000 rem | 50 % deaths occur within 4 days |
| 100,000 rem | Quick deaths |

## 1.6 Example of a Radiation-Level Calculation

The purpose of this section is to illustrate the concepts discussed in the previous sections with a simple example.

Let us assume a high-current implanter is running for four-milliamp boron beam at 80 kV, with an extraction current of 33 mA. Suppose in this case the reverse electron current is 30% of the extraction current, or 10 mA = $6.2 \times 10^{16}$ electrons/sec = $I_o$ (see Sec. 1.3).

Referring now to Table 4, we divide the energy range 0-80 keV into 5-keV intervals. We tabulate in the first column the probability of photon emission, $dP_x$, per incident electron for a molybdenum target, in each energy range (see Eq.(1), Sec.1.1). In the next column we list the linear absorption coefficient of lead in each range. Column 3 lists the quantity $I/I_o = \exp(-\mu_x x)$, from Eq.(3), Sec. 1.4 (x=0.3 cm). The entries in this column correspond to the number of photons transmitted through 3 mm of lead per incident photon. Finally, in column 4, we list the product dP $\exp(-\mu_x x)$, which gives the number of photons transmitted through the shielding for each electron incident on the target.

The last column of Table 4 shows clearly that the x-ray transmission through 3 mm of lead is insignificant below 60 keV. The total number of photons transmitted per incident electron ($N_{te}$) is obtained by summing all the entries in this column. To 10% accuracy, however, we need only consider the last four entries, giving $N_{te} = 2.2 \times 10^{7}$.

Multiplying $I_o$ by $N_{te}$ gives us the total number of photons transmitted per second through the shielding:

$$N_o = I_o \cdot N_{te} = 6.2 \times 10^{16} \times 2.2 \times 10^{-7} = 1.36 \times 10^{10} \text{ photons/sec.}$$

Table 4. See text of Sec. 1.6 for detailed discussion

| Energy range (keV) | $P_x$ (based on 5-keV energy intervals) | Linear absorption coefficient of lead ($cm^{-1}$) | Attenuation through 3 mm lead $\exp(-\mu_x x)$ | Photons transmitted per incident electron |
|---|---|---|---|---|
| 5-10 | $7.9 \times 10^{-3}$ | ---- | ---- | ---- |
| 10-15 | 3.7 | 912 | ---- | ---- |
| 15-20 | 2.3 | 1,370 | ---- | ---- |
| 20-25 | 1.6 | 798 | ---- | ---- |
| 25-30 | 1.2 | 399 | ---- | ---- |
| 30-35 | $8.8 \times 10^{-4}$ | 254 | ---- | ---- |
| 35-40 | 6.8 | 171 | ---- | ---- |
| 40-45 | 5.3 | 111 | $3.2 \times 10^{-15}$ | ---- |
| 45-50 | 4.1 | 82.1 | $2.0 \times 10^{-11}$ | ---- |
| 50-55 | 3.2 | 59.2 | $2.0 \times 10^{-8}$ | ---- |
| 55-60 | 2.4 | 46.7 | $8.1 \times 10^{-7}$ | ---- |
| 60-65 | 1.8 | 35.9 | $2.1 \times 10^{-5}$ | $3.8 \times 10^{-9}$ |
| 65-70 | 1.2 | 28.5 | $1.9 \times 10^{-4}$ | $2.3 \times 10^{-8}$ |
| 70-75 | $7.5 \times 10^{-5}$ | 22.8 | $1.1 \times 10^{-3}$ | $8.3 \times 10^{-8}$ |
| 75-80 | 3.5 | 19.4 | $3.0 \times 10^{-3}$ | $1.1 \times 10^{-7}$ |

Assuming these photons are emitted isotropically, the photon or x-ray flux one meter from the source is $1.1 \times 10^5$ photons per square centimeter per second.

According to Jaeger [7], the conversion from x-ray flux to Roentgens/hr in the energy range from 70 keV to 2 MeV is given by:

$$R/hr = 1.8 \times 10^{-9}\, IE,$$

where I is the x-ray flux (photons $cm^{-2}$ $sec^{-1}$) and E is the photon energy in keV. Using 80 keV for this energy, we obtain, finally:

$$R/hr = 1.8 \times 10^{-9} \times 1.1 \times 10^5 \times 80 \simeq 16 \text{ mR/hr}.$$

## 2. Poisonous Materials

Boron, phosphorus and arsenic are by far the most common dopants now being used in ion implanters.

The easiest way to generate ions in an implanter is to introduce a gas containing the dopant directly into the ion source, or by using a solid form of the dopant and heating it so that the vapor is introduced into the ion source.

Boron trifluoride ($BF_3$), arsine ($AsH_3$) and phosphine ($PH_3$) are the gases most commonly used in the ion source. Unfortunately, all three gases are <u>highly</u> toxic. Table 5 summarizes the toxic effects of all three gases.

Table 5. Summary of characteristics and toxic effects of arsine, phosphine and boron trifluoride.

| | Arsine ($AsH_3$) | Phosphine ($PH_3$) | Boron Trifluoride ($BF_3$) |
|---|---|---|---|
| Characteristics | Colorless, mild, garlic odor | "Decaying fish" odor | Pungent, irritating odor |
| Molecular Weight | 77.93 | 39.04 | 67.82 |
| Toxicity | | | |
| Acute Local | High | Moderate | High |
| Acute Systematic | High | High | Unknown |
| Chronic Local | Unknown | Unknown | High |
| Chronic Systematic | High | High | Unknown |
| Mechanism | Hemolytic action | Primarily a central nervous system depressant | Most effects due to hydrofluoric acid in lungs |
| Symptoms: Acute | Headache, dizziness, nausea, vomiting gastric pain, dark or bloody urine | Restlessness, followed by tremors, fatigue, drowsiness, nausea, vomiting | Powerful lung irritant |
| Symptoms: Chronic | Anemia, jaundice, edema of lungs, kidney damage, a known carcinogen | Anemia, bronchitis, gastro-intestinal disturbances; also visual, speech or motor disturbances | ------ |
| Fire Hazard | Moderate | High | ----- |
| Threshold Limit Value (TLV) | 0.05 ppm | 0.3 ppm | 1 ppm |

In most implanters, these gases will be stored in high-pressure metal bottles containing gas at pressures of several hundred psig. These bottles, along with the appropriate regulators and gas manifolds, reside within a hermetically sealed cabinet near the implanter's ion source. A toxic exhaust vent is brought down to the cabinet and maintained at a pressure slightly below atmosphere; in this way any slight gas leaks are routed away from operating personnel. When the access door is opened, the air velocity across the opening should exceed 100 ft/sec (30 m/s) [8].

It is the manufacturer's responsibility to provide gas fittings, regulators and manifolds which are appropriate for the gases being handled, and which allow safe and simple changing of the gas bottles. It is equally important for those personnel involved in gas-bottle changing to be thoroughly trained in the proper procedures. Such personnel must also be made aware of the extreme toxicity of these gases, as well as of emergency and first-aid procedures in the event of a gas leak.

The primary gas-safety system is the implanter's toxic vent. Flow or pressure sensors may be installed in this vent to monitor its operation, and an alarm sounded in the event that the negative pressure of the exhaust duct is not maintained. Devices are available which directly monitor the levels of arsine and phosphine in air; the Matheson Model 8040 is an example of such a device. In the event of a gas leak, all personnel in the vicinity should be instructed to leave immediately. Necessary corrective action should then be undertaken only by trained personnel using emergency air or oxygen equipment. Such equipment , along with a list of emergency phone numbers (local hospital, ambulance, toxicology department, etc.), should be stored and displayed in some conspicuous location outside the area in which the implanter resides.

As mentioned previously, some ion sources are designed to permit the use of solid materials. Elimination of bottles of toxic gas is a definite safety advantage. Solid source materials also generate fewer dissociation products , resulting in lower extraction current for a given beam current on target. The primary disadvantages of this type of operation are the added warm-up and cool-down times due to the thermal mass of the heater and crucible system. Furthermore, such operation tends to coat vacuum walls with films of toxic materials.

Unfortunately, toxic chemicals are not confined to the ion-source region. That portion of the extracted beam which does not land on the target wafer(s) will be deposited on some internal part of the implanter, or will end up in one of the implanter's vacuum pumps. Normal maintenance procedures involve cleaning and scrubbing of ion-source parts and other mechanical parts which have been in direct contact with the ion beam.

Contaminated parts should never be handled without plastic or rubber gloves. Furthermore, any scrubbing operation should be performed in a properly ventilated area (see Fig. 9) which ensures that particles are directed away from personnel. Fume hoods and "clean stations" are commercially available, and designed specifically to facilitate such operations.

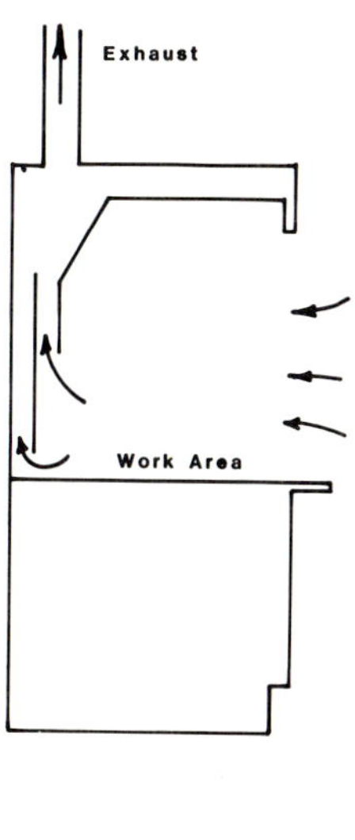


Fig. 9. Ventilated toxic-material-handling bench

Contaminated vacuum-pump oil is another potential hazard. Again, those who change oil in diffusion pumps or roughing pumps, or perform service on these pumps, should be wearing protective gloves. Contaminated oil and cleaning materials should be disposed of properly. If cryogenic pumps are used, the implanter's design must ensure safe venting during regeneration of the pump or in the event of a power failure. Because of their mode of operation (by condensing gases), cyropumps are not recommended for use in the source region of an implanter.

## 3. High Voltage

A typical ion implanter contains many electrical power supplies, ranging in voltage from several volts to many kilovolts. Table 6 lists some of the power supplies likely to be found in any modern implanter; many more are used to power control circuitry, valves, motors, and so on.

The implanter's design must be ensure that surfaces and circuitry at high potential are properly shielded from human contact under normal operating conditions. Maintenance procedures almost invariably expose personnel to some additional risk; in such situations there is no substitute for common sense and caution on the part of the user. Familiarity must not be allowed to breed carelessness.

A case in point is the extraction power supply. In most designs, we find that all of the electrical, mechanical and other support systems for the ion source are physically "grounded" to the positive output lead of this power supply. The operator or maintenance person who forgets this fact may be rudely, if not fatally, reminded.

In practice, the most effective safeguard against high voltages is to enclose them as completely as possible in a sturdy shell of conductive material. Occasionally such shells must be opened to allow access to the machinery and circuitry within them. For example, the shell surrounding the high-voltage terminal contains one or more doors on hinges (see Fig. 10a). A simple mechanical interlock - the drop bar - ensures that the terminal will be shorted to ground if any door is opened. Electrical interlocks should also be used to supplement this system (Fig. 10b). Microswitches can be installed on any door so that opening the door also opens the switch. All switches are wired in series; if any switch is opened, a relay or contactor will disconnect input power to the high-voltage power supply.

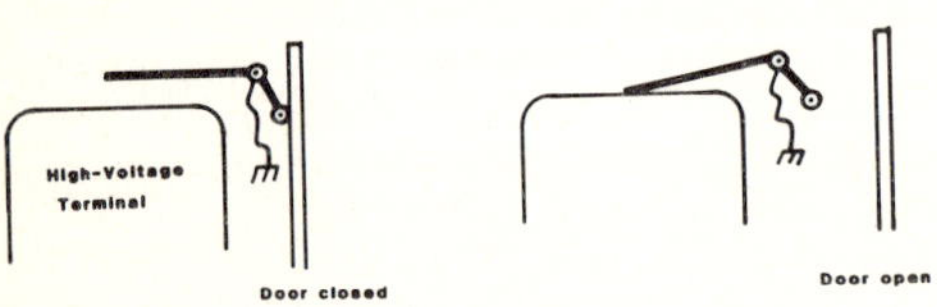


Fig. 10a. "Drop bar" arranged to short out surface at high voltage when enclosure door is opened

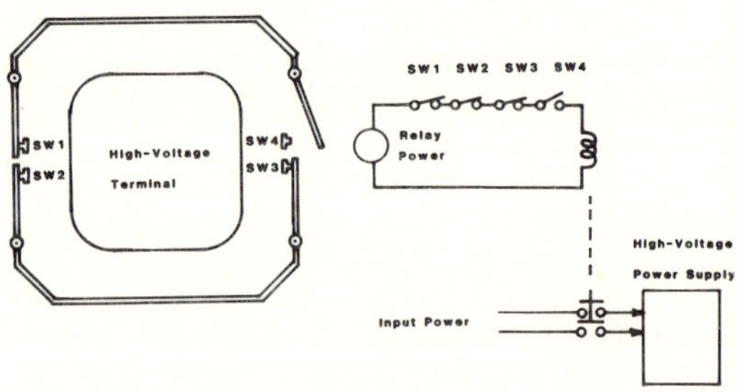


Fig. 10b. Doors in high-voltage enclosure showing placement of microswitches and corresponding interlock circuit

Table 6. Electrical power supplies and subsystems common to most types of commercial implantation machinery which could cause severe electrical shock

| Name and/or Function | AC or DC | Typical Voltage | Typical Current Rating |
|---|---|---|---|
| Arc Power Supply (maintains filament at positive potential with respect to arc chamber) | DC | 60 - 90 V | 0.1 - 10 A |
| Accel/Decl (maintains extraction electrode at negative potential with respect to beamline) | DC | 800-1600 V | 20 mA |
| Bias (faraday suppression voltage) | DC | 200-900 V | 20 mA |
| Scanning Plates | HF Saw tooth | 1 KV - 50 KV peak-to-peak | 5 mA |
| Quadrupole Lenses | DC | 1 KV - 20 KV | 5 mA |
| Grid Supply on Hot-cathode ionization-gauge tubes | DC | 160 V | 5 mA |
| Acceleration (High-current, single-stage implanter) | DC | 20 KV - 80 KV | 50 mA |
| Acceleration (medium-current extraction power supply) | DC | 25 KV | 5 - 10 mA |
| Acceleration (medium-current post-analysis) | DC | 180 KV | 2 - 5 mA |

As an additional safety feature, key-operated switches or door locks should be employed in order to keep non-authorized personnel from energizing the high-voltage supplies within the implanter. Individuals who need to work within these areas should remove the keys and keep them on their person while performing such work. Note: Jumping interlocks is dangerous and quite apart from high-voltage dangers, accidental exposure to x-rays is also feasible if the implanter is operated.

"DANGER" or "WARNING" labels should be applied to all surfaces at high potential. They should also appear on the outside surface of any assembly or enclosure which contains high voltages.

Numerous techniques are employed by designers of electrical equipment to safeguard the equipment itself and the users of the equipment. Such techniques include (but are not limited to) the following:

1. Fuses, circuit breakers and current-limiting resistors on the output of power supplies.

2. Bleeder resistors to discharge the filter capacitors on the output of D.C. power supplies.

3. "Varistors" and spark-gaps to short out transient over-voltage spikes.

4. Careful attention to current, voltage and power ratings of all components, wiring and connectors.

5. Attention to control-system design so that missing control signals, or control signals out of sight or sound range, do not create hazardous situations.

6. Emergency-off ("EMO") switches mounted conspicuously in several locations around the implanter.

## 4. Mechanical Hazards

Like other large machines which contain moving parts, ion implanters are capable of doing physical damage to careless or inattentive personnel. Again, the dangers to maintenance personnel are magnified.

Gate valves or in-line valves within the implanter's vacuum system often employ a guillotine-like motion, with forces large enough to severely injure fingers or hands. Where such valves need to be exposed for maintenance procedures, they should be deactivated, usually by removing the air lines to the air cylinders which drive them. Alternatively, a block of wood may be wedged into the valve to keep it from closing.

Implosion is always a potential hazard. The implanter's vacuum-control system should be interlocked so as to minimize this possibility. Anyone who operates vacuum valves while overriding these interlocks must pay attention to the pressure on either side of the valve, and should be very familiar with the machine's "vacuum map".

Hot-cathode ionization gauges are particularly fragile and tend to protrude awkwardly. Where possible, they should be mechanically protected against accidental contact.

Wafer-handling mechanisms at the implanter's end-station may present additional dangers. This is particularly true for high-current implanters which implant wafers in batches rather than one at a time, and employ wafer-scanning rather than beam-scanning techniques. Fortunately, the hazards are fairly obvious and not too difficult to deal with. Some of these are listed below:

1. Spinning Discs and Carousels: High rotational speeds and inertia, along with wafer-mounting hardware, present a serious hazard to fingers, hands and arms. In most cases, simple mechanical obstacles can be designed to keep operators out of harm's way.

2. Wafers Flying off Spinning Discs: Properly designed wafer clamps can virtually eliminate this possibility. Mechanical barriers should also protect the operator in this event.

3. Vacuum Chamber Doors: The vacuum chamber in which the disc or carousel resides generally has a large lid or cover which is opened to allow change of discs, and closed while implanting. Due to their size and weight, the actuators on such doors often employ rather large forces. Again, mechanical barriers should be designed to keep fingers and hands out of harm's way. "EMO" switches may be installed nearby so that the actuators may be instantly deenergized.

## 5. Fire, Flooding, Earthquake

### 5.1 Fire

Fire in an ion implanter is caused almost without exception by an electrical failure, which can range from a fire in an electric motor due to a mechanical fault or failure due to the overheating of an electronic component. Usually the fire is localized to the component which fails, and very often the circuit overloads and causes a system to blow a fuse or otherwise protect itself, and that is the extent of the problem. Occasionally the fire spreads to the surroundings, materials catch fire, and as a result there is damage to the implanter and the surrounding area.

The most significant fire hazard is posed by the presence of oil in high-voltage power supplies and in mechanical pumps. Designers must be aware of the fire danger created by the presence of oil, and of the different Federal and State regulations governing the use and location of the oil-filled units. Large oil-insulated power supplies may have to be located outside the building in which the implanter is located, and a long insulated cable must be used to couple the power supply to the implanter. Vacuum breakdowns which often occur sometimes can create significant overvoltages in such cables and cause a fire. To minimize risk, power supplies should be provided with a properly rated circuit breaker or fuse. Choosing the correct value is not so easy, as a power supply which shuts down too frequently can make an ion implanter unusable.

Since an implanter uses insulating material in many places such as in the high-voltage stack and in the vacuum insulators, a flame-retardant insulating material or ceramic should be used. Materials such as lucite, polyethylene and PVC are flammable and can catch fire quite easily. Smoke and fumes from all burning materials are dangerous, and the combustion of certain materials such as PVC creates some very toxic vapors. Further, an ion implanter usually has regions of forced draft for cooling or to ventilate regions where toxic gases are stored, and this increases the risk that the fire will spread rapidly.

Design Precautions:

- o Minimize the use of flammable materials.
- o Use Freon or $SF_6$ instead of oil as an insulating medium.
- o Use inorganic rather than plastic insulators.
- o Provide protective devices such as thermal switches.
- o Provide a fire or smoke alarm.

The safety of plastic material is usually measured by evaluating the rate at which a flame spreads over its surface when exposed to a test fire. This information is available for most materials. The flammability of a liquid is measured by determining its flash point, or the temperature at which it spontaneously bursts into flame. For transformer oil, this temperature is about 300°F(150°C). A surface coated with oil is more likely to heat up quickly, burst into flame and be the cause of a larger fire. Oily dirt should therefore not be allowed to collect in trays or corners.

### 5.2 Flooding

The least serious but certainly a very annoying hazard is the bursting of a water-coolant line in an implanter. This can be caused by the bursting of a tube made of an unsatisfactory material such as unreinforced rubber, or by a loose fitting. Corrosion can also in due time be the cause of a leak. Overheating in a clogged cooling line will also cause a plastic line to burst. When an ion implanter is being operated, a water leak is either observed directly or causes the element not being cooled to shut down, thus alerting the operator. However, most damage seems to occur during shutdown periods, weekends for example, when a burst may not be observed for a long time and considerable water damage can occur. (N.B.: In shutdown periods the water-line pressure often rises). There appears to be no satisfactory protection in use at the present time which gives an early warning of this type of event.

## 5.3 Earthquake

Mechanical shock can be transmitted to implanter structures by horizontal ground motions caused by truck or rail shipment, or by earthquakes. In either case, acceleration values seldom exceed 0.3 g. This suggests use of a horizontal shear-load criterion based on 0.5 g when sizing critical fasteners and structures, for example, analyzing magnet supports. It should be emphasized that much larger loads are possible, e.g. in the event of vehicular collision, but in these cases, damage is usually difficult to avoid. Cost of repair of such damage can be reduced, however, by providing adequate clearance around large masses so that adjacent equipment is not damaged by moderate movement; by ensuring that connections such as pipes and wires are more flexible than the mounting structure; and by fastening components that require critical alignment to each other rather than to auxiliary structures that will tend to move independently. Lastly, manufacturing controls that ensure use of correct fasteners are essential to avoid possible substitution of bolts that are too short or made of inadequate material, or that work loose due to the vibration occurring during shipment.

## References


1. F.S. Goulding, J.M. Jaklevic, Nucl. Instrum. and Methods 142, 323 (1979)

2. Beatty, Proc. R. Soc. London 89, 314 (1913)

3. H.A. Kramers, Philos. Mag. 46, 836 (1923)

4. J.F. Kircher, R.E. Bowman, Effects of Radiation on Materials and Components (Reinhold, NY, 1964), p. 13, 15

5. G. Grodstein, X-Ray Attenuation Coefficients from 10 keV to 100 MeV, N.B.S. Circular No. 583, p.19,34

6. M.R. Wehr, J.A. Richards Jr., Physics of the Atom (Addison-Wesley, Reading, MA, 1967), p. 211

7. T. Jaeger, Principles of Radiation Protection Engineering (McGraw-Hill, NY, 1965), p.55

8. N.I. Sax, Dangerous Properties of Industrial Materials, 4th Ed. (Van Nostrand-Reinhold, NY, 1975), p.73

# Part II

# Ion Ranges in Solids

# The Stopping and Range of Ions in Solids

J. P. Biersack

Hahn-Meitner-Institut, Glienicker-Strasse 100,
D-1000 Berlin 39, Fed. Rep. of Germany

J. F. Ziegler

IBM-Research, Yorktown Heights, NY 10598, USA

Abstract

The stopping and range of ions in matter is physically very complex, and there are few simple approximations which are accurate. However, if modern calculations are performed, the ion distributions can be calculated with good accuracy, typically better than 10%. This review will be in several sections:

a) A brief exposition of what can be determined by modern calculations.

b) A review of existing widely-cited tables of ion stopping and ranges.

c) A review of the calculation of accurate ion stopping powers.

## 1. Introduction

The range distribution for a particle injected through the surface of a target is defined as the probability of finding the particle at rest at a given position inside the target. Statistical notation enters this definition because a single ion history does not represent the experimental situation where a beam of ions is injected and only the mean history of many ions is observable.

In spite of the great interest in ranges, there exist surprisingly few actual experimental determinations of range distributions. Most of the experimental studies of ranges have been limited to certain characteristic parameters of the range distribution, such as the depth of the maximum of the distribution (most probable range), the mean value, and the full width at half maximum (FWHM) or the range straggling (standard deviation).

In range theory, the range distribution is regarded as a transport problem describing the motion of the ions during their slowing down to zero energy. The target is assumed to be undisturbed during the ion penetration, and after-effects (such as thermal or radiation-enhanced diffusion) are neglected. Final range distributions have been calculated using analytic transport theory, Monte Carlo calculation, and the direct simulation of ion trajectories. The common problem in these treatments has been to include reliable descriptions for the interactions between the moving ion and the surrounding target particles. The interactions are usually separated into binary elastic collisions between the ions and the screened nuclei of the target atoms, and into inelastic collisions by the ion with the entire electron system of the target.

The elastic interaction between the ion and target nuclei is described by means of a detailed interaction potential in direct simulations and by means of scattering cross sections in transport and Monte Carlo calculations. These elastic collisions cause both angular deflections and stopping (energy loss) of the ions. The inelastic interactions with the target electrons are conveniently collected into mean values such as the electronic stopping cross section and the electronic energy-loss straggling caused by the large variety of possible discrete collisions between the ion and the target electrons. Ion scattering by the target electrons is neglected because of the light electron mass. Because of this both the elastic as well as the inelastic collisions are statistical events. Even with the neglect of the angular deflections of the ion, the elastic collisions between the ion and target nuclei would result in a range distribution because of the spectrum of the elastic energy losses. A similar result holds true for the inelastic collisions between the ion and the target electrons, but this has far less quantitative importance at the usual implantation energies ($E \leqslant 10$ MeV/nucleon). Insufficient knowledge of the higher-order moments of the electron energy-loss spectrum has usually made it necessary to assume that the interaction of the ion with the target electrons is a simple energy loss, which depends only on the ion's energy. With this assumption, the electronic interaction alone leads to one specific range (a $\delta$-function distribution). The physics of ion slowing down can be summarized as follows:

<u>Interactions with target nuclei.</u> Binary collisions, energy loss, energy straggling and higher-order moments in the energy-loss spectrum, angular deflection, and elastic scattering cross section.

<u>Interactions with target electrons.</u> Collective interactions, energy loss, energy straggling, and electronic energy-loss distribution function.

Figure 1 shows typical results which can be obtained using modern calculations. The four plots are 2-dimensional maps of the Monte Carlo calculations of 100 keV boron ions into a silicon target. The upper left plot shows the final distribution of 1000 B ions injected into the center of the front edge of the plot. Since the spatial grid is fine, no more than a few ions end up in the same grid position. However, it is seen that the distribution is quite broad with the width of the distribution being about as broad as the depth distribution. Along the left and back edges of the plot are the summed longitudinal and lateral distributions. The two distributions are different, with the lateral distribution being peaked at the point of the incident beam, while the longitudinal distribution is skewed.

The lower left plot in Fig. 1 is the total energy-loss distribution of the 1000 injected ions. In this case the 2-dimensional plot appears to show that the energy loss is mostly in the first few grid spaces. But when the longitudinal distribution on the left side is analyzed, it is seen that the energy loss is almost constant for about half of the ion's track. It is the lateral scatter of the incident beam which makes it appear that the energy loss is concentrated in the surface layers, for once it penetrates past the surface atoms the incident beam is so widely scattered that the total energy loss is distributed into many spatial bins. In contrast to the longitudinal energy-loss distribution, the lateral distribution of the total energy loss is quite peaked.

The upper right plot shows the generation of vacancies in the target. The production of a vacancy is defined as a collision with a target atom which transfers more than some minimum amount of kinetic energy, usually

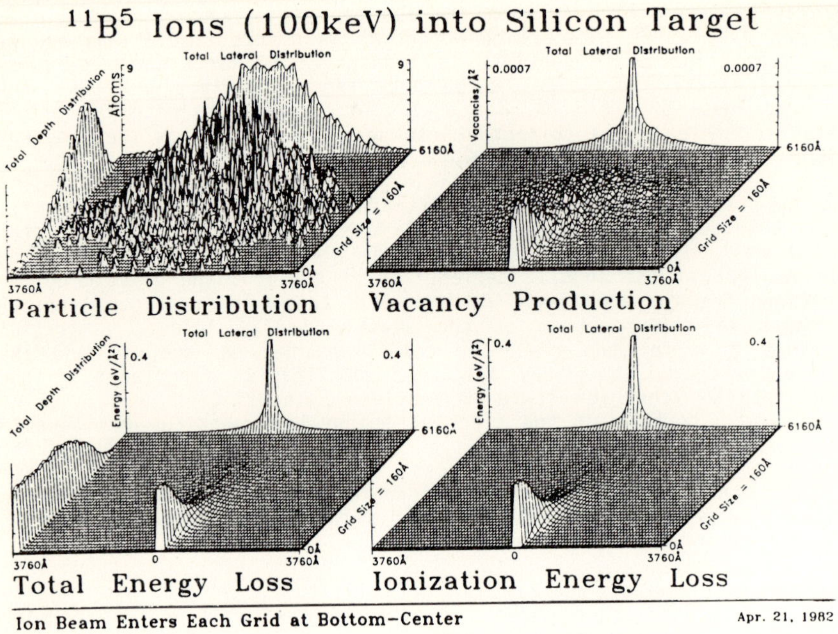


Fig. 1. Monte Carlo calculation of boron ions into silicon

called the displacement energy. If only a small amount of energy is transferred, e.g. about 10 eV, then the recoiling target atom will not have enough energy to be permanently separated from its vacancy, and the probability of recombination is almost unity. However, if more energy is transferred, e.g. more than 25 eV for the case of a Si target, then the probability is that the lattice atom will not recombine with its vacancy and a true vacancy will result. As a special additional note, if the collision is particularly violent, e.g. more than 60 eV of transferred kinetic energy, then the recoil itself may produce additional vacancies around its own vacancy. This effect is also included.

Finally, in the lower right plot in Fig. 1 is shown the energy distribution to ionization processes in the target. This type of distribution is important to certain classes of targets, especially insulators, which may be modified more by ionization effects than by vacancy production. For example, in the ion-beam exposure of photolithographic layers such as photoresist, it is the ion beam's ionization energy loss which does much of the chemical modification of the target.

## 2. Review of Some Stopping and Range Tables

During the last thirty years there have been published scores of tables and books evaluating the parameters of energetic ion penetration of matter. Rarely have the authors of these reference works included any evaluation of the accuracy of the tabulated numbers. We have chosen to show how, as theory developed through the years, various ideas have been incorporated into tables and increased their accuracy. This approach restricts our comments to those theoretical advances which have made significant contributions to the obtaining of practical ion stopping powers and range distributions. The tables reviewed were chosen because of their extensive citation in the literature.

Review of Ion Stopping Power Tables

1958 - Handbuch der Physik, W. Whaling [6]

1970 - Nuclear Data Tables, L.C. Northcliffe and R.F. Schilling [13]

1972 - A.I.P. Handbook, H. Bichsel [17]

1974 - Nuclear Data Tables, J.F. Ziegler and W.K. Chu [20]

1977 - Hydrogen Ion Stopping and Ranges, H.H. Andersen and J.F. Ziegler [27]

1978 - Helium Ion Stopping and Ranges, J.F. Ziegler [28]

1980 - Stopping Cross Sections for Energetic Ions, J.F. Ziegler [41]

The Current Accuracy of Stopping Tables

Review of Ion Range Distribution Tables

1970 - Nuclear Data Tables, L.C. Northcliffe and R.F. Schilling [13]

1970 - Ion Implantation Ranges, W.S. Johnson and J.F. Gibbons [15]

1975 - Ion Implantation Ranges, 2nd ed., J.F. Gibbons, W.S. Johnson and S.W. Mylroie [24]

1975 - Ion Implantation Ranges and Damage Distributions, Vol. 1, D.K. Brice [22]

1975 - Ion Implantation Ranges and Damage Distributions, Vol. 2, K.B. Winterbon [23]

1980 - Range Distributions for Energetic Ions, U. Littmark and J.F. Ziegler [43]

## 2.1 Stopping Power Tables

### 1958: The Whaling Table

The first of the widely used stopping power tables was that of Whaling in 1958 in the Handbuch der Physik, Volume 34 [6]. This table essentially presents experimental data for protons in various targets, with the curves extrapolated using the Bethe relation for high-energy stopping

$$S \sim (\ln E)/E \qquad (1)$$

where E is the particle's energy. Whaling stated that so little data was available for ions heavier than protons that no comprehensive tables could be made. So at this point, tables were based only on data, with possible extrapolation to high energy using Eq. (1), but no extrapolation could be made to lower energies or to heavy ion projectiles.

## 1970: The Northcliffe-Schilling Table

The next major table was the monumental effort by Northcliffe and Schilling published in Nuclear Data Tables in 1970 [13]. Any detailed analysis of this work is difficult because it is a judicious mixture of experiment, interpolation and a little theory. But in general, Northcliffe had spent a decade trying to organize in an empirical fashion the systematics of ion stopping in both solids and gases. This was done for solids by comparing the stopping of a wide variety of ions in Al, and from this array of data he established the variation of stopping with ion type. Then the stopping of a few ions was determined in a wide variety of solid targets, especially C, Al, Ni, Ag and Au. This established the target dependence of stopping; see Fig. 2. These two sets of curves were then used to parameterize the scaling laws, Eq. (1), so that any ion-target combination could be analyzed. The only basic theory introduced in this table was the use of LSS theory [17] for the low-energy stopping powers.

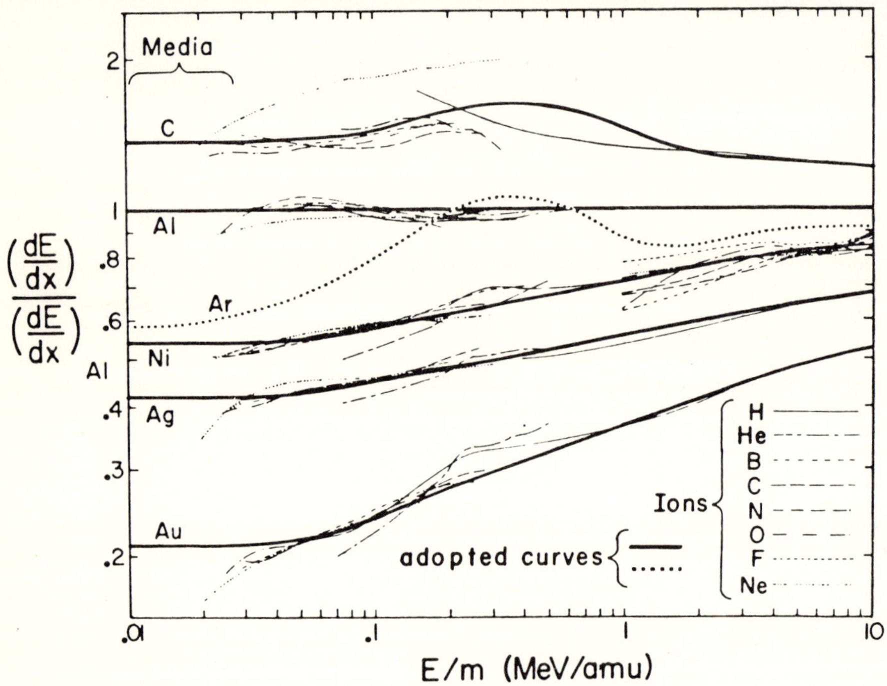


Figure 2. These curves show the primary stopping powers used for interpolation in the Northcliffe and Schilling Tables, 1970 [13]. The basic curve is for Al, which is a stable target which can be reliably made into thin pin-hole-free foils which do not significantly oxidize or corrode with time. The other primary targets similarly are corrosion resistant. The thin lines are experimental values for the ions noted in the lower right of the figure. At the time it was not appreciated that the chemical nature of these stable targets (tightly bound atoms without loose electrons to form oxides) made their stopping powers anomalously low, and not the best targets for linear interpolation (see Fig. 3). The one curve for Ar indicates that the stopping power of gas targets is quite different from solids

Northcliffe's work brought the scaling laws which were developed between 1927-1941 to their most advanced state. The stopping powers predicted for high-energy ions, E ⩾ 10 MeV/amu, were quite good and in general were accurate to better than 20 %. This is because Bethe-Bloch theory is quite accurate if the ion is moving much faster than most of the electrons in the target. For this reason we quote energy in the units of MeV/amu. (A communications problem arises because nuclear physicists usually think in particle energies, while stopping theorists think in velocities, especially ion velocities relative to target electron velocities. The compromise usually made is to discuss non-relativistic stopping in units of "energy-/mass".)

As we consider ions of energy below 10 MeV/amu we find the N-S scaling laws begin to break down in special cases and this breakdown generates large errors of up to 400% at the lowest energies of the tables, 0.0125 MeV/amu. The reason for this breakdown is most easily explained by analyzing a plot such as in Fig. 3. In this plot are shown the N-S stopping

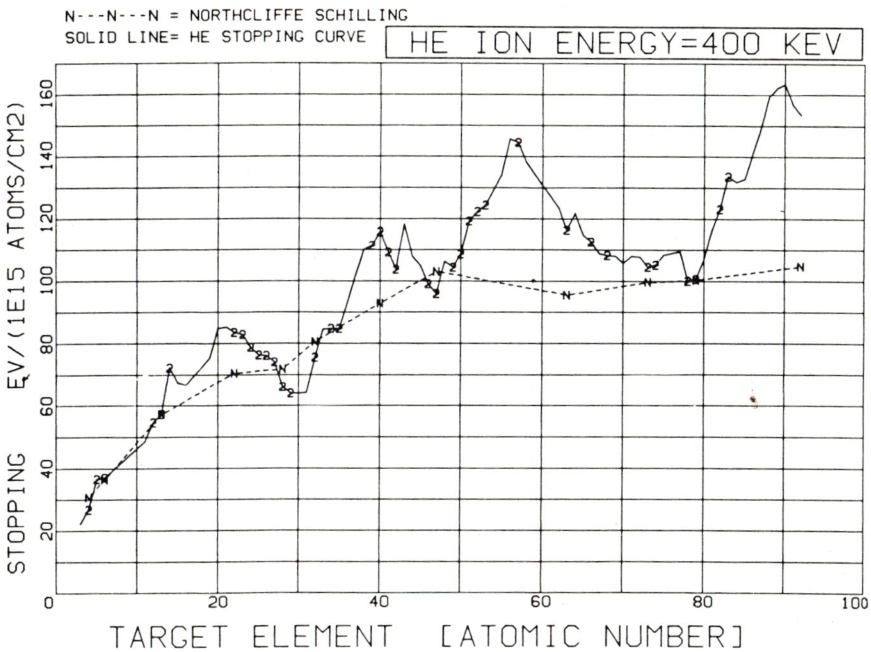


Figure 3. This figure [38], produced ten years after the N-S tables, shows the N-S table stopping values for He ions at 400 keV in various solids as small "N" symbols. Also shown with the plot character "2" are experimental values for 38 solids, most of which are averaged values from three to six different papers. It can be seen that the stopping of ions in targets which do not oxidize are near the minima of these stopping values, and this can be directly related to the tightly bound nature of the outer electrons in these targets (see Figs. 14 and 15). The use of non-oxidizing targets for the N-S standards appears to be the main source of error in this magnificent work. The solid line is from a table made in 1977, Ref. [28]

values for He at 0.4 MeV in various targets. Their values change relatively smoothly from target to target as would be expected from the N-S scaling laws. Also shown with small number 2's are experimental results for He stopping in 38 elemental targets (almost all of these experimental values were published after the N-S tables). Each value shown is usually an average of three to six published experiments. These experimental values show wide oscillations with the target atomic number, and the N-S values tend to occur at the minima of the oscillations. We shall show that this occurs because of an unfortunate choice of normaliziation targets for the N-S scaling. Northcliffe concentrated his experimental efforts on targets which did not easily oxidize or corrode. Since the experiments extended over years, it was convenient to have targets which did not degrade with time. But the very reason these targets (Au, Ag, Ni, Al and C) are materials which are chemically stable is the same reason they lie at the stopping minima! They do not have lightly bound outer-shell electrons. We shall see that any lightly bound electrons absorb a disproportionate amount of energy from a penetrating ion, and materials which oxidize readily have large stopping powers when compared to nearby tightly bound atoms.

Hence, since the N-S tables are based on stable non-oxidizing targets, their stopping powers should be a lower limit or the stopping value for any ion-target combination, unless the target is one of those actually measured. As seen in Fig. 3, for these targets, $Z_2$ = 79, 47, 28, 13 and 6; the values are quite good. The solid line in Fig. 3 represents the values of a 1977 table discussed below which is largely based on theoretical calculations.

## 1972: The Bichsel Table

Two years after the N-S tables were published, Bichsel published an abbreviated table in the American Institute of Physics Handbook [17]. This contribution is noteworthy because it includes for the first time details of atomic structure in the calculations. Bichsel had published many papers in the previous decade suggesting corrections to the Born approximation of the Bethe theory, which assumed the ion was moving much faster than the target electrons. Bichsel calculated the correction necessary to account for reduced absorption of energy by inner-shell electrons which often had classical velocities greater than the ion velocities (e.g. the innermost electrons of very heavy atoms such as Au have relativistic velocities, equivalent to ion velocities over 20 MeV/amu). He also introduced similar corrections to the mean ionization potential of atoms, which introduces into the stopping calculation the excited-state level structure of the target atoms. His corrections are essential for the correct calculation of the stopping of high-energy particles, and his tables are quite accurate, better than 10%, for energies from 5 MeV/amu to ~ 2 GeV/amu. This upper energy limit of accuracy is caused by extreme relativistic effects which were not included in his formalism.

The decade of the seventies produced two new user groups which greatly expanded the interest and experiments in stopping theory. Ion beam analysis of materials proved to be a quantitative technique which gave detailed views of surfaces and thin films. Examples of this field are the nuclear backscattering analysis of integrated circuit technology, ion-induced x-ray analysis of environmental pollution, nuclear reaction analysis of the hydrogen embrittlement of metals, and a wide variety of age dating techniques. Most of these techniques require stopping powers of greater accuracy than was required by experiments in basic nuclear physics. Especially important was the stopping of He ions, and over 100 experimental papers for this one ion were published during the short span of 1972-1980.

A second major set of new users occurred in the field of ion implantation. In 1970 there were less than 10 accelerators in industry, while there were about 400 in academia. In 1980 the statistics were reversed, with about 1000 ion accelerators in industry, with those in academia shrinking to fewer than 200. Ion implantation provided the semiconductor industry with a new way to create materials. It provided an accurate and reproducible way to introduce impurities into a solid in a cold process. Each of these advantages - accurate, reproducible and cold process - helped to make ion implantation into an indispensable manufacturing tool, and once accelerators were introduced into development laboratories they were used for many other materials besides semiconductors.

These new user groups substantially increased the amount of experimental data on ion stopping and range. Whereas in the sixties there were only about forty papers a year published in this field, by 1980 there were over 200 about ion implantation profiles in silicon alone, and over 600 per year on ion implantation in general [31].

## 1974: Ziegler and Chu Tables

The needs of this new community directly stimulated tables of He ion stopping powers published by Ziegler and Chu in the Atomic and Nuclear Data Tables in 1974 [20]. He ions were used in much of ion-beam analysis, and a major subject of such analysis in those years was ion implantation profiles. As with Northcliffe, Ziegler and Chu collected experimental data (from 11 papers) which established approximate stopping values for 47 elements over the narrow He ion energy range of 0.4-4 MeV (almost half of these 47 elements were represented by a single datum at one energy, so the experimental values were not really comprehensive).

The major new contribution of this work was to interpolate stopping powers using elaborate calculations based on Hartree-Fock atomic structure and using the detailed formalism of the Lindhard theory of electronic stopping [1]. The inclusion of these two concepts created a significant leap in accuracy over previous tables. Figure 4 shows a comparison of this new theoretical approach (originally demonstrated by Rousseau, Chu and Powers [14]), and a wide variety of experimental data points for He ions at 1 MeV. There is remarkable agreement between pure theory and experimental values as shown in Fig. 4, and it is evident that only a slight normalization was necessary to create the semi-empirical curves used in the tables. This figure can be compared with the similar Fig. 3 to see how dramatically stopping accuracy was improved by changing from an interpolation scheme based on scaling laws to detailed quantal calculations.

During the mid-seventies the outpouring of new experimental data overshadowed all previous data, both in quantity and more importantly in the increased understanding of ion penetration physics. It was shown that a large amount of the stopping and range data taken in the previous two decades was inaccurate, as the experimental apparatus used did not measure average phenomena, but anomalous or unusual phenomena, or just plain garbage.

As an example, there is the phenomenon of ion channeling in crystals. This type of ion penetration takes place down major crystalline axes, and

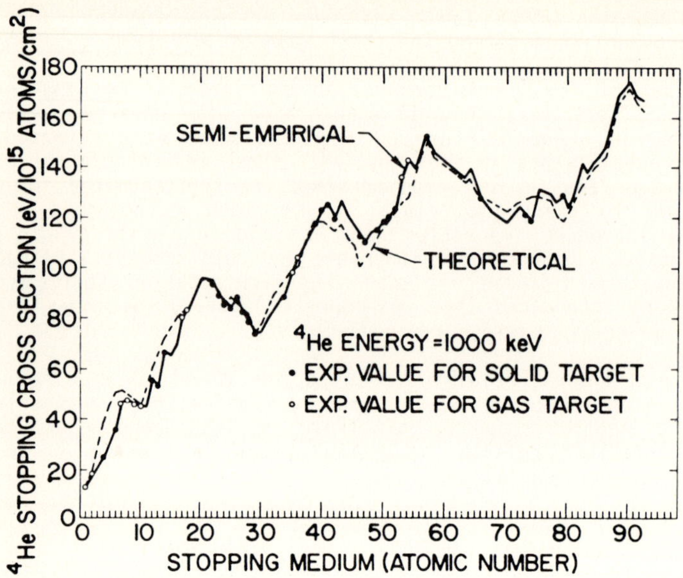


Fig. 4. Extensive theoretical calculations were included in tables of the stopping of medium-energy ions (~ 0.25 MeV/amu) for the first time in 1974. These calculations reproduce the large changes in stopping from one target to another as shown in Fig. 3 (this effect was called the "$Z_2$ oscillation effect"). The dashed line shows the theoretical calculations [28] based on the local density approximation and using Lindhard's theory [1] of particle stopping in a free-electron gas. The dots show experimental measurements for MeV He, and the solid line gives the semi-empirical tabulated values. The theoretical curve has been used to interpolate between the experimental values, giving accuracies of better than 10%

it has been shown that, as an extreme case, ions penetrating down loose salt structures can have stopping powers reduced by over 90%! Channeling was known since 1962, but another aspect of metal films (which was known by metallurgists) was not to penetrate to stopping power measurements until 1978. This was the 'target texture' quality of metals in which polycrystalline metals do not have their micro-crystallites randomly oriented, but in fact over 90% may be aligned within a degree or two about a common axis. This effect is particularly true of evaporated films, which are built up layer by layer, and atoms have time to move about the surface before they are buried. The stopping power experimenters had turned to evaporated targets in the sixties because of their inability to make pin-hole-free rolled thin films. But the problem of target texture took a decade to catch up to the users. In Fig. 5 is shown a typical result. This figure shows the collected stopping powers of H ions in Au. As can be seen, there appear to be two independent curves which can be drawn through the data. One is much lower than the other at the stopping power peak. This is also the energy at which a large difference occurs between the energy loss of channeled and random-path ions (the maximum difference in random and channeled stopping powers occurs when the ion is at the same velocity as the electrons in the channel). An analysis of the experimental details of the 14 papers reported in Fig. 5 is partially convincing. Three of the higher data sets were from experiments in which the ion beam was

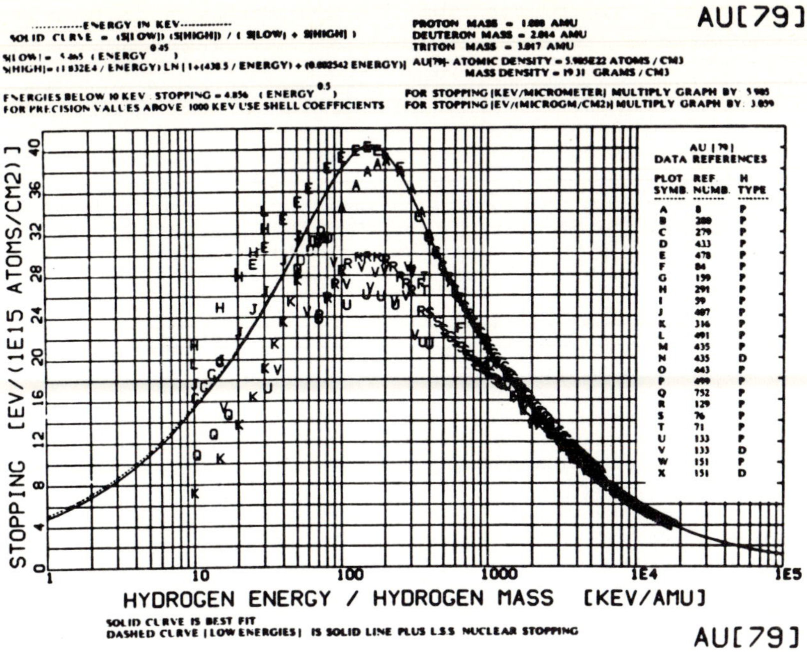


Fig. 5. One of the unappreciated problems of early stopping power measurements was 'target texture'. This is the phenomenon especially found in evaporated films where the separate crystallites of the polycrystalline film have one axis in marked alignment, usually perpendicular to the substrate surface. For example, over 95 % of the crystallites may have their <111> axis within 1° of each other, and this can lead to ion channeling and an apparent reduction of stopping powers. It has been suggested that the above H ion stopping plot shows two distinct stopping curves, with the lower one being from experiments where the beam was inadvertently aligned with the target texture. Later theoretical work affirms that the upper measurements are probably correct. The subject of target texture is treated at length in Ref. [37]

not perpendicularly incident to the target; three of the lower sets had perpendicularly incident beams; and 8 of the papers did not report the experimental conditions (which is typical of papers in Physics Letters or Soviet journals where the author has severe space limitations on his text).

In their paper on texture effects in ion penetration phenomena, Andersen, Tu and Ziegler [37] pointed out how Au had significant texture problems for almost all techniques of preparation, and the major Au axis of the polycrystalline films was always within 1° of being perpendicular to the target plane.

A second problem in stopping power measurements prior to the 70's was the lack of appreciation of straggling effects, especially the angular scatter of the ion beam from its incident trajectory. During the 50's and 60's it was common to use a magnetic spectrometer to determine a particle's energy. Because of its high energy resolution, this was a natural way to measure energy loss. And so a technique was developed and widely used to simultaneously measure stopping powers and ion ranges. This was called the foil-stack method. An incident beam was made to directly enter the small entrance aperture of a magnetic spectrometer. Then thin foils were placed in the beam, and for each foil the energy loss was measured by using the magnet. This was continued until the energy spectrum of the emerging beams was so wide in energy distribution that its mean was hard to find. This kind of experiment produced a plot of $\Delta E$ vs x, where $\Delta E$ is the energy loss in a film thickness x. A second curve was also made showing $\Delta I$ vs x, where $\Delta I$ is the amount of current lost in the foil stack. The differential of this plot was the final ion distribution, for each incremental $\Delta I$ was presumed to have stopped in the last added foil.

This type of experiment became one of the standard techniques of the 60's; then it disappeared as more advanced methods were developed. But the data stood in the literature, and it bugged the hell out of the theorists who were calculating good numbers which were sometimes in sharp disagreement with the "classic experiments". Finally, Biersack was persuaded to set up a Monte Carlo program to replicate one of the more famous experiments, and his results are shown in Fig. 6 [43]. He found in his simulation that the current detected was profoundly changed as the detector solid angle was increased (typical magnetic spectrometer entrance angles were less than 3°). When the data from various solid angles were differentiated, as shown in the lower figure, the profiles showed little resemblance to each other. Note the shaded area in the lower figure - this is the actual implantation profile from a full Monte Carlo calculation, and all the other solid lines indicate the relative errors from this ideal depending on the experimental set-up.

A second problem with this experiment is very subtle and was overlooked by the scientists, but it led to erroneous conclusions. The authors felt that the foil thickness which caused the transmitted fraction of the incident current to reach 50% was the mean depth of penetration of the ion beam. Half of the beam was stopped in the foil stack. But the mean range of an ion beam is defined as the mean depth in a target much thicker than the mean range. A little reflection may convince one that if a plane is inserted parallel to a solid's surface at the depth of the true mean range, almost an equal number of ions will be found scattering back towards the surface as penetrating deeper. This is especially true for light particles in heavy substrates. At this point of mean range the average vector momentum of the ions is zero. There is no way that this fact can be incorporated in a foil-stack experiment, for here there can be no back reflection. So the experiment fails to be meaningful in determining implantation distribution. (Further problems with earlier measurements are found in Ref.[43], pages 19-24.)

1977: Andersen and Ziegler: H Tables

Because of all these problems, the next step in stopping tables was actually a necessary step back to the Whaling concept in 1957 - just collect data and try to separate the good from the bad. This was aided by Andersen in his massive bibliography [32] (done before computer retrieval systems became wide-spread), and Andersen and Ziegler [27] published in 1977 their book on the stopping of H ions in all elements. By this time over 70 papers had been published on H stopping powers, and the authors made a data base of over 7000 experimental values. The special advance of this table was in the evaluation of high-energy stopping ($E \geqslant 1$ MeV/amu).

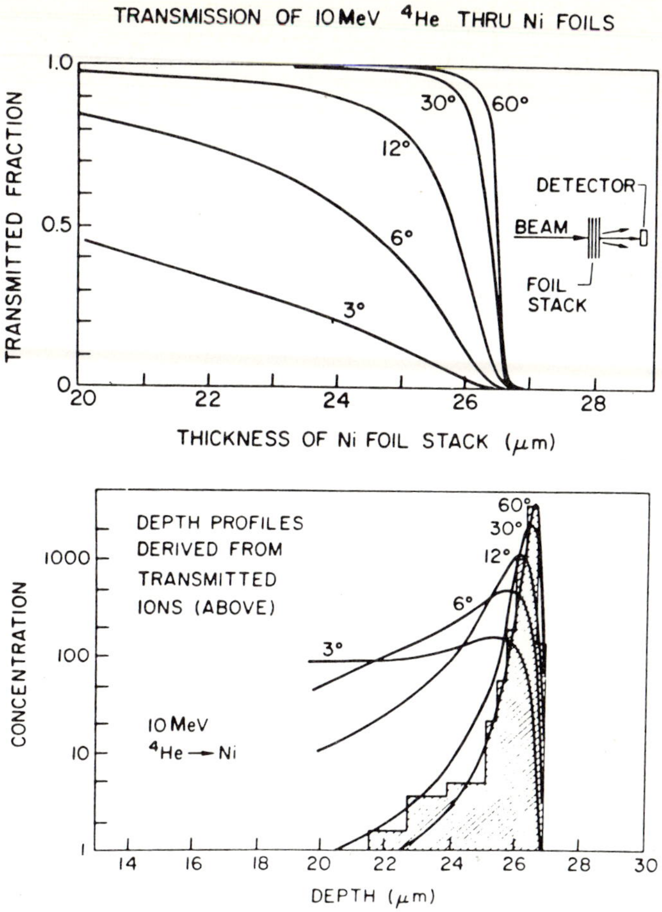


Fig. 6. Two-dimensional Monte Carlo calculations of ion-beam ranges determined by transmission through foil stacks. This technique is the most popular method of measuring high-energy ion ranges. The upper plot shows the transmitted fraction of the beam for foil stacks of various thicknesses. The calculation shows how this fraction significantly depends on the solid angle subtended by the detector (the plot shows the detector half-angle in degrees). The lower plot shows the "range profile" obtained by differentiating the spectra of the upper plot. Also shown as a shaded area is the calculated final ion distribution of ions into a solid target. Large errors can occur unless the detector is arranged to collect all the transmitted ions. The common assumption that the 'mean ion range' is where the transmitted beam falls to one-half of the initial beam current is shown to be somewhat in error, and the estimated stragglings would be very erroneous

Previously, we commented how Bichsel introduced theoretical mean ionization potentials (which brings in the atomic excitation structure) and also corrections for the fact that inner-shell electrons are not correctly treated in an impulse approximation like the Bethe theory (this correction is usually called a "shell correction"). Two other scientists, Janni [10] and Bonderup [12], each also had produced significant papers on these two subjects. Andersen and Ziegler took these three works and normalized them to the data. Since the data base was comprehensive (36 elements were covered over wide energy spans) they were able to produce high-energy stopping curves which were believed to be accurate to better than 10 % at 0.5 MeV/amu and better than 5 % above 5 MeV/amu. Only a few experiments have proven to be reproducibly more accurate than this (although accuracies to 0.1% have been estimated by enthusiastic experimentalists).

For energies below ~ 0.5 MeV/amu these authors made only coarse estimates of interpolated stopping powers, basing their values mostly on the wiggles shown in the Ziegler-Chu tables (see Fig. 4).

## 1978: Ziegler: He Tables

At this point, about 1978, stopping power theory and experiment begin to come together. In the development of the next major table, "He Stopping Powers and Ranges in All Elements" [28], a total of 121 experimental papers were included in the data base, and these were amplified by using scaling laws so that the H stopping powers of the previous book could be scaled to equivalent He values, as shown in Fig. 7. A total of over 12,000

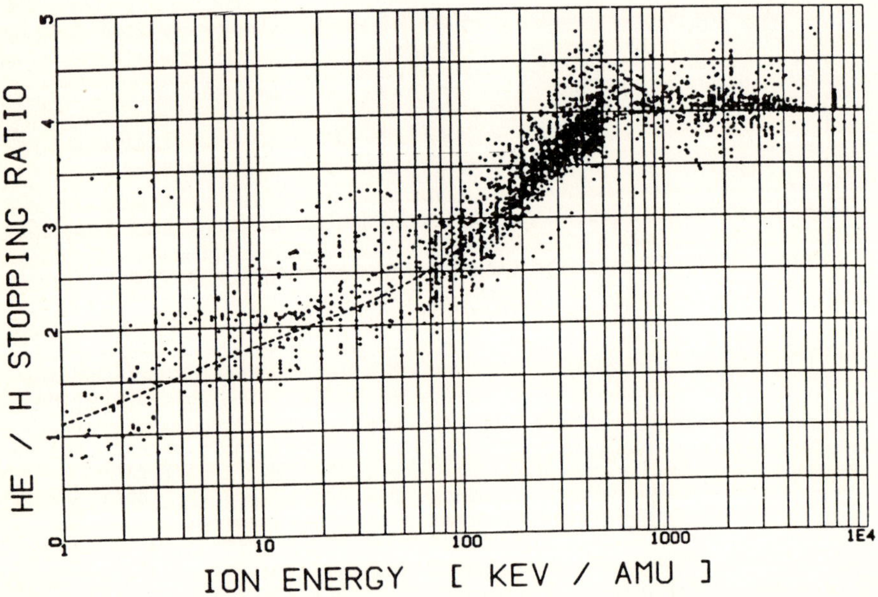


Fig. 7. This plot [28] shows the ratio of the experimental stopping of He ions in elemental matter to the stopping of H ions at the same velocity and in the same material. Because of the extensive data available, a curve could be drawn (dashed line) which allowed the production of an accurate "master stopping" curve upon which scaling predictions for heavy-ion stopping could be based (see Fig. 8). This single curve presumes there is no target dependence of the He/H ratio and above about 80 keV/amu this appears to be accurate. Below this value this presumption breaks down

experimental data points thus became available to allow finally the weeding of the good data from the inaccurate. Using this data base, the shell corrections and mean ionization potentials for high-velocity stopping were further refined. The accuracy of theoretical models of electronic stopping was further enhanced by the inclusion, for the first time, of actual solid-state charge distributions for solid targets (previously the charge distribution for isolated atoms was used). This type of calculation proved to be quite accurate for the energy span from where the Bethe theory broke down, ≈ 1 MeV/amu, down to where a new problem arose, ≃0.1 MeV/amu. Below this energy the He ion began to neutralize, and the target electrons had time to move and coalesce about the slow-moving ion (called target polarization), and these effects were not included in the theoretical approach.

## 1980: Ziegler Energetic Ion Tables

In 1980, the next step in stopping tables was the publication of "Handbook of Stopping Cross Sections for Energetic Ions" [41]. This book used all the previous advances, and added one more improvement. First it was built on the H and He stopping books, and by the time this table was published there were over 15,000 experimental data points for these two ions alone. As noted before, considerable success was achieved in scaling He to H stopping powers, so together they formed what was called a master stopping set. The next step was to scale this master set to obtain the electronic stopping for heavy ions. Considerable theory had been done on this subject from 1941-1961, using statistical models of the atom for which scaling from one element to another was done by a simple multiplication factor. There was no shell structure in these atoms, but as long as they were concerned with only the stripping of outer-shell electrons these approaches were adequate. In general, they lead to equations such as

$$\frac{Z^*}{Z} \sim 1 - \exp[-(v/v_o)]^{2/3} \qquad (2)$$

where $Z^*$ is the 'effective charge' of an atom of atomic number $Z$ with a velocity $v$. This velocity is related to the Bohr velocity $v$, about $2 \times 10^8$ cm/sec. At very low velocities the atom is unstripped, but as its velocity increases beyond the Bohr velocity (a typical velocity of the ion's outer-shell electrons) the ion's electrons begin to be stripped off. Another theory is to assume all the ion's electrons which have velocities below the ion's velocity will be stripped off (called the Bohr criterion).

The first extensive application of this approach for stopping power scaling was by Northcliffe [8] in 1962, who used the form:

$$Z^*_{HI}/Z^*_H \sim 1 - \exp[-(v/v_o Z_{HI}{}^{2/3})] \qquad (3)$$

which describes how the effective charge of a heavy ion, $Z_{HI}$, is related to the effective charge of a H ion (at the same velocity). This expression can be combined with the scaling law for stopping power S:

$$S_{HI} = S_H \left(\frac{Z^*_{HI}}{Z^*_H}\right)^2 \qquad (4)$$

From the experimental "master stopping" set, Fig. 7, which can give a $S_H$ based on 12,000 data points, it is only necessary to establish the best form of Eq. (4) to scale to heavy-ion stopping powers. Scores of papers have been written on this, but except for Northcliffe and Schilling [13] the previous authors used limited data bases for $S_H$. For the preparation of the Handbook, about 4000 heavy-ion data points were compiled and a final parameterized variation was established. The accuracy is shown in Fig. 8 where the scaling law is tested by comparing actual experimental values to those predicted by using the variation of Eq. (4). For plotting convenience, what is shown is reduced stopping, defined as $S_{HI}/S_H Z_{HI}^2$ for both

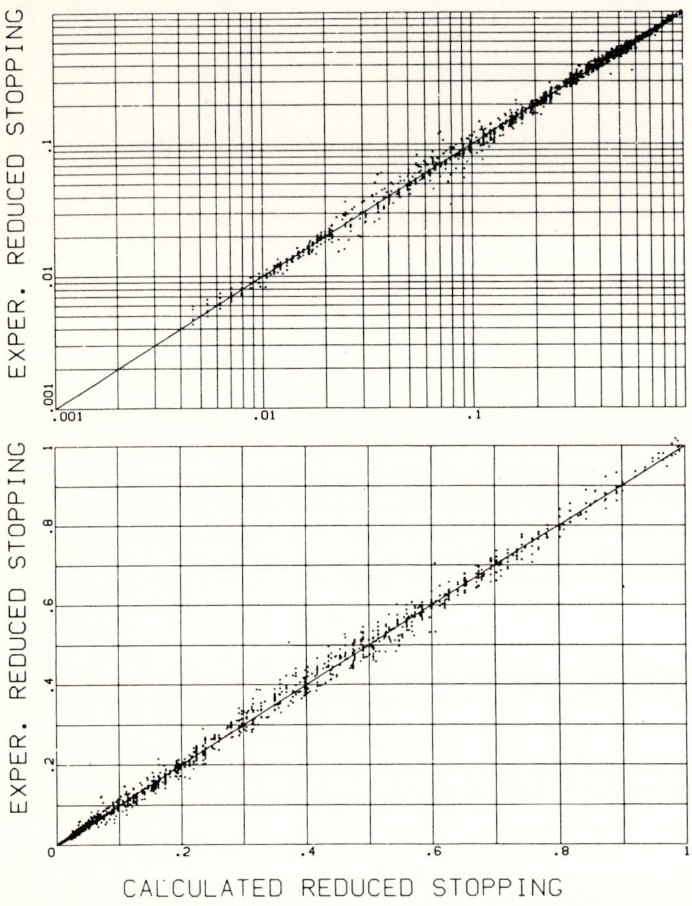


Fig. 8. Shown are experimental versus calculated values of heavy-ion stopping to hydrogen stopping. The experimental values are from 127 ion-target combinations ranging from ions of C to U, in both gaseous and solid targets ranging from Be to Au, at ion energies from 0.22-22 MeV/amu. These experimental heavy-ion stopping values are reduced by dividing by equivalent hydrogen ion stopping as in Eq. (4). These reduced experimental values are plotted versus calculated values. The experimental and calculated values agree to about 5%

the experimental and the predicted values. Thus, if the heavy ion is totally stripped, and stopping increases with the square of the ion's atomic number, then the reduced stopping value is one. For lower velocities where the heavy ion is only partially stripped, the reduced stopping must be less than one. The agreement in Fig. 8 is about 5%.

The Handbook thus introduced advanced heavy-ion scaling of electronic stopping powers based on a huge experimental data base. It also made a significant step in upgrading the calculations of nuclear stopping powers. Finally, it shows not only the final values, but plots all existing experimental stopping powers (see Fig.9) so users could review relevant data.

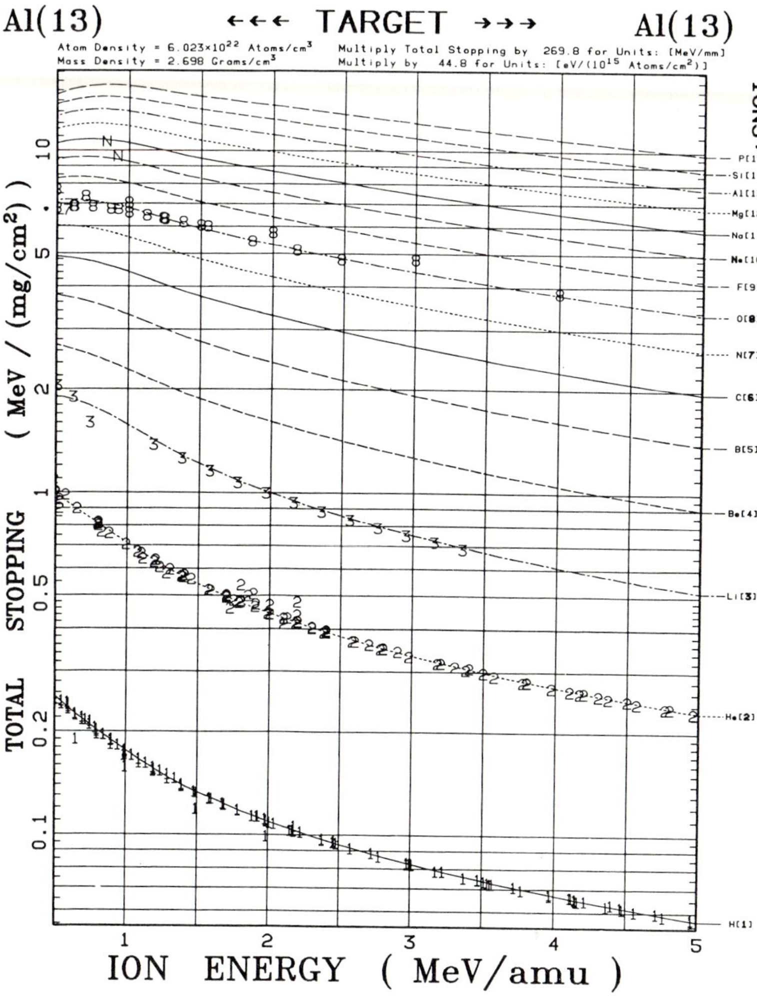


Fig. 9. Shown is a page of Ref. [41] which displays for a target of Al the calculated stopping of 15 ions. Also shown are the experimental results for each so that the accuracy of the table may be judged. The plot symbol for each ion, e.g. 7=nitrogen ions, is indicated by printing that symbol in bold type in the right margin

The book is limited by not giving stopping values for ion velocities below 0.2 MeV/amu. Below this velocity there is a direct interaction between the ion $Z_1$ and the target $Z_2$, and all scaling laws begin to break down.

The Current Accuracy of Stopping Tables

$E > 10$ MeV/amu: Well-developed theoretical models with accuracy better than 5%.

$E > 1$ MeV/amu: Reasonably good combination of theory and scaling laws. Accuracy better than 10%.

$E > 0.2$ MeV/amu: Tables based on theory, scaling, and extensive data. Accuracy better than 20%.

$0 \leq E \leq 0.2$ MeV/amu: Pure theory [9] is accurate to a factor of two. No published tables are any better except for H and He ions where extensive experimental data is available.

## 2.2 Range Tables

There was less interest in tables of ion range distributions than in stopping powers until the industrial development of ion implantation. The first extensive table was in Northcliffe and Schilling [13] in 1970. After 1970 followed a flood of books all related to ion implantation. These books [15, 22-24, 33, 43] were basically the same in both theoretical approach and accuracy. Range calculations can be broken into two parts: the basic physics of energy loss and scatter from atom-atom collisions, and the mathematics to follow these processes until the ion stops. Except for small points (noted later) all these books are based on the theories of Lindhard et al., with particular importance being placed on what is called the LSS theory of ion ranges [9]. This theory develops analytic expressions for both the electronic and nuclear stopping of ions. It then takes these and sets up a Boltzmann transport equation to solve the statistical problem of the final ion distribution. For the numerical solution of the transport equations, expansions are made and thus what is obtained are the moments of the distribution with the first moment being defined as the mean range, the second being the straggling; the third is the skewness, etc. Given this theory, it is a direct computer step to go from the universal stopping concepts to the final moments.

What is the accuracy of LSS calculations? First, the range distribution will be as accurate as the collision physics, and we have discussed before that when one used universal atoms such as the Thomas-Fermi atom used in LSS, the collisional accuracy is usually within a factor of two. However, the first moment is usually better than this since the mean range is the integral of the stopping at all energies down from the initial energy, and, if the worst deviation is a factor of two, it will possibly be compensated at other energies giving a more accurate range than stopping power. But the other moments are increasingly sensitive to errors in the interaction, and they are more inaccurate.

A second problem is equally important. The users of range tables do not wish to have moments of a range distribution, they wish to have the distribution itself, i.e. the shape of the concentration vs depth. Given only two or four moments, can a shape be uniquely described? The answer is yes, sometimes, but lurking at innocent ion-target combinations can be hidden devils!

The usual method of obtaining an ion distribution shape from calculated moments is to fit an analytic expression which has a single peak in its shape. For example, a distribution used to give a peak beneath the surface, x=0, with a distribution skewed toward the surface is

$$P(x) = ax^b \exp(-cx) \tag{5}$$

with a, b, and c being parameters which can be adjusted to give the correct first three moments. Usually, the distribution is a rising quadratic which is cut off by an exponential factor, forming a peak with an abrupt final edge. This type of shape is typical of light ions in heavy substrates where most of the ions go the full range with little deflection, but a few undergo a strong collision on the inward path and are distributed between the surface and the peak.

Winterbon considered the problem of converting moments to shapes and unearthed a significant worm. He took several typical ion-target combinations and solved for the moments. He then took all the popular shapes (analytic expressions) and found parameters for them which gave them the correct moments.

The problem arises when a three-parameter distribution like Eq. (5) is used to fit a four-moment distribution. Winterbon took various distributions used by various authors and showed how much variation could be obtained; see Fig. 10. It is clear that wild errors may occur when a shape is forced to fit a series of moments. The most suitable - and recommendable - distribution functions may be the Pearson distributions using 4 moments.

Some of the major range tables up to now are:

1970: Northcliffe and Schilling, "Range and Stopping Power Tables" [13]

The ranges were calculated by the integration of stopping powers,

$$R(E) = \int_0^E (-dE/dx)^{-1} dE \tag{6}$$

where R is the range of an ion with energy E, with a stopping power dE/dx, which was calculated from the N-S electronic stopping powers. A correction was applied for nuclear stopping and deflection using LSS theory [9]. Only mean ranges are tabulated. The results are good for energetic ions ($E \geqslant 0.1$ MeV/amu) and for heavy ions in light substrates ($M_1/M_2 \geqslant 3$). The results are rather inaccurate for low energies and where the ion is not heavy ($M_1/M_2 \leqslant 3$), which is where the ion penetration is not a straight path and more detailed calculations are necessary. As an example, for 100 keV He ions in Al ($M_1/M_2$=0.15), the difference between the ion path length and its final mean range below the surface is about 43 % [25]. The authors do not indicate what are the typical corrections which they apply to their ranges found using Eq. (6), and the actual procedure they use is too complex to evaluate. For the case of He in Al, the N-S results are about 35% higher than several measured ranges, and for 100 keV H ions into Al, their results are about 60% higher than experiments.

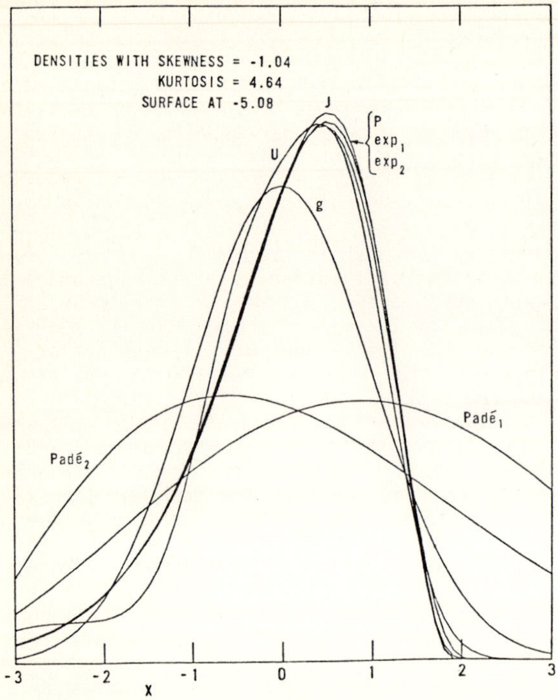


Fig. 10. This plot shows the results of attempting to produce an ion range distribution from a series of distribution moments as calculated by a transport equation approach to range theory. The four moments are 5.08, 1, -1.04 and 4.64 with the x-axis in units in which the standard deviation (the second moment) is one. Each of the curves comes from an analytic function with parameters which can make a non-negative distribution with a single peak below the surface. Each function has been fitted up to four moments. This figure illustrates that there is not a unique range distribution, and without experimental data no clean conclusions can be reached about details of the distribution; from Winterbon, Ref. [52] (g = Gaussian, U = Egdeworth, J = Johnson, P = Pearson, exp = Exponential).

## 1970: Johnson and Gibbons LSS Range Tables

The next major step in range tables was by Johnson and Gibbons [15] in 1970 where they produced the first distribution calculations. Their book was specifically oriented for ion implantation in semiconductors. They developed a computer code to calculate LSS theory [9], and the authors were very generous in that they published the code in the book so it could be used by others for different ion-target combinations. Unfortunately, the code had a serious error, and only the calculation of the first distribution moment (the mean range) was reliable [33].

By using LSS theory, the authors included a full statistical treatment of ion penetration of matter. The LSS theory incorporates an electronic stopping of the form

$$S_e = kE^{1/2} \qquad (7)$$

where k is a constant and the stopping is proportional to the ion velocity. This stopping may approximate low-velocity stopping, but since electronic stopping goes through a peak at about 90 keV/amu, this limits the applicability of LSS theory to energies below this value. The nuclear stopping is based on the collision of Thomas-Fermi atoms and so is accurate to about a factor of two as discussed before. The result of this major step forward in range tables may be indicated by comparing the results to the previous N-S tables. Recall that for 100 keV He in Al the N-S tables predict a range of 0.8 μm. The J-G tables show about the same, 0.89 μm, for the ion's total path length, but when the statistics of the penetration process are included the J-G tables predict a mean range of 0.56 μm. The difference is in the multiple deflection of the He from its original trajectory. This result is reasonably close to the experimental data scattered from 0.5 to 0.6 μm.

He ions were used in the above example because the nuclear stopping component is small compared to the electronic component. For heavy ions, which have large nuclear stopping at low energies, the LSS theory is well established to be accurate to a factor of two, and although many papers have found better agreement these are coincidental in that some ranges will agree well, while others will be a factor of two away. Examples of this are shown in Figs. 11 and 12. Figure 11 [25] shows range and straggling measurements for 800 keV $N^{14}$ ions in targets of atomic number 22 through 32. The solid lines are the LSS prediction and the dashed lines go

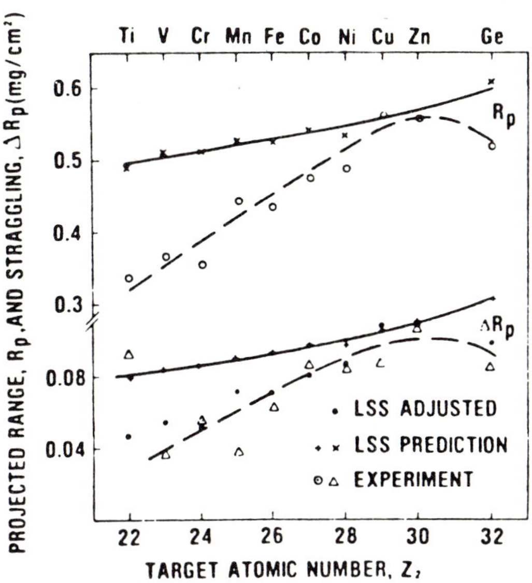


Fig. 11. A comparison of LSS range theory with experimental results for N ions at 800 keV in various targets [25]. The solid lines are LSS predictions for the projected range $R_p$ and the projected straggling $\Delta R_p$. The dashed lines are drawn through the experimental results shown as open symbols. For some targets such as Zn there is excellent agreement, while for the more loosely bound solids of Ti and V there is an error approaching a factor of two. Similar results are shown in Fig. 12 for many ions in a single target

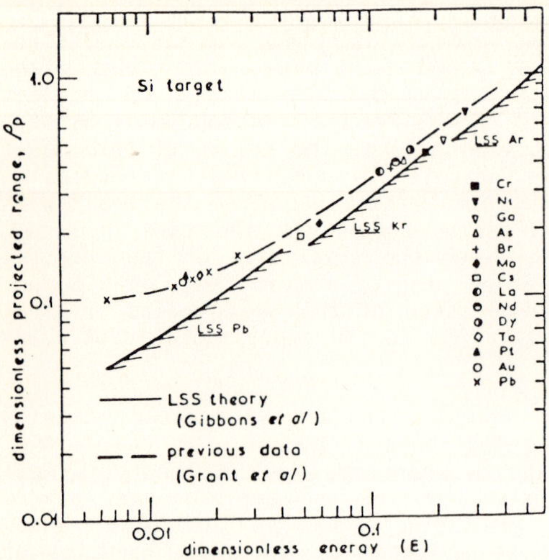


Fig. 12. A comparison is shown between LSS range predictions [9] and the experimental results for many ions in silicon targets [34]. Various ion energies were used, but this factor can be mostly eliminated by plotting in LSS-reduced coordinates $\varepsilon$ and $\rho$, as shown on the plot. The LSS values are shown as the solid lines, with a possible range of values depending on the ion mass being shown by the narrow cross-hatching. The dashed line shows previously published experimental values, while the dots show new measurements which essentially agree with the previous ones. The accuracy of LSS predictions for this single target ranges from 15% for high energy $\varepsilon$, to about a factor of two for low values of $\varepsilon$

through the experimental points. For nitrogen ions into a Cu target the agreement is excellent, while for a target of V ($Z_2$ = 23) the experimental values are about 1/2 the LSS prediction. Similar results are shown [34] for many ions into a single target, Si, in Fig. 12. This plot shows the ranges of 14 ions with energies from 10-400 keV into silicon. The plot has coordinates which are called LSS-reduced coordinates with the abscissa corresponding to energy and the ordinate being related to range. In these units the LSS range-energy predictions for all heavy ions falls about on a single line, shown with slight hatch-marks on the figure. Extensive prior experimental range data are shown as a dashed line and the results of Ref. [34] are shown as various plot symbols. For the lowest energies there is a deviation between LSS theory and experiment by about a factor of two, while at the higher energies the agreement is good, about 15%.

## 1975: Gibbons, Johnson and Mylroie Range Tables

The calculation errors in the above J-G tables were corrected by the authors in the 2nd edition: Gibbons, Johnson and Mylroie, "Projected Range Statistics", Dowden, Hutchison and Ross, Inc., USA (1975)[24]. This edition also included two major advances over their previous tables. First they realized that the LSS electronic stopping powers $S_e$ could be improved by using experimentally determined parameters to create a similar power form of $S_e$:

$$S_e = aE^b \tag{8}$$

where a and b were determined from experiments. This significantly increased the accuracy of the light-ion range tables. Secondly, they used a fit to the nuclear collisions of Thomas-Fermi atoms as suggested by Winterbon, Sigmund and Sanders [16, 18]. Since this correction remained in the spirit of Thomas-Fermi atoms, the basic nature of the nuclear stopping did not change, and the results do not significantly modify their previous results. However, GJM made the point that correcting $S_e$ and $S_n$ were the most important ways to improve the accuracy of LSS range calculations, and it is along these lines that the major improvements of later tables were accomplished.

1975: Brice and Winterbon Range Tables

In the same year as the JGM tables were published, 1975, there were also published the books of Brice and Winterbon called "Ion Implantation Range and Energy Deposition Distributions" [22, 23].

The Brice book, "Vol. 1 - High Ion Energies", also indicated that $S_e$ and $S_n$ as developed by LSS in 1963 could be improved on, and Brice developed his own parameterized form of $S_e$ so that he could calculate ranges for energies above the stopping power peak. His $S_e$ equation is too complex to review, but it contained 3 parameters which allowed control over the initial (low-energy) slope of Se, the energy of the peak of Se, and the down-slope of the high-energy Se. The parameters were found by fitting the Northcliffe-Schilling [13] values of Se for all ions except He. Since we have noted that the N-S tables were empirical improvements over the LSS values, Brice used values which were probably better than the GJM values which were based on very limited experimental data. Brice used the Thomas-Fermi atom nuclear scattering of LSS for $S_n$.

But the major advance of this book was the inclusion of damage distributions of the ions. Brice had previously extended LSS theory to obtain the transport equations which govern the ion spatial distributions at all energies as the ions slow to a stop in the target. That is, if the incident ions are 100 keV boron atoms, then the distribution of 100 keV ions is a delta function at the target surface. As the ions penetrate the target the spatial distribution of ions at, say, 90 keV will be wider as shown in Fig. 13. Brice showed how to construct spatial distributions within the target for ions at any energy below the incident energy, with the distribution at E = 0 being the final implantation distribution. From these spatial distributions Brice developed a direct procedure to calculate the energy deposition distributions of ions in targets. The nuclear stopping component of these energy deposition distributions were felt to describe where target damage was created. Brice's innovative technique is still a fundamental way which is used to calculate damage profiles, and it has only two main problems. First, there is no account made of the fact that solids have a displacement energy, $E_d$, which is defined as the minimum energy necessary to displace an atom from its lattice site, usually about 25 eV. If the collision transfers less than this energy there will probably be no permanent damage (atom displacement) and the energy may be dissipated as phonons. Since nuclear energy transfers may vary as $E^{-2}$ these low-energy collisions will lead to overestimates of damage. Secondly, Brice's transport equation method does not account for what happens to the recoiling target atom after it is displaced. Some collisions will transfer significant energy to the target atom which can recoil and knock out other target atoms, or it can dissipate its energy in electronic processes or phonons. Neglect of recoil cascades usually leads to overestimates of target damage.

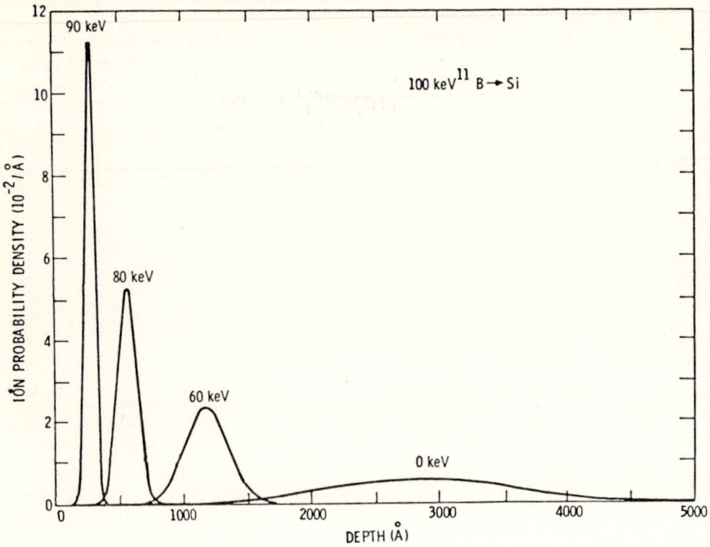


Fig. 13. Illustration of the Brice technique to calculate the damage introduced in a target by an ion beam (see Ref. [22] and citations within). Each curve shows the spatial distribution of ions at successively lower energies as 100 keV B ions penetrate into silicon. The original distribution is a delta function at zero depth, and the final ion distribution is shown for zero energy. The damage distribution is calculated by the difference between two successive energy distributions. No account is made of displacement energy thresholds or secondary cascades as discussed in the text.

Volume 2 by Winterbon, "Low Incident Ion Energies", is different theoretically and in the items calculated. Winterbon was the first to include calculations which were basic to other phenomena of ion implantation: (a) the fraction of the ion distribution which is outside the target surface, i.e. the fraction of the ions which are backscattered, (b) the energy which is in the backscattered ions and hence not injected into the target, (c) the damage distribution at the target surface which determines the sputtering yield, and (d) the number of sputtered target atoms per incoming projectile ion. This book is clearly the most sophisticated of the ones we have reviewed, for it covers many of the phenomena of low-energy ion impact of solids. For the electronic and nuclear stopping Winterbon chose to use the traditional LSS formula, and the errors introduced by this choice have been previously discussed.

1981: Littmark and Ziegler, Energetic Ion Range Tables

Finally, Littmark and Ziegler published in 1980 the "Handbook of Range Distributions for Energetic Ions in All Elements" [43]. This book contains the most advanced form to date of solving ion range distributions using transport equations. Recall that LSS theory in 1963 used a closed form of nuclear and electronic interactions between the ions and the target atoms. The authors in the range books reviewed above basically maintained the LSS formula approach. In the L-Z book a new expansion was made to the transport equations which allowed relatively complete freedom in using whatever nuclear and electronic interactions were appropriate. Since great strides had been made in calculating stopping powers (as reviewed before), this new range solution allowed these to be used to obtain equally accurate range distributions (in general, the mean range will be more accurate than the accuracy of the stopping powers on which it is based).

The prime limitation in the L-Z book was that only the first two moments of the final range distribution were given. They also only considered ranges for ion energies above 200 keV/amu, which was the lower limit of the accurate stopping powers of Ref. [41].

## 3. Stopping Powers for Ions in Solids

We shall review the three primary subjects which have led to a simple accurate code for the stopping of ions in solids at all energies. First there will be a review of the development of a universal nuclear stopping calculation based on the work of Gombas [53] and the enhancements introduced by Wilson, Haggmark and Biersack [26]. Then a review will be made of a universal low-energy electronic stopping calculation based on the ideas of Lindhard [1], and Brandt and collaborators [21, 45-48]. Finally, these two stopping powers are combined and compared with the experimental results of 900 papers, including those showing the pronounced $Z_1$ (ion) and $Z_2$ (target) oscillations in stopping power at constant ion velocity.

### 3.1 Nuclear Stopping Powers

About 1960 there was extensive theoretical work on the energy loss of an energetic ion to target atoms based on the work of Firsov [2-5, 7] and Lindhard et al. [9]. These works laid the ground work for most later nuclear stopping calculations. These original studies were based on the Thomas-Fermi model of the atom, which led to erroneous results for collisions in solids between widely separated atoms since the Thomas-Fermi model has an unrealistic long tail of charge. This leads to the calculation of too great stopping powers, and of too shallow ion ranges in solids. A more realistic diatomic potential in a solid would drop to zero rather sharply for atomic separations much greater than about one Ångstrom, thus reducing atomic scattering at larger impact parameters.

Extensive work has been done in the last ten years on realistic Hartree-Fock atomic distributions and we have used those calculated by Moruzzi et al. [36]. In this book the authors treat only 26 solids and we have used their approach to create charge distributions for all 92 elements in their normal crystallographic form. For normal gases, we have used their most thermodynamically stable solid structure, for these shapes will be useful as a first approximation in calculating the stopping power of gaseous compounds such as $Al_2O_3$. This procedure has been described in Ref. [41].

For the charge distribution of the ions, it was decided to use the same charge distribution as for its elemental solid. This was done because (a) target polarization by the ion reduces the effective spatial charge distribution of the ion, and this is better accounted for by solid-state distributions than isolated atom distributions, and (b) this makes the calculation reversible and, for example, the interatomic potential of Ni on Ag is the same as Ag on Ni.

The interaction between atoms can be evaluated reasonably well by introducing corrections based on physical data such as phonon distribution curves, elastic constants, compressibility, etc. as reviewed by Johnson [19]. This approach is not universally applicable, in part because of lack of data, and further it is not very accurate for rather hard collisions. On the other hand, there is the quite complex approach which is known as the multiconfigurational self-consistent-field method which is quite accurate, but which consumes so much computer power that it is prohibitively expensive. However, Wilson et al. showed [26] that an accurate approximation to this elaborate calculation is the method of Wedepohl [11] which is called the free-electron method of interatomic potentials. This consists of first calculating the Coulomb interaction energy $V_c$ of the collision for any given separation (allowing no change in the electron spatial distribution during the collision). Then, in the volume of space in which the two atoms' electron distributions overlap, the electrons are allowed to absorb energy $V_k$ in excitation. This is done by considering this volume to be a free-electron gas, and calculating the difference in energy between the two isolated overlap distributions, and in the same number of electrons compressed together into the overlap volume. For this overlap region, the free-electron gas energy must be decreased because of exchange and correlation energy $V_a$, because the electrons are not randomly distributed, since their relative separations are affected by the exclusion principle regarding spin and repulsive Coulomb forces among themselves.

The results of this type of calculation of interatomic potentials are shown in Fig. 14, for the case of $U^{92}$ and $U^{92}$. In the upper left are shown the values of the various potentials for increasing nuclear separation

Atoms = $U^{92}$ and $U^{92}$

| $r_{12}$ | $V_c$ | $V_k$ | $V_a$ | V | Phi | $r_{12}/a_{TF}$ |
|---|---|---|---|---|---|---|
| .05 | 1037103.46 | 306223.22 | -4143.93 | 1339182.75 | .5494 | .7650 |
| .10 | 262790.31 | 164139.36 | -2945.61 | 423984.06 | .3479 | 1.5300 |
| .15 | 89553.22 | 102657.03 | -2204.94 | 190005.32 | .2338 | 2.2950 |
| .20 | 35572.78 | 66453.83 | -1666.97 | 100359.64 | .1647 | 3.0600 |
| .40 | 1193.09 | 15175.67 | -641.43 | 15727.33 | .0516 | 6.1199 |
| .60 | -824.07 | 5326.43 | -319.09 | 4183.27 | .0206 | 9.1799 |
| .80 | -709.13 | 2322.64 | -179.83 | 1433.68 | .0094 | 12.2399 |
| 1.00 | -527.59 | 1205.13 | -111.16 | 566.38 | .0046 | 15.2999 |
| 1.20 | -355.10 | 670.57 | -71.81 | 243.65 | .0024 | 18.3598 |
| 1.40 | -235.39 | 399.06 | -47.27 | 116.40 | .0013 | 21.4198 |
| 1.60 | -155.16 | 246.98 | -30.82 | 61.01 | .0008 | 24.4798 |
| 1.80 | -58.80 | 112.09 | -18.60 | 34.68 | .0005 | 27.5397 |
| 2.00 | -22.45 | 51.45 | -10.84 | 18.17 | .0003 | 30.5997 |
| 2.20 | -8.64 | 24.80 | -6.16 | 10.00 | .0002 | 33.6597 |
| 2.40 | -3.11 | 12.22 | -3.40 | 5.70 | .0001 | 36.7197 |
| 2.60 | -.98 | 5.79 | -1.77 | 3.03 | .0001 | 39.7796 |
| 2.80 | -.24 | 2.48 | -.81 | 1.43 | .0000 | 42.8396 |
| 3.00 | -.03 | .83 | -.28 | .52 | .0000 | 45.8996 |

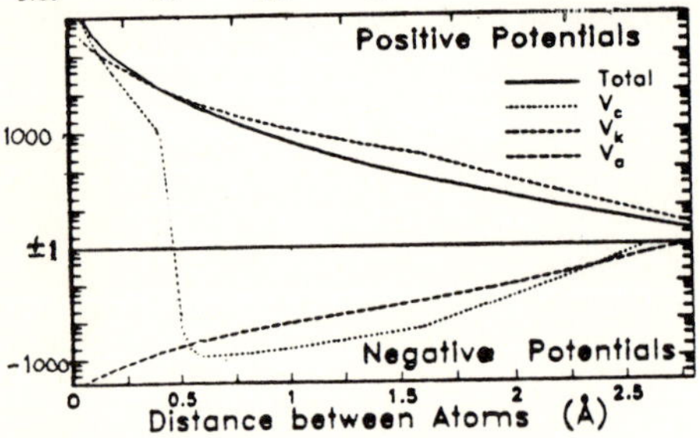


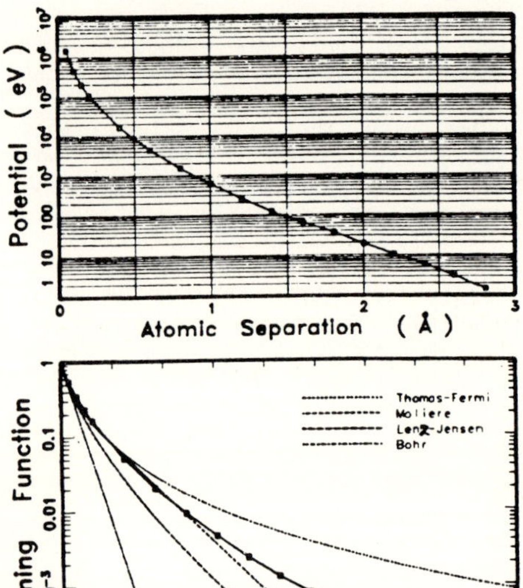


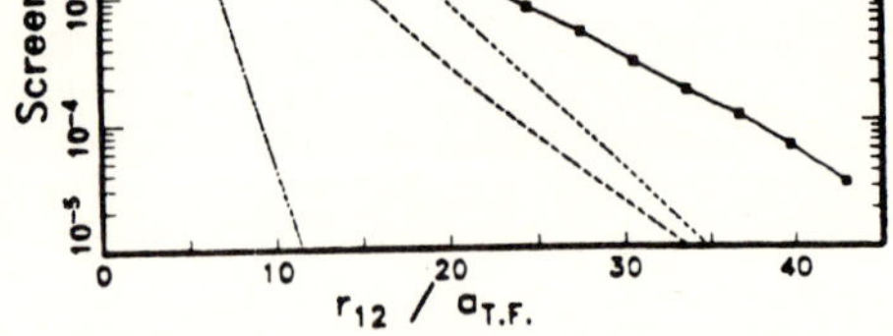


Fig. 14. Results of free-electron gas calculations of the interaction potential for two colliding atoms (U-U)

distances, $r_{12}$, in units of Ångstroms. The various potentials are described in the text and are known as $V_C$, the Coulomb interaction; $V_k$, the electron excitation energy; $V_a$, the exchange energy between electrons; and V, the total interaction, $V_C + V_k + V_a$, in units of eV. The next column labeled "Phi" is called the screening function, and is the summed potential divided by the potential of the two bare nuclei at the same separation without any electron screening. This screening function goes from unity (at zero separation) to zero at the point where both atoms are totally separate and completely screened from each other. The final column of tabulated numbers is labeled $r_{12}/a_{TF}$, which is the atom separation $r_{12}$, reduced by the Thomas-Fermi-Firsov screening distance of the two atoms. This final unit is useful in comparing these results with traditional potentials based upon statistical models of the atoms such as the Thomas-Fermi atom.

The lower left plot shows the four potentials tabulated above. Of particular interest is how the Coulomb energy goes from repulsive to attractive as a function of interatomic separation. Note further that the total potential V, shown as a solid curve, is greater than any of its components for small separations, but once the Coulomb potential goes negative it becomes less than $V_C$. This plot also shows the relative magnitudes of the various contributions, and how they all are significant. The upper right figure shows the final interatomic potential in the physical units of eV versus Ångstroms. The lower right figure compares this potential with traditional potentials such as the Thomas-Fermi, the Moliere, the Lens-Jensen and the Bohr. The radial separation units are the reduced units based on LSS theory as described above, and it can be seen that the U-U atom potential lies about half-way between the Thomas-Fermi and the Moliere potentials. Other potentials are usually less than this, and they range all the way down to lying between the Bohr and Lens-Jensen potentials for He on He.

The calculation of interatomic potentials was completed for about 500 of the possible 8100 ion-target combinations, chosen at random. The purpose of our interatomic potential calculations was to try to attempt to find a new, more accurate way to calculate interatomic potentials in a simple algorithm, since each of these individual calculations consumed significant computer time, and it would be prohibitively expensive to repeat the calculation every time it was needed. And merely doing them once for all ion-target combinations and fitting the resultant potentials would yield an unwieldy data base of about 50,000 numbers. Successful attempts were made to fit the individual interatomic potential components [51] but there exists also a simpler and sufficiently accurate method based on the Bohr concept of introducing a reduced radius. This is shown in Fig. 15, where all 500 interatomic screening functions are plotted in the reduced radial coordinates shown. The rms average of these potentials can be fitted by the equation:

$$\phi_u = 0.1818e^{-3.2x} + 0.5099e^{-0.9423x} + 0.2802e^{-0.4029x} + 0.02817e^{-0.2016x} \quad (9)$$

where the reduced radial coordinate x is defined as $r/a_I$ with

$$a_I = 0.8854a_0/(Z_1^{0.23} + Z_2^{0.23}), \quad (10)$$

where $a_0$ is the Bohr radius, 0.529 Ångstroms. This algorithm has an rms accuracy of about 16% for all potentials above 2 eV.

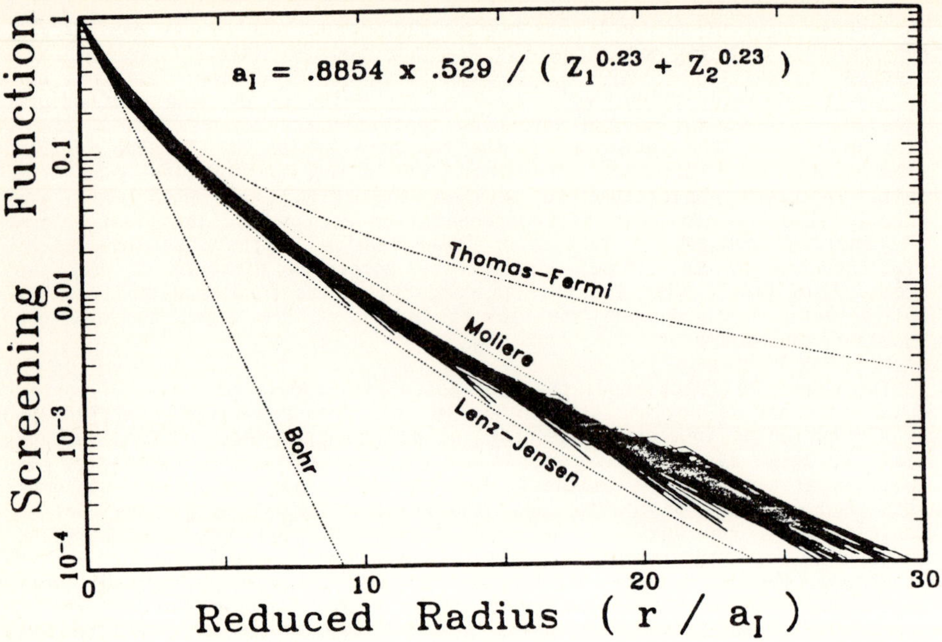


Fig. 15. Interatomic screening functions for 500 individual calculations. By applying a least-squares fit on $a_I$ they could be shifted into close agreement

This universal interatomic potential can now be used to evaluate the nuclear stopping power of an ion in a solid. The energy loss is integrated over all impact parameters, and the result is shown in Fig. 16 using LSS reduced coordinates of energy loss versus ion energy. The result is somewhat similar to that of Wilson et al. [26] which they identified as the result of the collision of the carbon-krypton system. In the upper part of the figure is shown the form of the screening potential used in the calculation. The nuclear stopping results are shown as small dots on the figure. Through these values a curve was fitted, and it is shown as the solid line. The accuracy of the fit was better than 1 %. The formula used is shown in the lower part of the figure. Also shown are the nuclear stopping curves based on various statistical models of the atom.

The universal interatomic potential can also be used in the Magic Formula as constructed by Biersack and Haggmark [42]. This formula allows the quick calculation of the angle of deflection of an energetic screened atom colliding with another, based on the incident energy and the impact parameter. This Magic Formula is invaluable in Monte Carlo calculations in that it speeds up this calculation by large factors. We have used the universal screening potential to calculate by complete numerical integration of the orbit equations the final angle of scatter, and these are shown in Fig. 17 as dots for the values of reduced energy indicated, ranging from 0.00001 to 10. These have been fitted with the Magic Formula and the coefficients are shown at the top of the figure. The values of the scattering angle vs impact parameter have then been calculated using the Magic Formula and the results are shown as solid lines. The accuracy of the Magic Formula to the detailed scattering calculations is about 2 %.

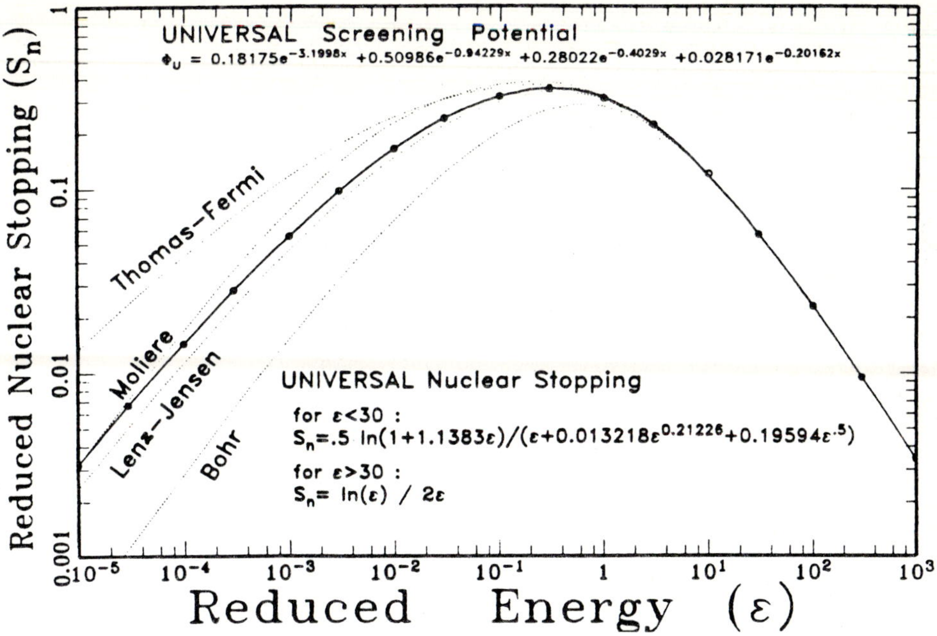


Fig. 16. Universal reduced nuclear stopping for the new universal potential of Eq. (9)as compared to previously used stopping powers

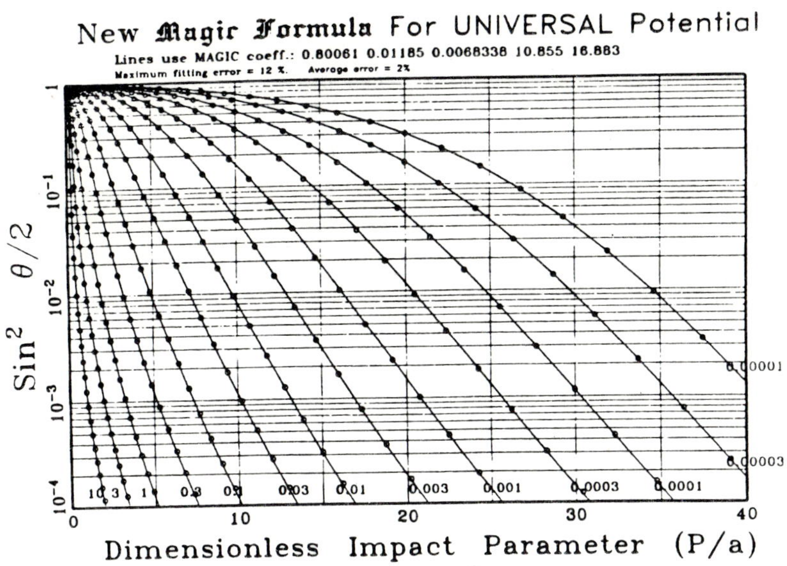


Fig. 17. New "magic formula" for $\sin^2\theta/2$ using the new universal potential. The curves using the coefficients 0.80061, 0.01185, 0.0068338, 10.855, 16.883 are the analytically calculated values; the dots are the exact evaluations of the scattering integral [42]

## 3.2 The Electronic Stopping of Ions in Solids

Our approach for the electronic stopping of ions in solids is based fundamentally on the Lindhard formalism of the stopping of a particle by free-electron gas [1]. This approach then must be modified by the screening of the ion by its electrons and by its ionization into large charge states by the interaction of the solid on the ion's electrons. Bohr suggested that this stripping of the projectile's electrons could be approximated by assuming that all electrons with classical velocities below the ion's velocity would be stripped. Lindhard suggested that the shielding of the ion by its electrons could be resolved in the Thomas-Fermi model using the TF screening distance.

Recently, Brandt has suggested that the Bohr concept should be revised to include the fact that the electron gas of the solid has itself an intrinsic velocity, and that the electrons which are stripped should be relative to the Fermi velocity of the solid. He has detailed this concept in a series of papers [45-48]. Ferrill and Ritchie have recently considered the screened potential of an ion and its effect on the ion-solid interaction. They have developed a formalism which is more general than the previous Lindhard approach in that it is not as specific to a given atomic model, such as the Thomas-Fermi in the Lindhard case. This new approach allows for the screening to be used on more general electron distributions, especially that of the shape of $e^{-r}/r$.

Brandt has also proposed the new concept that the charge fraction of an ion is dependent only on the ion's velocity, and not on its atomic number. That is, the fraction of the charge left on a Ne ion is the same as that on a Pb ion at the same velocity in the same target. Although this is hard to swallow as a prime condition, it then allows the development of a new formalism which is very powerful in predicting the electronic stopping powers. For Brandt has pointed out that this "charge fraction" of an ion is not physically important; what is important is the screening of the ion. And each ion will be screened differently by its electrons even at the same velocity. And so it is the screened ion which gives the concept of "effective charge", not the charge fraction.

To clarify these points, consider a large ion penetrating through a target. It has been stripped of some electrons, and those remaining are distributed in space, screening the nuclear charge. However, the theorist is most comfortable with a point charge, penetrating through a free-electron gas medium. To go from the pure point charge to the lumbering broad partially stripped ion we need a formalism of an effective charge. This effective charge concept is developed so that we may use this charge as an equivalent point charge in the well-developed point charge approach to get stopping powers. So the effective charge includes in it both the ionization level of the ion and the screening of the nucleus by the electrons.

The effective charge as developed by Brandt first assumes that all ions are ionized to the same fraction at the same velocity. Then approximating the ion by an exponential screening function, he can extract a screening length. From this, the effective charge is determined by considering the effect of the screening on the energy loss to the electron sea of the solid.

In practice, a factor must be adjusted for the ionization level of the projectile ion. This factor is small, and does not change much in nearby atomic numbers, and ranges from about 1 to 1.5. Since this is the only free parameter in the calculation of electronic stopping powers, it heroically must swallow all the other approximations also.

## 3.3 Empirical Ion Stopping Powers

The nuclear stopping power of ions in solids is taken directly from the preceding discussion of universal interatomic potentials. There are no free parameters in these values.

The electronic stopping powers are developed using the framework of Bohr, Lindhard, Brandt and Ritchie as discussed before. The main steps are shown in Figs. 18-20. First, it was necessary to develop a method of calculating the Fermi velocities of all solids. This is not rigorously possible, since either real or Hartree-Fock solids do not have a single, well–defined Fermi-velocity, and this concept is most useful only in statistical models of the atom. But one may redefine the Fermi velocity both theoretically and experimentally. If one looks at the transmission of electrons through a thin foil, the energy loss of the beam will have a large peak at certain electron energies. One may define the Fermi velocity of the solid as that corresponding to a free-electron gas which has its maximum interaction at that energy. This approach was used by Isaakson and Brandt [39] to deduce Fermi velocities of many solids from photon and electron absorption experiments. From a theoretical view, one may take a Hartree-Fock solid and isolate the interstitial electrons, i.e. those outside the muffin-tin spherically symmetric distributions. Then it can be hypothesized that these electrons constitute the free electrons of the solid, and by considering their local density, the Fermi-velocity of the solid can be determined. Both of these approaches are shown in Fig. 18.

The Fermi velocities of all elemental solids are shown in Fig. 18 in units of the Bohr velocity. The solid dots are the values derived by Isaakson and Brandt [39] from photon and electron transmission experiments (see text). The solid line indicates the values we have calculated from Hartree-Fock solids considering only the interstitial electrons. The two results are remarkably consistent, with a mean deviation of 9 %. In our

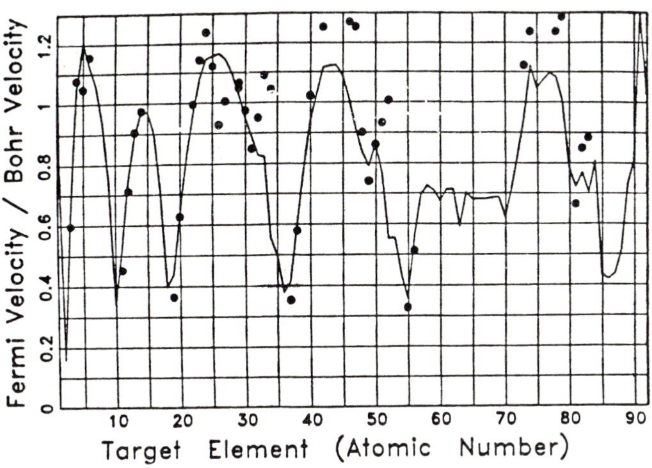


Fig. 18. Reduced Fermi velocity

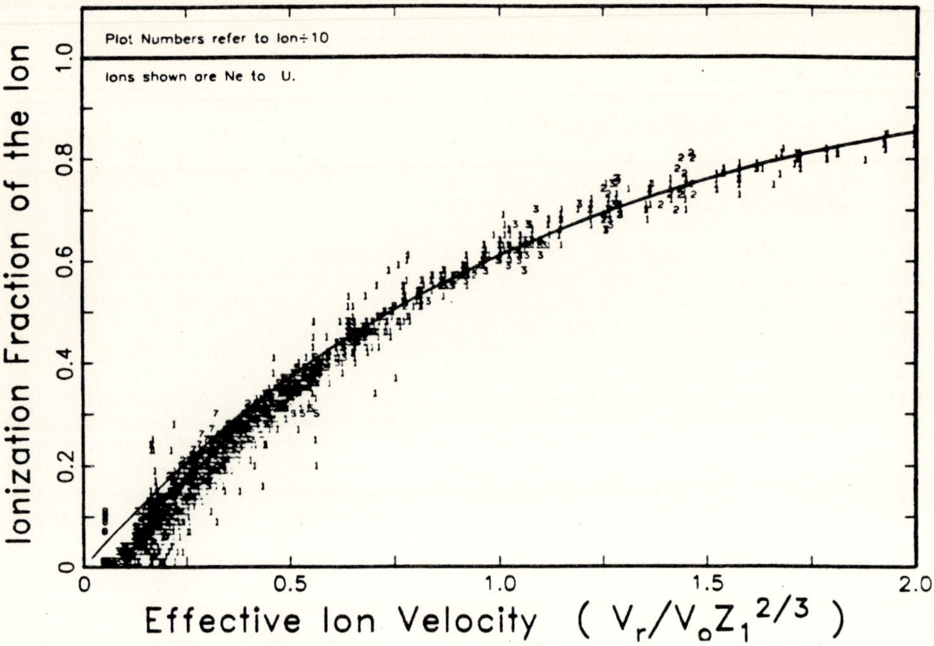


Fig. 19. Fractional ionization level of ions in solids

work we have used the experimental numbers where they are available, and the theoretical numbers elsewhere.

Figure 19 shows the ionization fraction of ions in solids as suggested by Brandt et al. vs the effective ion velocity (the ion velocity relative to the Fermi velocity of the solid). Using the Brandt approach it is possible to go from experimental stopping powers to the ionization fraction of the ions, which he states is independent of ion. He has published similar curves, but this figure contains many more data points. The solid line is Brandt's suggested curve.

For ions lighter than Ne (excluded from this plot) the tight bunching of the data points disappears. For ions of atomic number of 10-92, there is a tight bunching which reasonably follows Brandt's curve above about 0.7. Below this there is a small divergence. The divergence can be accounted for by adjustments of the ion screening length, which is called lambda in Brandt's papers. In order to adjust this screening length, it is first necessary to obtain a new ionization fraction curve which goes through the data points. Once this is done it is now possible to adjust the ion screening length.

In Fig. 20 is shown the screening length, lambda, versus the ionization fraction for ions of oxygen in all solids. The data show various solids by plotting their atomic number divided by ten for the data point. The solid line is the Brandt screening factor, multiplied by the factor of 1.3 as noted on the figure's left margin. The bump in the solid line is the additional screening due to inner-shell electrons.

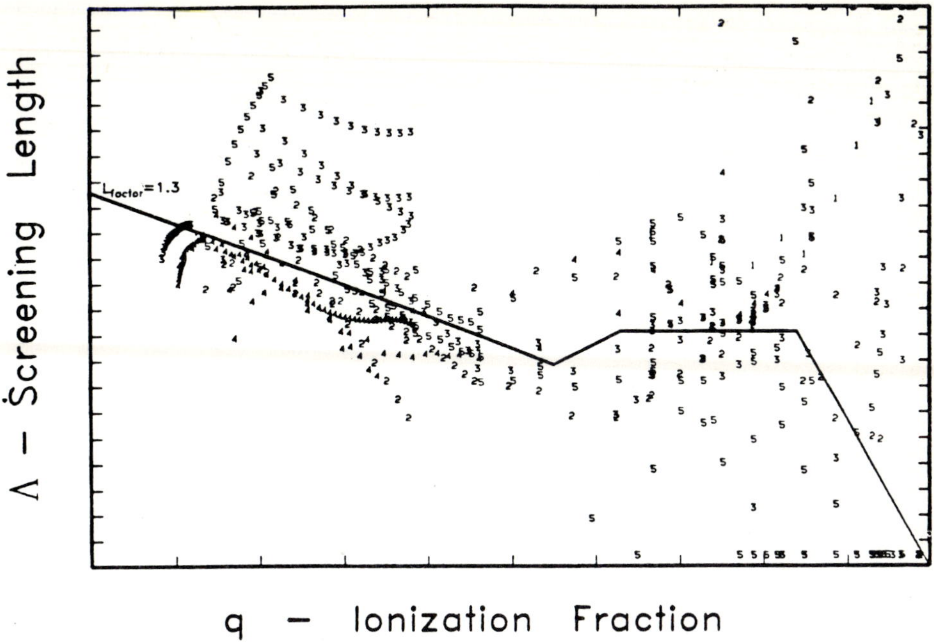


Fig. 20. Screening lengths Λ for oxygen ions in all solids

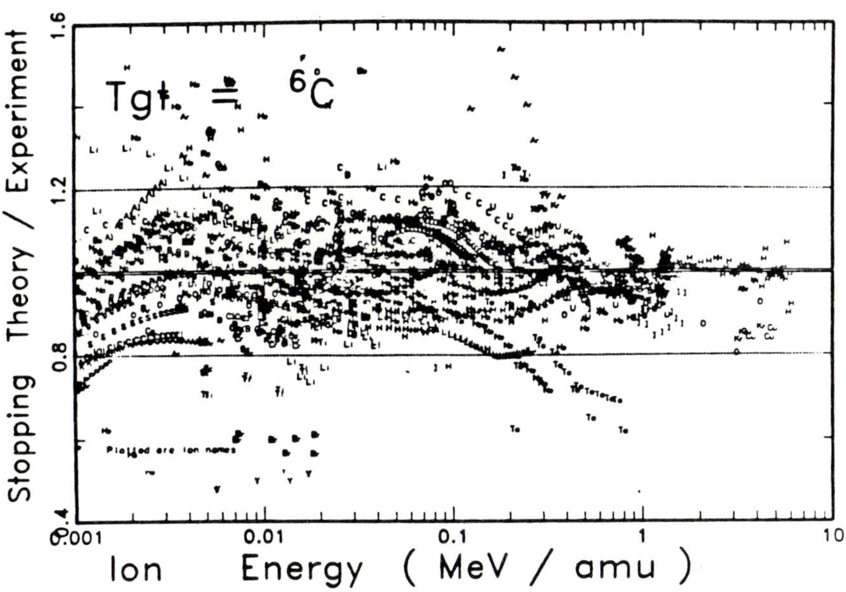


Fig. 21. Comparison of predicted and experimental stopping powers for all ions in carbon

A comment is necessary on Figs. 19 and 20. The parameter which is found is the screening length of the ion, which is independent of the target. However, this length comes from an assumption of the ionization fraction vs velocity as shown in Fig. 19. These two things circle each other, and either can be varied and accounted for by the other. Thus, the data can be brought up to the original Brandt curve by suitable adjustment of the correction factor on the screening length. Or the reverse can be adjusted: the ionization versus velocity curve could be adjusted to reduce the magnitude of the screening correction factor.

The final empirical stopping powers of ions in solids are now possible to assemble from the various components. First one starts with a data base of proton stopping powers in all elements. These have been assembled over several years by techniques outlined in Ref. [28], where the data base of hydrogen and helium experimental stopping powers has been combined with calculations based on the Lindhard stopping in a free-electron gas to get stopping powers accurate to about 10% above the energy of 80 keV/amu. These were then extended to 20 keV/amu by using values predicted by the ENR theory [49] and any experimental data. Below 20 keV/amu the stopping powers were assumed to go as $E^{0.45}$ as suggested by an analysis of a great number of stopping measurements [27, 28].

The second item which is calculated is the effective charge of the ion, which is calculated using the Brandt approach, modified as shown in Figs. 19 and 20. Finally, the nuclear stopping power is calculated.

Typical results are shown in Fig. 21, which shows the ratio of the predicted stopping power to the experimental stopping power for all ions in carbon. A carbon target is shown because there is more data for ion stopping in carbon than in any other material because thin target foils of carbon are readily obtainable in accelerator laboratories. Plotted are the ion names. The accuracy of the predicted stopping powers is about 10%, which is remarkable since this is about the variation of stopping in carbon just due to its method of fabrication. And this source of error is

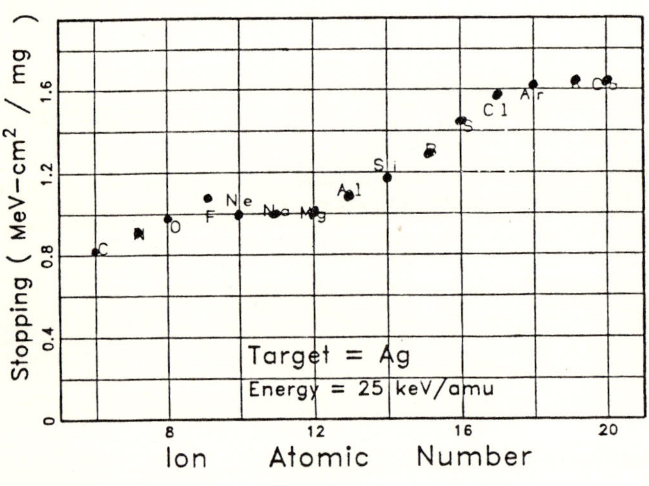


Fig. 22. Electronic stopping of various ions at the Bohr velocity in silver

combined with the fact that carbon greatly absorbs gases, especially water vapor, which can cause erroneous stopping power measurements.

Finally, in Fig. 22 is shown the stopping of various ions in Ag at the Bohr velocity [30]. The predicted values are shown as letters identifying the ions. The values of Ward et al. are shown as dots. The agreement is good. The work on stopping power oscillations of ions in solids appears to be a measurement of the screening function of the ion. Similarly, the change in the degree of oscillation from target to target is a measure of how much interaction there is between the conduction electrons of the solid on the degree of ionization of the ions.

## References


1. J. Lindhard, K. Dan. Vidensk. Selsk. Mat. Fys. Medd. 28 (1945).
2. O.B. Firsov, Zh. Eksp. Teor. Fiz. 32, 1464 (1957).
3. ibid, 33, 696 (1957).
4. ibid, 34, 447 (1958).
5. ibid, JETP, 7, 308 (1958).
6. W. Whaling, Handbuch der Physik, Bd. XXXIV, 13, Springer-Verlag, Berlin (1958).
7. ibid, JETP, 9, 1076 (1959).
8. F.W. Martin and L.C. Northcliffe, Phys. Rev. 128, 1166 (1962).
9. J. Lindhard, M. Scharff, and H.E. Schiøtt, K. Dan. Vid. Selsk., Mat. Phys. Medd., 33, 1 (1963).
10. J.F. Janni, AFWL-TR 65-150 (1966), now out of print.
11. P.T. Wedepohl, Proc. Phys. Soc., 92, 79 (1967).
12. E. Bonderup, K. Dan. Vid. Selsk., Mat. Fys. Medd., 35 No. 17 (1967).
13. L.C. Northcliffe and R.F. Schilling, Nucl. Data Tables, 7, 233 (1970).
14. C.C. Rousseau, W. K. Chu, and D. Powers, Phys. Rev. A, 4, 1066 (1970).
15. W.S. Johnson and J.F. Gibbons, Projected Range Statistics in Semiconductors, Stanford University Bookstore, Stanford, CA, 1970 (now out of print).
16. K.B. Winterbon, P. Sigmund, and J.B. Sanders, K.Dan. Vid. Selsk. Mat. Fys. Medd. 37, No. 14 (1970).
17. H. Bichsel, American Institute of Physics Handbook, 8-142, McGraw-Hill, New York (1972).
18. K.B. Winterbon, Rad. Effects 13, 215 (1972).
19. R.A. Johnson, J. Phys. F 3, 295 (1973).
20. J.F. Ziegler and W.K. Chu, Atomic Data and Nucl. Data Tables, 13, 463 (1974).
21. W. Brandt, Atomic Collisions in Solids, Plenum Press, 1, 261 (1975).
22. D.K. Brice, Ion Implantation Range and Energy Deposition Distributions, 1, High Energies, Plenum Press, New York (1975).
23. K.B. Winterbon, Ion Implantation Range and Energy Deposition Distributions, 2, Low Energies, Plenum Press, New York (1975).
24. J.F. Gibbons, W.S. Johnson, and S.W. Mylroie, Projected Range Statistics: Semiconductors and Related Materials, 2nd Edition, Halsted Press Stroudsbury, PA, USA (1975).
25. D.G. Simons, D.J. Land, J.G. Brennan, and M.D. Brown, Phys. Rev. A, 12, 2383 (1975).
26. W.D. Wilson, L.G. Haggmark, and J.P. Biersack, Phys. Rev. B, 15, 2458 (1977).
27. H.H. Andersen and J.F. Ziegler, Hydrogen Stopping Powers and Ranges in All Elements, Pergamon Press (1977)
28. J.F. Ziegler, Helium Stopping Powers and Ranges in All Elemental Matter, Pergamon Press (1977).

29. T.L. Ferrill and R.H. Ritchie, Phys. Rev. B, 16, 115 (1977).
30. D. Ward, H.R. Andrews, I.V. Mitchell, W.N. Lennard, and R.B. Walker, Can. J. Phys.,57, 645 (1979).
31. A.H. Agajanian, Ion Implantation: A Bibliography, 6 (Covering 1977-1980), IBM Technical Comm., Rpt. No. TR 22.2374, East Fishkill, New York (1981).
32. H.H. Andersen, Bibliography and Index of Experimental Range and Stopping Power Data, Pergamon Press, New York (1977).
33. B. Smith, Ion Implantation Range Data for Silicon and Germanium Device Technologies, Research Studies Press, Forest Grove, Oregon, USA (1977).
34. W.A. Grant, D. Dodds, J.S. Williams, G.E. Christodoulides, R.A. Baragiola, and D. Chivers, Proc. 5th Int. Conf. on Ion Implantation in Semicond. and Other Materials, Plenum Press (1977).
35. J.F. Ziegler, Appl. Phys. Lett. 31, 544 (1977).
36. V.L. Moruzzi, J.F. Janak, and A.R. Williams, Calculated Electronic Properties of Metals, Pergamon Press (1978).
37. H.H. Andersen, K.N. Tu, and J.F. Ziegler, Nucl. Inst. and Meth., 149, 247 (1978).
38. J.F. Ziegler, Nucl. Inst. and Meth. 149, 129 (1978).
39. W.K. Chu, Proceedings of the 1978 International Conference on Ion Beam Modification of Materials, Hungarian Academy of Sciences, Budapest, Hungary (1979), p. 179.
40. S. Kalbitzer, ibid, page 3.
41. J.F. Ziegler, Handbook of Stopping Cross-Sections for Energetic Ions in All Elements, Pergamon Press (1980).
42. J.P. Biersack and L.G. Haggmark, Nucl. Inst. and Meth., 174, 257 (1980).
43. U.Littmark and J.F. Ziegler, Handbook of Range Distributions for Energetic Ions in All Elements, 6, Pergamon Press, New York (1980).
44. K.B. Winterbon, Chalk River Laboratory Rpt. No. CR-1817 (unpublished).
45. S. Kreussler, C. Varelas, and W. Brandt, Phys. Rev. B, 23, 82 (1981).
46. A. Mann and W. Brandt, Phys. Rev. B, 24, 4999 (1981).
47. M. Kitagawa and W. Brandt (unpublished).
48. W. Brandt, to be published in Nucl. Inst. and Meth. 191 (1982).
49. P.M. Echenique and R.H. Ritchie, Sol. St. Comm. 37, 779 (1981).
50. U. Littmark and J.F. Ziegler, Phys. Rev. A, 23, 64 (1981).
51. J.P. Biersack and J.F. Ziegler, Nucl. Instr. and Meth. 194, 93 (1982).
52. K.B. Winterbon, Chalk River R.N.L. Rep. # 1817 (1978).
53. P. Gombas, Die statistische Theorie des Atoms, Springer-Verlag, Wien (1949).

# The Calculation of Ion Ranges in Solids with Analytic Solutions

J. P. Biersack

Hahn-Meitner-Institut, Glienicker-Strasse 100
D-1000 Berlin 39, Fed. Rep. of Germany

J. F. Ziegler

IBM-Research, Yorktown Heights, NY 10598, USA

## 1. Introduction

There exist numerous approaches [1-7] to solve the transport equation for ions slowing down in amorphous matter. They use different expansions and iterations, but have in common the use of an approximation to the scattering cross section in combining the two variables $\varepsilon$(energy) and sin $\Theta/2$ ($\Theta$ scattering angle) into one single variable $t^{1/2}=\varepsilon \sin\Theta/2$. As previously noted [8-10], this leads to deviations from exact values, particularly at low energies, where the effective potential becomes steeper and large deflection angles become more frequent. Some of the former approaches do not account for the straggling in nuclear energy loss, all neglect electronic straggling, and an exact treatment of compound targets often has difficulties. Recently, a new projected range theory was developed, which is based on well-known stopping powers and energy-loss stragglings, thus avoiding any uncertainties of prescribing differential scattering cross sections. It turned out that this can be accomplished by connecting directional angular spread to the nuclear energy loss directly, and that this approach yields good agreement with other existing theories and experimental results on projected ranges. The method has already been applied for tabulations of ion ranges in semiconductors, including multi-atomic composite materials [11,12].

## 2. The Basic Ideas of the Model

### 2.1 Directional Spread of Ion Motion During the Slowing-Down Process

We want to obtain the (mean) projected range of ions by summing the (mean) projections of each path length element. To accomplish this, we need to know the mean directional cosine of the ion motion during the slowing-down process, i.e. we have to consider the distribution of the directions of ion motion as a function of the ion energy. The general picture is simply that with each collision with a target atom, the ion loses energy and at the same time changes direction, where the so-called "nuclear" energy loss T (due to momentum transfer to the target atom) is directly related to the deflection angle $\Theta$ of the ion:

$$T = \frac{4M_1M_2E}{(M_1+M_2)^2} \sin^2 \frac{\Theta}{2} , \tag{1}$$

where $M_1$ and $M_2$ denote the masses of ion and target atom, respectively. Consequently, with increasing nuclear energy loss the ion will deviate - on average - more and more from its original direction. We may represent the directions of ion motion by the polar and azimuthal angles $\varphi$ and $\vartheta$, and depict them as points on the unit sphere, Fig. 1a, with the initial direction of ion motion on the pole ($\varphi$=0), corresponding to the x axis, along which we want to measure the projected range. As the direction of motion changes at random with each collision, with no correlation between successive deflections, the point on our directional sphere performs a random walk, like a Brownian motion, as indicated in Fig. 1a. If we project the point of this unit sphere onto the x axis (polar axis), we obtain values between -1 and +1, corresponding to the directional cosine of the ion motion, Fig. 1b.

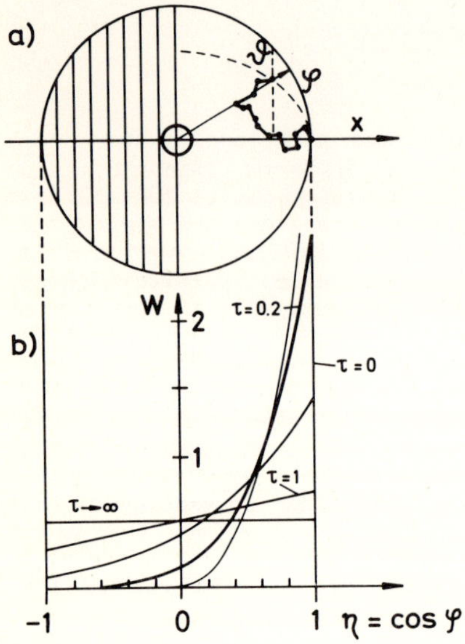


Fig. 1. Directional changes of ion motion are depicted as a Brownian motion on a unit sphere. Initially ($\tau \leq 1$), the ion motion is preferentially directed forward. Towards the end of the slowing-down process ($\tau \geq 1$), all directions become equally probable, resulting in a uniform distribution on the sphere (a) or in the directional cosines (b). In (b) intermediate distributions are also shown, according to the solution of Eq. (3) (cf. Appendix)

If we now consider the probability distribution function W either of the polar angle $\varphi$ or of the directional cosine $\eta = \cos\varphi$, we observe that (i) originally it is a delta function centered at $\varphi$=0 or $\eta$=1, (ii) the probability distribution then spreads out in a diffusion-like manner, while the ion slows down, until (iii) finally the ion has completely "forgotten" its initial direction, i.e. at the end of the trajectory, all directions of motion are equally probable. At this stage, the probability distribution is constant over the $\varphi,\theta$ sphere, and hence also constant over the range -1, +1 for $\eta$ (slices indicated in Fig. 1a have equal surface areas).

The qualitative picture can easily be converted into a more rigorous mathematical description: The stochastic motion on our unit sphere is governed by the diffusion equation:

$$\frac{\partial W}{\partial \tau} = \frac{\partial}{\partial \eta}\left[(1 - \eta^2)\frac{\partial W}{\partial \eta}\right] , \qquad (2)$$

where $\eta = \cos\varphi$, and $\tau$ is equivalent to Dt in ordinary diffusion. Solutions of this equation are readily obtained, and yield the curves shown in Fig. 1b.

In assessing projected ranges, we actually need not know the distribution functions $w(\eta,t)$ explicitly. For our purpose it is sufficient to find the average value of the directional cosine $\bar{\eta}(\tau)$ which is obtained from Eq. (2), yielding the simple expression:

$$\bar{\eta}(\tau) = \int_{-1}^{+1} \eta w(\eta,\tau)d\eta = e^{-2\tau} \; . \qquad (3)$$

This result shows again - now in a quantitative way - how the initial forward motion, $\bar{\eta} = \langle\cos\psi\rangle = 1$, changes in the course of slowing down, i.e. with increasing $\tau$, to the finally isotropic motion, $\bar{\eta} = \langle\cos\psi\rangle \to 0$, where no net forward motion but rather diffusional spread of the ions occurs: Forward, $\eta \geqslant 0$, and backward motion, $\eta \leqslant 0$, become equally likely towards the end of the ion trajectory.

## 2.2 Connection Between Angular Spread Parameter $\tau$ and Energy Loss

We still have to establish the relation between $\tau$ and the nuclear energy loss. From the way $\tau$ was placed in the diffusion Eqs. (2) or (3), $\tau$ is seen to correspond to Dt or $\int D dt$ in ordinary space-time diffusion. In our case of a surface diffusion on the unit sphere, we may evaluate $\tau$ by using the two-dimensional Einstein relation:

$$\delta\tau = \frac{1}{4}\,\delta\phi^2 \; , \qquad (4)$$

which is valid in the plane, and - to a good approximation - also in small surface regions of our unit sphere. $\delta\phi^2$ is the (mean) square distance on the sphere connected to the increment $\delta\tau$ ( or $D\delta t$). If in that interval a number n of collisions occurs, we may write:

$$\delta\tau = \frac{1}{4}\delta\phi^2 = \frac{1}{4}\sum_1^n \phi_i^2 \;\; * $$

$$\approx \frac{1}{4}\sum_1^n \frac{\Theta_i^2}{(1 + M_1/M_2)} \approx \frac{1}{4}\frac{M_2}{M_1}\frac{\sum T_i}{E} \; . \qquad (5)$$

Here the symbol "≈" was used twice to indicate that small-angle approximations were used in both cases, (i) for converting the lab angles $\phi_i$ into center-of-mass angles $\Theta_i$ and (ii) to relate the scattering angle $\Theta_i$ to the transferred energy $T_i$, according to Eq. (1).

Equation (5) may be rewritten in the shorter form:

$$\delta\tau = \frac{\mu}{4}\,\frac{\delta E_n}{E} \; , \qquad (6)$$

which most clearly indicates the natural connection between angular spread and nuclear energy loss $\delta E_n$ ($\mu = M_2/M_1$). Using the standard notations S (E), $S_t(E)$ for nuclear and total stopping powers, and $\delta E$ for the change in projectile energy, one finally obtains:

---

*In a random walk, the squares of the distance are additive.

$$\delta\tau = -\frac{\mu}{4}\frac{S_n}{S_t}\frac{\delta E}{E}, \qquad \tau(E_o,E) = -\frac{\mu}{4}\int_{E_o}^{E}\frac{S_n}{S_t}\frac{dE}{E}, \tag{7}$$

where $E_o$ is the initial energy of the ion.

As mentioned above, our treatment has up to now been based on small-angle approximations corresponding to using only first terms of power series expansions which lead to $\delta\tau \simeq \Sigma T_i \simeq S_n$. Thus, $\tau$ depends up to now on $S_n(E)$ only, which of course contains no more information than a differential cross section depending only on a single variable, as in previous theories. In order to improve the precision beyond this level, one has to maintain higher terms of the various expansions in Eq.(5). For example, the last step in Eq. (5):

$$\frac{\Theta^2}{4} = \sin^2\frac{\Theta}{2} + \frac{1}{3}\sin^4\frac{\Theta}{2} + \frac{8}{45}\sin^6\frac{\Theta}{2} + \ldots \tag{8}$$

would lead - with Eq. (1) - to the terms $\Sigma T_i$, $\Sigma T_i^2$, $\Sigma T_i^3$,.... and finally to $S_n$, $Q_n$, and higher moments in nuclear energy loss. Evidently, the set of energy-loss moments contains the full information on the differential cross section. The effect of using such higher terms in $\tau$ will be discussed separately in Sect. 5.

## 2.3 Calculation of the Mean Projected Range

The mean projected range $\bar{x}$ is now - by application of Eqs. (3) and (7) - directly accessible, as easily as the total path length s:

$$s = \int ds = \int_0^{E_o} \frac{dE}{S_t(E)}, \tag{9}$$

$$\bar{x} = \int \langle\cos\varphi\rangle ds = \int_0^{E_o} \exp(-2\tau(E_o,E))\frac{dE}{S_t(E)} \tag{10}$$

Each path length element $\delta s$ of the ion trajectory is projected on the x axis by multiplying with the corresponding directional cosine. Although the mean projected range for ions of an energy $E_o$ may be calculated directly by first obtaining $\tau(E_o,E)$ through Eq. (7), and inserting this into Eq. (10), it should be mentioned that more practical methods for obtaining tables of $\bar{x}$ vs E will be developed in Sect. 3.

## 3. Application of the Model

### 3.1 Heavy-Ion Ranges as an Example of Possible Analytic Treatment

For heavy ions of low energy, the nuclear stopping is the dominant slowing-down process. If we use $S_t \approx S_n$ in Eq. (7), we obtain:

$$\tau = \frac{\mu}{4}\ln\frac{E_o}{E},$$

and the mean projected range becomes, through Eq. (10):

$$\bar{x} = \int_0^{E_0} \frac{e^{-2\tau} dE}{S_t} = \int_0^{E_0} \underbrace{\left(\frac{E}{E_0}\right)^{\mu/2}}_{u} \underbrace{\frac{dE}{S_t(E)}}_{v'}$$

$$= \left(\frac{E}{E_0}\right)^{\mu/2} s(E) \Big|_0^{E_0} - \int_0^{E_0} \frac{\mu}{2E_0} \left(\frac{E}{E_0}\right)^{\mu/2-1} s(E) dE \quad . \qquad (11)$$

We have applied partial integration in the last step, as symbolized by u and v'. In this interim result (11), the first term yields only s(E), the total path length itself, while the second term indicates the reduction due to angular deviations from a straight forward flight path.

In the low-energy regime, stopping powers and ranges follow power laws over wide energy regions; e.g. it was shown in Ref. [10] that the expression $d\varepsilon/d\rho = 1/2 \cdot \ln(1+\varepsilon)/(\varepsilon + A\varepsilon^B)$ gives the best fit to realistic stopping powers. This expression leads for low energies, $\varepsilon \ll 1$, to $d\varepsilon/d\rho \approx \varepsilon^{1-B}/2A$, and hence to a total range $\rho \sim (2A/B)\varepsilon^B$, where B is a parameter of the order of 1/2. If we use $s \sim E^B$ in Eq. (11), we obtain:

$$\bar{x} = s(E_0) - \frac{\mu s(E_0)}{2B + \mu} = \frac{s}{1+\mu/2B} \quad . \qquad (12)$$

Correspondingly, the second moment (range straggling) can be obtained through:

$$\overline{x^2} = \frac{s^2}{1 + \mu/B} \; ,$$

which yields the variance of the projected range distribution:

$$\sigma_x^2 = \overline{x^2} - \bar{x}^2 = \frac{(\bar{x}\mu/2B)^2}{1+\mu/B} \quad .$$

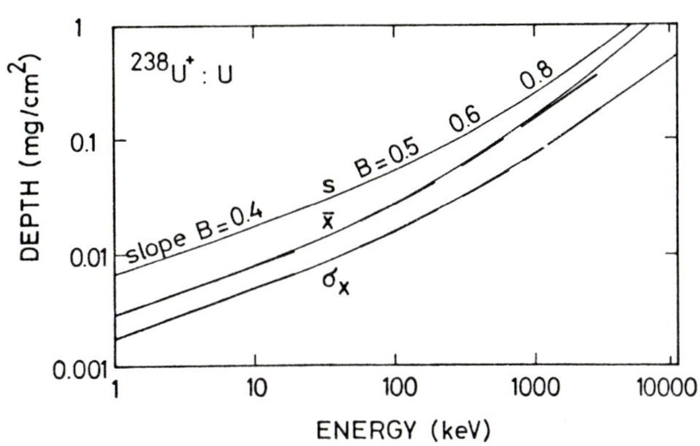


Fig. 2. Simple analytical range predictions for heavy ions, where nuclear stopping is predominant: The expressions $s/x = 1 + \mu/2B$, and $\sigma/x = (1 + \mu/B)^{-1/2}\,\mu/2B$, are applied to deduce x and σ from the total path length s(E). B is the slope of s(E) in the log-log plot, and $\mu = M_2/M_1$

These simple expressions yield quite satisfactory results as compared with other more sophisticated calculations. An example is shown in Fig. 2, where our results (heavy lines) are compared to solutions of the transport equation as obtained by Littmark and Ziegler [7,13] (light lines).

## 3.2 Differential Equation and Universal Algorithm for Projected Ranges

In some cases, the user wants to apply arbitrarily chosen functions for $S_e$ and $S_n$, other than the previously used ones, or he has to deal with composed targets, containing more than one element (cf. Section 4). For such more general calculations of projected ranges, the theory can be developed further to provide $\bar{x}(E + \Delta E)$ from a given $\bar{x}(E)$ in an iterative way, starting with $\bar{x}(0) = 0$. This is most easily accomplished by differentiating Eq. (10) with respect to $E_o$, which appears once in the integral boundary and once in the function $\tau(E_o,E)$, thus yielding the following two terms:

$$\frac{d\bar{x}}{dE_0} = \frac{e^{-2\tau(E_o,E)}}{S_t(E_0)} + \int_o^{E_o} \frac{\partial e^{-2\tau(E_o,E)}}{\partial E_o} \frac{dE}{S_t(E)} \ . \tag{13}$$

From Eq. (7) for $\tau(E_0, E)$ one obtains $\tau(E_0, E_0) = 0$, and:

$$\frac{\partial \tau}{\partial E_o} = \frac{\mu}{4} \frac{S_n(E_0)}{S_t(E_o)E_o} \ ,$$

which is used in Eq. (13) to yield:

$$\frac{d\bar{x}}{dE_0} = \frac{1}{S_t(E_0)} - 2\frac{\mu S_n(E_0)}{4S_t(E_o)E_o} \int_o^{E_0} e^{-2\tau(E_o,E)} \frac{dE}{S_t(E)} \ ,$$

$$\frac{d\bar{x}}{dE_o} = \frac{1}{S_t} - \frac{\mu S_n}{2S_t E_o} \bar{x} \ . \tag{14}$$

This is a linear differential equation for $\bar{x}(E_0)$ with the initial condition $\bar{x}(0) = 0$, which is easily solved by standard computer routines. For home applications, it was found practical to replace $d\bar{x}/dE_0$ by $[\bar{x}(E_o + \Delta E_o) - \bar{x}(E_o)]/\Delta E_o$ for obtaining the range algorithm:

$$\bar{x}(E_0 + \Delta E_0) = \bar{x}(E_0) + \left(1 - \frac{\mu S_n \bar{x}(E_o)}{2\,E_0}\right) \frac{\Delta E_o}{S_t} \ . \tag{15}$$

This iteration has the advantage of using the least number of program steps and memory locations, thus making feasible the calculation of projected ranges on pocket calculators (cf. Appendix).

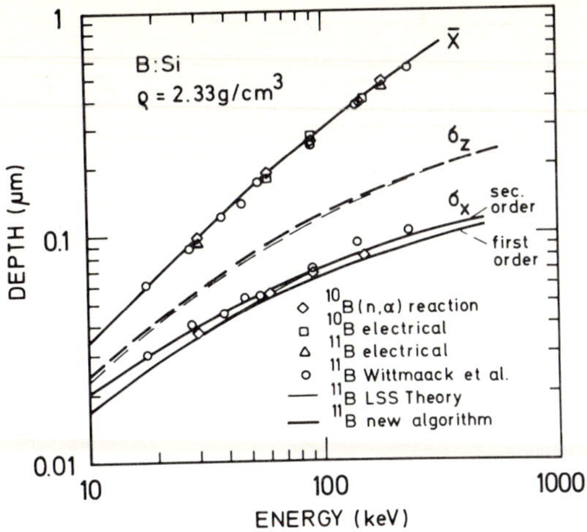


Fig. 3. Comparison of the present range algorithm with transport theory [14] and precision measurements [15], for boron ranges in silicon. Projected ranges x(E) from the LSS theory, and from the present calculation to the first order or to a higher order, fall all on one line. Range stragglings longitudinal $\sigma_x(E)$ and lateral $\sigma_z(E)$ are depicted by heavy lines (present predictions) and by light lines (LSS predictions). For electronic stopping, the standard value of $k = 1.59\ k_{LS}$ was used

An example for applying Eq. (15) is shown in Fig. 3 for the case of boron implantation in silicon, which has been experimentally well studied. Besides the projected range $\bar{x}$, the projected range straggling $\sigma_x$ is now also accessible in experiment and theory, and is depicted in Fig. 3 as well. In the present theoretical model, the variance

$$\sigma_x^2 = \overline{x^2} - \bar{x}^2 \tag{16}$$

is obtained through the following equations:

$$\frac{d\overline{x^2}}{dE} = (\overline{z^2} - \overline{x^2})\frac{\mu S_n}{ES_t} + \frac{2\bar{x}}{S_t} \quad \text{and} \tag{17}$$

$$\frac{d\overline{z^2}}{dE} = (\overline{x^2} - \overline{z^2})\frac{\mu S_n}{ES_t} , \tag{18}$$

where $\overline{z^2}$ denotes the lateral straggling, measured along a Cartesian axis normal to the x axis. The lateral straggling $\sigma_z^2 = \overline{z^2}$ is here obtained as a by-product and is depicted as a dashed line in Fig. 3. The agreement of the present results with LSS results [14] and with experiments [15] is good, including $\sigma_z$-values. Equations (17), (18) can be applied in various ways, e.g., summing the two equations yields a simpler equation:

$$\frac{d}{dE}(\overline{x^2} + \overline{z^2}) = \frac{2\bar{x}}{S_+} , \tag{19}$$

which together with Eq. (18) leads to the following algorithm, using the notation $\xi = \overline{x^2} + \overline{z^2}$:

$$\xi(E + \Delta E) = \xi(E) + \frac{2\bar{x}(E)}{S_t(E)} \Delta E, \tag{20}$$

$$\overline{z^2}(E + \Delta E) = \overline{z^2}(E) + [\xi(E) - 2\overline{z^2}(E)] \frac{\mu S_n(E)}{E S_t(E)} \Delta E. \tag{21}$$

Equations (15), (20), and (21), in that sequence, provide an efficient method for calculating projected ranges ( see Appendix 6.1).

## 3.3 Universal Analytic Approximation for Projected Ranges

In the energy region where most ion implantations are performed, i.e. below 100 keV/amu, the electronic stopping is velocity proportional, $S_e = K\sqrt{E}$, which leads to considerable simplifications in the actual range calculation and finally to an analytical expression for the projected range. In this case, we may use the Lindhard-Scharff equation for the electronic stopping power:

$$S_e^* = \left(\frac{d\varepsilon}{d\rho}\right)_e = \frac{Z_1^{2/3} Z_2^{1/2} (1 + \mu)^{3/2}}{12.6 \, (Z_1^{2/3} + Z_2^{2/3})^{3/4} M_2^{1/2}} \sqrt{\varepsilon} = k_{LS}\sqrt{\varepsilon}, \tag{22}$$

$$\varepsilon = \frac{E/(1 + M_1/M_2)}{Z_1 Z_2 e^2/a} \tag{23}$$

and a new nuclear stopping power approximation:

$$S_n^* = \left(\frac{d\varepsilon}{d\rho}\right)_n = 1/\left(\frac{a}{\sqrt{\varepsilon}} + b\sqrt{\varepsilon}\right), \; a = 0.6, \; b = 2.7 \tag{24}$$

which is accurate within ± 10 % in the energy range $10^{-4} \leq \varepsilon \leq 10$ as compared to the "universal" or Kr-C potentials. They are considered the most realistic general potentials, when used with the new screening length:

$$a = 0.4685\,\text{Å}/(Z_1^{0.23} + Z_2^{0.23}) . \tag{25}$$

The electronic stopping power, Eq. (22), may be improved by applying a correction factor CK, if known.

With the above stopping powers, Eqs. (22) and (24), the differential equation (14) - written in dimensionless units $\varepsilon,\rho$:

$$S_t^* \frac{d\rho_p}{d\varepsilon} = 1 - \frac{\mu}{2} \frac{S_n^*}{\varepsilon} \rho_p \tag{26}$$

allows approximate solutions. For $\varepsilon \ll a/b$, one obtains - by assuming $a+b\varepsilon$ to be constant in this low-energy region:

$$\rho_p = \frac{2\sqrt{\varepsilon}}{k} \frac{1}{1 + \frac{(1+\mu)/k}{a + b\varepsilon}} = \rho_{el} \frac{1}{1 + \frac{\kappa}{a + b\varepsilon}} , \tag{27}$$

where the abbreviations:

$$\rho_{el} = 2\sqrt{\varepsilon}/k \quad \text{and} \tag{28a}$$

$$\kappa = \frac{1 + \mu}{k} \tag{28b}$$

were introduced. For high energies, $\varepsilon \gg a/b$, the differential equation (26) reduces to

$$(1 + bk\varepsilon)\cdot \frac{d\rho_p}{d\varepsilon} = b\sqrt{\varepsilon} - \frac{\mu}{2\varepsilon}\rho_p$$

with the solution:

$$\rho_p = \frac{2\sqrt{\varepsilon}}{k} \left(1 - \sqrt{\frac{\pi\kappa}{2b\varepsilon}} \exp \frac{\kappa}{2b\varepsilon}\cdot \mathrm{erfc}\sqrt{\frac{\kappa}{2b\varepsilon}}\right) . \tag{29}$$

In the range of validity, $\rho_p$ may be approximated by the simpler expression:

$$\rho_p = \rho_{el}\,(1 - 1.2\chi e^{-\chi}), \quad \chi = \sqrt{\frac{2\kappa}{\pi b\varepsilon}} . \tag{30}$$

It seems noteworthy that both high- and low-energy solutions contain only one parameter $\kappa$ which combines the $\mu$ and $k$ dependence. The other parameters, $a$ and $b$, are universal constants in our chosen system $\varepsilon, \rho$.

As both solutions are valid in different energy regions and approach the value $\rho_{el}$ in the other (non-applicable) region, they have little mutual interference and can be combined in one formula:

$$\rho_p = \rho_{el} \frac{1 - 1.2\chi e^{-\chi}}{1 + \frac{\kappa}{a + b\varepsilon}} . \tag{31}$$

As this calculation was performed in the dimensionless units $\varepsilon$ and $\rho$, the result, Eq.(31), has to be converted to regular units, e.g. to Ångstroms:

$$\bar{x} = \frac{(M_1 + M_2)^2}{4\pi a^2 N M_1 M_2}\,\rho_p , \tag{32}$$

where $N=0.602\ \rho(g/cm^3)/M_2$ is the atomic density of the target (atoms/Å$^3$). The precision of the final result is about ±10%, which appears satisfactorily accurate in view of the present uncertainties in the stopping powers used.

## 4. Projected Ranges in Compound Targets

The present theory naturally expands when used for targets which are composed of more than one atomic species. The previous Sub-Sect. 2.2, in particular the Eqs. (5), (6), and (7), could have been written equally well for any number of components. From Eq. (5), it is apparent that each species would contribute to the angular spread $\delta\tau = 1/4\ \delta\phi^2$ proportional to the product of its mass ratio:

$$\mu_j = \frac{M_{2j}}{M_1}$$

and its nuclear energy transfer:

$$\delta E_{nj} = -\frac{S_{nj}}{S_t}\,\delta E,$$

where $S_t = \Sigma j(S_{nj}+S_{ej})$ is the total stopping power. The atomic composition is implicitly accounted for by using the proper atomic densities in the stopping powers. Thus, Eqs. (6) and (7) can be rewritten as:

$$\delta\tau = \sum_j \frac{\mu_j}{4}\,\frac{\delta E_{nj}}{E} = -\frac{\sum \mu_j S_{nj}}{4E}\,\frac{\delta E}{S_t(E)},$$

$$\tau = -\int_{E_0}^{E} \frac{\sum_j \mu_j S_{nj}(E)}{4ES_t(E)}\,dE, \qquad (33)$$

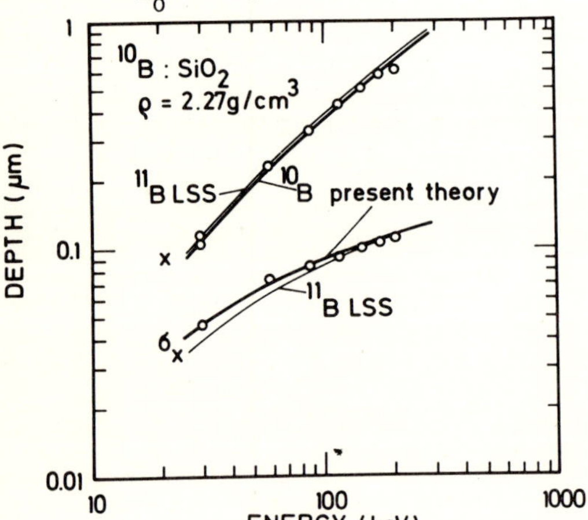


Fig. 4. Example of range predictions for a compound target: $^{10}$B implantation in $SiO_2$. Experimental data from Jahnel [16]

which actually amounts to using a properly weighted average $<\mu S_n> = \sum_j \mu_j S_{nj}$ in all equations, particularly in the final results, Eqs. (14), (17) and (18). An example of ranges in a compound target is given in Fig. 4, comparing results of the present algorithm with LSS results and recent measurements [16].

## 5. Higher Terms and Precision

In order to estimate the precision of our projected range predictions, we need to include higher terms of the expansions used in Eqs. (6) and (7), and to determine their contributions. In Eqs. (6) and (7), the small-angle approximations were used in all places where the symbol "≈" was written instead of the equal sign. If all these approximations were developed to higher terms, then - besides T - $T^2$, $T^3$,...would also enter the equations, leading finally to $S_n$, $Q_n$ and higher energy-loss moments in nuclear stopping. Including for example the second moment in nuclear energy loss, $Q_n = d(E_n^2)/dx$ (but still neglecting electronic energy-loss straggling as in LSS range calculations), our range equations would read - instead of Eqs. (14), (18) and (19):

$$\left(S_t - \frac{\mu Q_n}{2E}\right)\frac{d\bar{x}}{dE} = 1 - \left(\frac{\mu S_n}{2E} + \frac{\mu Q_n}{4E^2}\right)\bar{x}, \tag{34}$$

$$S_t\,\frac{d\overline{(x^2 + z^2)}}{dE} = 2\bar{x}\,, \tag{35}$$

$$\left(S_t - \frac{\mu Q_n}{E}\right)\frac{d\overline{z^2}}{dE} + \frac{\mu Q_n}{E}\,\frac{d\overline{x^2}}{dE} = \left(\frac{\mu S_n}{E} - \frac{(1-\mu)^2 Q_n}{4E^2}\right)(\overline{x^2}-\overline{z^2})\ . \tag{36}$$

Results of this set of equations, however, were found to differ only slightly from the above results obtained by the first-order treatment. E.g., differences of only a few percent were observed in $\bar{x}$. This good precision inherent in the original description can best be understood when considering that, at low energies, $\bar{x} \sim E^B$ with $B \approx 1/2$ holds true (cf. Fig. 2) and that for the case B = 1/2, the two terms in Eq. (34) containing $\mu Q_n$ cancel each other exactly. At high energies, the $\mu Q_n$ terms become small enough to be neglected. In calculating the second range moments $\overline{x^2}$ and $\overline{z^2}$, it is somewhat more significant to maintain the $Q_n$ terms in Eqs. (35) and (36). In applying Eqs. (34), (35) and (36) for multi-atomic targets, one has to use the properly weighted averages $<\mu Q_n> = \Sigma_j \mu_j Q_{nj}$ and $<(1 - \mu)^2 Q_n> = \Sigma_j (1 - \mu_j)^2 Q_{nj}$ in addition to the above-mentioned $<\mu S_n>$. Useful analytic expressions for $Q_n$ can be found in Refs. [9] or [10].

## 6. Appendix

### 6.1 Numerical Evaluation of Projected Ranges on Programmable Pocket Calculators (PRAL)

In recent years, a number of pocket calculators have appeared on the market which allow at least 25 items of data, and some 200 program steps, to be stored. Permanent memory or magnetic cards (for the program) and a printer (for the results) would be helpful. With this limited capacity and the low speed of a pocket calculator, at least the iterations for $\bar{x}$,

$\sigma_x, \sigma_z$, according to Eqs. (15), (20),(21), can be executed for an elementary (mono-atomic) target in the velocity-proportional electronic stopping region, i.e. for $E \leq 100$keV/amu.

The program consists of a preliminary part "A" which determines and stores all necessary parameters. It requires the

INPUT: $Z_1$, $Z_2$, $M_1$, $M_2$, $\rho$ (g/cm³), and calculates and stores in this sequence:

$a = 0.468\ (Z_1^{0.23} + Z_2^{0.23})$ screening length

$N = 0.602\ \rho/M_2$ atomic density

$f = \dfrac{69.4\ a\ M_2}{Z_1\ Z_2\ (M_1 + M_2)}$ conversion factor $\varepsilon = fE$(keV)

$\mu = M_2/M_1$ mass ratio

$k_{LS} = \dfrac{38.8\ Z_1^{7/6}\ Z_2\ N}{(Z_1^{2/3} + Z_2^{2/3})^{3/2}\ \sqrt{M_1}}$ electronic stopping factor for $S_e$(eV/Å) $= k_{LS}\sqrt{E}$(keV)

$F = \dfrac{181 Z_1\ Z_2 N\ a}{1 + \mu}$ conversion factor $S_n$(eV/Å) $= F\ S_n^*$ (dimensionless)

After these constants are determined, one may multiply the correction factor CK into the memory location of $k_{LS}$, or alter the exponent 1/2 to other powers of E in the program steps which determine the electronic stopping $S_e = k\ E^{1/2}$. Part "A" of the program should also initialize the following variables:

$$s;\ \bar{x};\ \overline{z^2};\ \xi = \overline{x^2} + \overline{z^2} = 0 \text{ and } E = 0.01 \text{ (keV)} .$$

The main program, called "B", iterates all variables stepwise from their value at E to their next value at E + $\Delta$E or , if the final energy $E_F$ is reached, prints the results:

$$E,\ S_n,\ S_e,\ s,\ \bar{x},\ \sigma_x = \sqrt{\xi - \overline{z^2} - \bar{x}^2},\ \sigma_z = \sqrt{\overline{z^2}} .$$

It requires only the

INPUT: $E_F$(keV)

and determines $\Delta E$ as 1% of $E_F$.

The iteration consists of the following steps:

LABEL B

$E = E + \Delta E$

$S_e = k\ \sqrt{E}$

$\varepsilon = fE$

$$S_n = F \frac{0.5 \ln (1 + \varepsilon)}{\varepsilon + 0.107\ \varepsilon^{0.3754}}$$

$$S_t = S_e + S_n$$

$$\Delta s = \Delta E / S_t$$

$$\Delta\tau = 0.5\ \mu\ S_n\ \Delta s / E$$

$$s = s + \Delta s$$

$$\bar{x} = \bar{x} + \Delta s - \bar{x}\Delta\tau$$

$$\xi = \xi + 2\ \bar{x}\ \Delta s$$

$$\overline{z^2} = \overline{z^2} + 2(\xi - 2\overline{z^2})\Delta\tau$$

if $E<E_F$ go to LABEL B

$$\sigma_x = \sqrt{\xi - \overline{z^2} - \bar{x}^2}$$

$$\sigma_z = \sqrt{\overline{z^2}}$$

print $E, S_e, S_n, s, x, \sigma_x, \sigma_z$

The last step would consist of determining the next $E_F$ and $\Delta E = 0.01\ E_F$ (according to personal taste), and then returning to LABEL B in the program. If the computer capacity allows, one may add the next higher terms (containing $Q_n$) according to Eqs. (34) and (36), where $Q_n$ is given in an analytic form - similarly to $S_n$ - in Ref. [10], table II, listed under the name W, Kr - C.

## 6.2 Numerical Evaluation of Projected Ranges on Computers (DIMUS)

In addition to the pocket calculator version (PRAL), the Fortran IV version (DIMUS) includes the following features: The correct electronic and nuclear stopping power are used particularly beyond the electronic stopping maximum, but excluding relativistic corrections, i.e. up to about 10 MeV/amu. The Fortran version is extended to account for higher-order energy-loss moments in both nuclear and electronic stopping. The treatment of multiatomic targets is also feasible with this program.

In the following, a printout of the program is given, together with a short explanation of the operation of the program and several examples. To start the program one has to log on the computer and type:

run dimus

The computer now asks for the required information:

TARGET

arbitrary target description

RHO, NTA, (ZT(I), MT(I), N(I))

```
<target density>, <number of different elements in the compound>
<atomic number>, <atomic mass>, <number of atoms of the first element>
<..............>.<...........>, <...of second element>.......
```

```
ION
```

arbitrary ion specification

```
ZI, MI, COR
<atomic number>,<atomic mass>,<correction factor for the LSS-
                                  stopping power, default 1>
```

A sample session is shown for krypton in PMMA ($C_5H_8O_2$, $\rho = 1.19$ g cm$^{-3}$)

```
run dimus
TARGET
polymethylmethacrylate
RHO, NTA, (ZT(I), MT(I), N(I))
1.19. 3
6, 12.010,5
1, 1.008, 8
8, 15.999, 2
ION
kr
ZI, MI, COR
36, 83.912,1
ION

<either a new ion specification for the same target or ctrl Z>
```

The program is terminated by typing ctrl Z. To obtain a print-out of the range listing one has to type "print for 002.dat".

References


1. J. Lindhard, M. Scharff, and H.E. Schiøtt, K.Dan. Vid. Selsk. Mat. Fys. Medd. 33, (1963) 14.
2. H.E. Schiøtt, K. Dan. Vid. Selsk. Mat. Fys. Medd. 35, (1966) 9.
3. W.S. Johnson and J.F. Gibbons, Appl. Phys. Lett. 9, 321 (1966).
4. J.B. Sanders, Can. J.Phys.46, 445 (1968).
5. D.K. Brice, Ion Implantation Range and Energy Deposition Distributions (IFI/Plenum, New York 1975).
6. K.B. Winterbon, Rad. Effects 13, 215 (1972).
7. U. Littmark and J.F. Ziegler, Phys. Rev. A 23, 64 (1981)
8. J.P. Biersack, Hahn-Meitner-Report, HMI B-37 (1964)
9. J.P. Biersack, Z. Physik 211, 595 (1968).
10. W.D. Wilson, L.G. Haggmark, and J.P. Biersack, Phys. Rev. B15, 2458 (1977).
11. J.P. Biersack and W. Krüger, Tables in the book: Ionenimplantation, by H. Ryssel and G. Ruge (B.G. Teubner, Stuttgart 1978).
12. H. Ryssel, G. Lang, J.P. Biersack, K. Müller, and W. Krüger, IEEE Transactions on Electron Devices, ED-27, 58 (1980).
13. U. Littmark, J.F. Ziegler, and R.A. Fiorio, IBM, unpublished.
14. J.F. Gibbons, W.S. Johnson, and S.W. Mylroie, Projected Range Statistics, (Dowden, Hutchinson and Ross, Stroudsburg, Pa. 1975).
15. H. Ryssel, H. Kranz, K. Müller, R.A. Henkelmann, and J.P. Biersack, Appl. Phys. Lett. 30, 399 (1977).
16. F. Jahnel, H. Ryssel, G. Prinke, K. Hoffmann, K. Müller, J.P. Biersack, and R. Henkelmann, Nucl. Instr. Meth. 182/183 223 (1981).

```
FORTRAN IV-PLUS V3.0              11:46:00     5-Jul-82            Page 1
DIMUS.FTN;63               /TR:BLOCKS

0001                PROGRAM DIMUS
0002                DIMENSION X(100),ITARG(20)
0003                COMMON /WPAR/NTA,D(10),CA(10),F(10),RK(10),EPSK(10),SBK(10)
                    1 ,OT(10),QMT(10),ZT(10)
0004                REAL MY,MI,KL,IO,N,K,
                    1 M111,M112,M021,M022,M221,M222
0005                DOUBLEPRECISION ION,SCR
0006                DATA EE/1001./,COR/1./
                 * ,X
                 * / 10.,  20.,  30.,  40.,  50.,  60.,  70.,  80.,  90., 100.,
                 *  110., 120., 130., 140., 150., 160., 170., 180., 190., 200.,
                 *  220., 240., 260., 280., 300., 320., 340., 360., 380., 400.,
                 *  420., 440., 460., 480., 500., 550., 600., 650., 700., 750.,
                 *  800., 850., 900., 950.,1000.,1500.,2000.,2500.,3000.,3500.,
                 * 4000.,4500.,5000.,47*0./
                 * ,ION/'BOR'/, RHO/2.32/
                 * ,NTA,ZT(1),QMT(1),OT(1)  /1,14.,28.,1./
                 * ,ZI,MI,COR/5.,11.,1.59/
0007                CALL ASNLUN(6,'TI',0)
0008      500       WRITE(6,*)'TARGET'
0009                READ(5,1005,END=9999)ITARG
0010      1005      FORMAT(20A2)
0011                WRITE(6,*)'RHO,NTA,(ZT(I),MT(I),N(1))'
0012                READ(5,*,END=9999)RHO,NTA,(ZT(I),QMT(1),OT(I),1=1,NTA)
0013      600       WRITE(6,*)'ION'
0014                READ(5,1010,END=9998)II,ION
0015                IF(II.EQ.0)GOTO 500
0016      1010      FORMAT(Q,A8)
0017                WRITE(6,*)'ZI,MI,COR'
0018                READ(5,*,END=9998)ZI,MI,COR
0019                WRITE(2,2001)
0020      2001      FORMAT('1')
0021                WRITE(2,2000)
0022      2000      FORMAT(5X,'I',51(1H-),'I',51(1H-),'I')
0023                WRITE(2,2010)
0024      2010      FORMAT(5X,'I',51X,'I',51X,'I')
0025                WRITE(2,2020)ITARG,ION
0026      2020      FORMAT(5X,'1',3X,'TARGET :',20A2,'1',2X,
                    1 'ION :',A8,36X,'I')
0027                WRITE(2,2030)INT(.5!ZT(1)),QMT(1),RHO,OT(1),
                    1 INT(.5+ZI),MI,COR
0028      2030      FORMAT(5X,'1',3X,'ZT :',I3,4X,'MT :',F7.3,4X,'RHO :',F7.3,
                    1 2X,F7.3,1X,'I',3X,'ZI :',I3,5X,'MI :',F7.3,10X,'COR :'
                    1 ,F7.3,3X,'1')
0029                IF(NTA.LE.1)GOTO 550
0030                DO 540  1=2,NTA
0031      540       WRITE(2,2035)INT(.5!ZT(I)),QMT(1),OT(I)
0032      2035      FORMAT(5X,'1',3X,'ZT :',I3,4X,'MT :',F7.3,5X,'NT :',F7.3,
                    1 10X,'I',51X,'I')
0033      550       WRITE(2,2010)
0034                WRITE(2,2000)
0035                WRITE(2,2040)
0036      2040      FORMAT(5X,'I',6(1H-),'I',3(14(1H-),'I'),
                    1 4(12(1H-),'I'))
0037                WRITE(2,2050)
0038                WRITE(2,2055)
```

```
FORTRAN IV-PLUS V3.0            11:46:00      5-Jul-82            Page 2
DIMUS.FTN;63              /TR:BLOCKS

0039      2050      FORMAT(5X,'I',2X,'E ',2X,'I',6X,'SE',6X,'I',
                    1 6X,'SN',6X,'I',6X,'W2',6X,'I',5X,' R',5X,'I',
                    2 5X,'DR',5X,'I',5X,'RP',5X,'I',5X,'DRP',4X,'I')
0040      2055      FORMAT(5X,'I','[KEV]',1X,'I',3X,'[KEV/MY]',3X,'I'
                    1 3X,'[KEV/MY]',3X,'I',1X,'[KEV**2/MY]',2X,'I',4X,'[MY]',4X,'I'
                    2 4X,'[MY]',4X,'I',4X,'[MY]',4X,'I',4X,'[MY]',4X,'I')
0041                WRITE(2,2040)
          C
          C****     RECHENPROGRAMM - ANFANG
          C
0042                ELZ=0.
0043                QT=0.
0044                DO 10 I=1,NTA
0045                QT=QT+QMT(I)*OT(I)
0046      10        ELZ=ELZ+OT(I)
0047                QT=QT/ELZ
0048                N=.6022*RHO/QT
0049                DO 20 I=1,NTA
0050                CALL COEF(COR,N,MI,ZI,QMT(I),ZT(I),GA(I),D(I),F(I)
                    1 ,QK(I),EPSK(I),SBK(I))
0051      20        OT(I)=OT(I)/ELZ
0052                MY=QT/MI
0053                B=.1
0054                E1=B
0055                SN1=WI(0,1,E1)
0056                W21=WI(0,2,E1)
0057                W31=WI(0,3,E1)
0058                W41=WI(0,4,E1)
0059                W51=WI(0,5,E1)
0060                W61=WI(0,6,E1)
0061                SE1=WI(1,0,E1)
0062                S1=SE1+SN1
          C
0063                P11=.5*MY*SN1/E1-.25*(.5-MY)*W21/E1**2-.125*(1.-1.5*MY)*W31
                    1 /E1**3-5./32.*(.75-MY)*W41/E1**4-7./64.*(1.-1.25*MY)*W51
                    2 /E1**5-105./1024.*(1.-1.2*MY)*W61/E1**6
          C
0064                P21=1.5*MY*SN1/E1-.375*(1.-MY)**2*(W21/E1**2+W31/E1**3
                    1 +W41/E1**4+W51/E1**5+W61/E1**6)
          C
0065                R1=.5*B/S1
0066                DR1=SQRT(.5*B*W21/S1**3)
0067                EX11=EXP(-.5*B*P11/S1)
0068                EX21=EXP(-.5*B*P21/S1)
0069                M111=R1
0070                M021=B*M111/S1
0071                M221=B*M111/S1
          C
0072                DO 100 I=1,100
0073      3         C=10.-E1
          C
0074                Q=(10.-B)/16.
          C
0075                DE=5.
0076                IF(I.LE.45)DE=1.
0077                IF(I.LE.35)DE=.5
```

```
FORTRAN IV-PLUS V3.0               11:46:00      5-Jul-82              Page 3
DIMUS.FTN;63                /TR:BLOCKS

0078                 IF(I.EQ.1)DE=AMIN1(B,R,C)
0079                 E2=E1+DE
          C...
0080                 IF(E2.GT.EE)GOTO 600
          C...
0081                 SN2=WI(0,1,E2)
          C
0082                 W22=WI(0,2,E2)
0083                 W32=WI(0,3,E2)
0084                 W42=WI(0,4,E2)
0085                 W52=WI(0,5,E2)
0086                 W62=WI(0,6,E2)
          C
0087                 SE2=WI(1,0,E2)
          C
0088                 S2=SE2+SN2
          C
0089                 P12=.5*MY*SN2/E2-.25*(.5-MY)*W22/E2**2-.125*(1.-1.5*MY)*W32
                     1 /E2**3-5./32.*(.75-MY)*W42/E2**4-7./64.*(1.-1.25*MY)*W52
                     2 /E2**5-105./1024.*(1.-1.2*MY)*W62/E2**6
          C
0090                 P22=1.5*MY*SN2/E2-.375*(1.-MY)**2*(W22/E2**2+W32/E2**3
                     1 +W42/E2**4+W52/E2**5+W62/E2**6)
          C
0091                 R2=.5*DE*(1./S2+1./S1)+R1
0092                 DR2=SQRT(.5*DE*(W22/S2**3+W21/S1**3)+DR1**2)
0093                 EX12=EXP(-.5*DE*(P11/S1+P12/S2))
0094                 EX22=EXP(-.5*DE*(P21/S1+P22/S2))
          C
0095                 M112=M111*EX12+.5*DE*(1./S2+EX12/S1)
0096                 GM11=.5*DR2**2*((P12**2+S2*(P12-P11)/DE)*M112-P12)
0097                 RP=M112+GM11
0098                 M022=M021+DE*(M112/S2+M111/S1)
0099                 GM02=DR2**2*(1.-P12*M112)
0100                 RM02=M022+GM02
          C
0101                 M222=M221*EX22+DE*(M112/S2+M111*EX22/S1)
          C
0102                 GM22=.5*DR2**2*((P22**2+S2*(P22-P21)/DE)*M222-2.*(P12+P22)
                     1 *M112+2.)
          C
0103                 RM22=M222+GM22
0104                 DRP=SQRT(1./3.*RM02+2./3.*RM22-RP**2)
0105                 E1=E2
0106                 SE1=SE2
0107                 SN1=SN2
0108                 W21=W22
0109                 W31=W32
0110                 W41=W42
0111                 W51=W52
0112                 W61=W62
          C
0113                 S1=S2
0114                 P11=P12
0115                 P21=P22
0116                 R1=R2
```

```
FORTRAN IV-PLUS V3.0                 11:46:00      5 Jul-82            Page 4
DIMUS.FTN;63               /TR:BLOCKS

0117                DR1=DR2
0118                M111=M112
0119                M021=M022
0120                M221=M222
          C
0121                IF(E1.NE.X(I))GOTO 3
          C
0122                WRITE(2,2060)INT(.5+E1),SE1,SN1,W21,R1,DR1,RP,DRP
0123      2060      FORMAT(5X,'I',1X,I4,1X,'I',3(1X,1PE12.3,1X,'I'),
                    1 OP,4(1X,F10.4,1X,'I'))
          C
0124                IF(I.EQ.20.OR.I.EQ.35.OR.I.EQ.45.OR.I.EQ.53)
                    1 WRITE(2,2040)
          C
0125      100       CONTINUE
0126                GOTO 600
0127      9999      STOP 'EOF AFTER TARGET'
0128      9998      STOP 'EOF AFTER ION'
0129                END

PROGRAM SECTIONS

 Name        Size                      Attributes

$CODE1   005576   1471                 RW,I,CON,LCL
$PDATA   001340    368                 RW,D,CON,LCL
$IDATA   000042     17                 RW,D,CON,LCL
$VARS    001270    348                 RW,D,CON,LCL
$TEMPS   000026     11                 RW,D,CON,LCL
WPAR     000552    181                 RW,D,OVR,GBL

Total Space Allocated = 011270   2396
```

```
FORTRAN IV-PLUS V3.0               11:46:35      5 Jul 82              Page 5
DIMUS.FTN;63                  /TR:BLOCKS

0001                FUNCTION WI(IE,I,EN)
0002                COMMON /WPAR/NTA,D(10),GA(10),F(10),QK(10),EPSK(10),SBK(10)
                    1 ,OT(10),QMT(10),ZT(10)
          C
          C****     POTENTIALE
          C
0003                DIMENSION WPA(4,6)
0004                DATA
                    1 WPA /4*0.,.18696,-1.6828,6.9825,-.94342,
                    2        11.348, .9569 ,.1833 ,-1.738 ,
                    3        15.8  ,-.9512 ,.29   ,-1.71  ,
                    4        20.495,-.9538 ,.2697 ,-1.752 ,
                    5        24.95 ,-.957  ,.2989 ,-1.76    /
0005                WI=0.
0006                IF(IE.NE.0)GOTO 10
0007                IF(I.LE.2)GOTO 20
0008                DO 1 J=1,NTA
0009      1         WI=WI+OT(J)*D(J)*GA(J)*(GA(J)*EN)**(I-2)/(F(J)**2*
                    1 (4*I-4+WPA(1,I)*(F(J)*EN)**WPA(2,I)+
                    2         WPA(3,I)*(F(J)*EN)**WPA(4,I)))
0010                RETURN
0011      20        IF(I.EQ.1)GOTO 30
0012                DO 2 J=1,NTA
0013      2         WI=WI+OT(J)*D(J)*GA(J)/(F(J)**2*
                    1 (4+WPA(1,2)*(F(J)*EN)**WPA(2,2)+
                    2    WPA(3,2)*(F(J)*EN)**WPA(4,2)))
0014                RETURN
0015      10        DO 3 J=1,NTA
0016      3         WI=WI+OT(J)/(1./(QK(J)*SQRT(EN))+EPSK(J)*EN/(SBK(J)*
                    1 ALOG(1.+EPSK(J)*EN+(5./(EPSK(J)*EN)))))
0017                RETURN
0018      30        DO 4 J=1,NTA
0019      4         WI=WI+OT(J)*D(J)*.5*ALOG(1.+F(J)*EN)/(F(J)*(F(J)*EN+.051953*
                    1 (F(J)*EN)**.32011))
          C
          C****     ENDE DER POTENTIALE
          C
0020                RETURN
0021                END
```

PROGRAM SECTIONS

| Name | Size | | Attributes |
|---|---|---|---|
| $CODE1 | 001256 | 343 | RW,I,CON,LCL |
| $PDATA | 000010 | 4 | RW,D,CON,LCL |
| $IDATA | 000004 | 2 | RW,D,CON,LCL |
| $VARS | 000142 | 49 | RW,D,CON,LCL |
| $TEMPS | 000016 | 7 | RW,D,CON,LCL |
| WPAR | 000552 | 181 | RW,D,OVR,GBL |

Total Space Allocated = 002224   586

```
FORTRAN IV-PLUS V3.0                 11:46:44      5 Jul 82             Page 6
DIMUS.FTN;63               /TR:BLOCKS

                SUBROUTINE COEF(COR,N,MI,ZI,MT,ZT,GA,D,F,K,EPSK,SBK)
                REAL MY,MI,MT,KL,N,K,IO
                A=.8853*.529/(SQRT(ZT)+SQRT(ZI))**(2./3.)
                GA=4.*MI*MT/(MI+MT)**2
                D=3.1415927*A**2*N*GA*1E4
                F=A*MT/(ZT*ZI*14.41*(MI+MT))*1E3
                KL=38.33*ZI**(7./6.)*ZT/(ZI**(2./3.)+ZT**(2./3.))**1.5/SQRT(MI)
                K=COR*KL*N/1000.*1E4
                IO=12.+7./ZI
                IF(ZI.GE.13.)IO=9.76+58.5*ZI**(-1.19)
                EPSK=1./(.456*ZI*MI*IO)
                SBK=8.*3.1415927*N*(ZT*14.41)**2/(1000.*IO)*1E4
                RETURN
                END

PROGRAM SECTIONS

 Name       Size                      Attributes

$CODE1   000632   205                 RW,I,CON,LCL
$PDATA   000044    18                 RW,D,CON,LCL
$VARS    000020     8                 RW,D,CON,LCL
$TEMPS   000004     2                 RW,D,CON,LCL

Total Space Allocated = 000722   233
```

# Range Distributions

Heiner Ryssel

Fraunhofer-Institut für Festkörpertechnologie, Paul-Gerhardt-Allee 42
D-8000 München 60, Fed. Rep. of Germany

## 1. Introduction

According to the classical LSS theory as well as the alternative Biersack approach [1, 2], the range profiles of implanted impurity distributions are given by symmetrical Gaussian distributions. Experiments have shown, however, that profiles are skewed and possess tails, not only in crystalline semiconductors where channeling of ions may occur. These experimental findings are confirmed by numerical Monte Carlo simulations. Both theories can be extended to calculate higher moments of the range distributions. Winterbon [3] calculated 4 moments of range distributions for many ion-target combinations; however, no easily accessible tables or formulas (such as those by Gibbons [4], Smith [5], and Biersack [29]) are available. Only Gibbons gives an estimate of the third moment.

In the following section, the most widely used analytical expressions for range profiles will be described. Since all implantations for practical applications are performed into tilted samples, the profile description for amorphous targets is sufficient. Residual channeling tails can be modeled by a slight adjustment of the parameters.

## 2. Gaussian Profiles

A Gaussian distribution is described by two moments, the projected range $R_p$ and the projected standard deviation or straggling $\Delta R_p$. Together with the implanted dose $N_\square$, they describe the implanted profile by:

$$C(x) = \frac{N_\square}{\sqrt{2\pi}\,\Delta R_p} \exp\left[-(x-R_p)^2/(2\,\Delta R_p^2)\right] \qquad (1)$$

where x is the distance measured along the axis of incidence. Gaussian distributions are very useful for fast estimation of the range distribution of implanted ions, or for calculating the thickness of masking layers.

In deriving Eq. (1), it has been assumed that the ions come to rest in a volume extending from $-\infty$ to $+\infty$. The semiconductor, however, extends only from 0 to $+\infty$. If the fraction of the ions which is backscattered is neglected, one arrives at a more exact expression:

$$C(x) = \frac{2\,N_\square}{\sqrt{2\pi}\,\Delta R_p\left[(1+\mathrm{erf}\,R_p/\sqrt{2}\,\Delta R_p)\right]} \exp\left[-(x-R_p)^2/(2\,\Delta R_p)^2\right] \ . \qquad (2)$$

The error in using Eq. (1) is small, and is usually below 1%. Only for very low energy implants can it be appreciable. Moreover, backscattered particles have been completely neglected in Eqs. (1) and (2). Calculations and measurement by Bøttiger et al. [31] show that the backscattered particles are those lying in the part of the profile which lies in front of the sample surface. Therefore, the use of Eq. (1) could lead to better results.

If the profile is only slightly asymmetrical, the third moment is sufficient to obtain a good profile description [6]. In this case, the profile is given by two joint half-Gaussian profiles with range $R_M$ and range straggling $\Delta R_{p1}$ and $\Delta R_{p2}$:

$$C(x) = \frac{2N_{\square}}{(\Delta R_{p1}+\Delta R_{p2})\sqrt{2\pi}} \exp\left[-\frac{(x-R_M)^2}{2\Delta R_{p1}^2}\right] \qquad x \geq R_M$$

$$C(x) = \frac{2N_{\square}}{(\Delta R_{p1}+\Delta R_{p2})\sqrt{2\pi}} \exp\left[-\frac{(x-R_M)^2}{2\Delta R_{p2}^2}\right] \qquad 0 \leq x \leq R_M \; . \tag{3}$$

$R_M$, $\Delta R_{p1}$, and $\Delta R_{p2}$ can be calculated, e.g. from the tabulated data of Gibbons [4], according to:

$$R_p = R_M + 0.8(\Delta R_{p2}-\Delta R_{p1})$$

$$\Delta R_p = -0.64\,(\Delta R_{p2}-\Delta R_{p1})^2 + (\Delta R_{p1}-\Delta R_{p1}^2\Delta R_{p2}+\Delta R_{p2}^2) \tag{4}$$

$$\gamma = \Delta R_p^{-3}(\Delta R_{p2}-\Delta R_{p1})(0.218\;\Delta R_{p1}^2+0.362\;\Delta R_{p1}\Delta R_{p2}+0.218\Delta R_{p2}^2)\;;$$

$\gamma$ is the skewness of the distribution (for its definition, see next section). Approximate values for $\Delta R_{p1}$ and $\Delta R_{p2}$ are obtained by interpolation using Table 1 [4]. For a negative skewness, $\Delta R_{p1}$ and $\Delta R_{p2}$ have to be exchanged.

Table 1. Standard deviations of the joint half-Gaussian distribution

| $\gamma$ | 0 | 0.1 | 0.2 | 0.3 | 0.4 | 0.5 | 0.6 | 0.7 | 0.8 | 0.9 | 1.0 |
|---|---|---|---|---|---|---|---|---|---|---|---|
| $\Delta R_{p1}/\Delta R_p$ | 1 | 1.062 | 1.123 | 1.182 | 1.241 | 1.301 | 1.360 | 1.422 | 1.486 | 1.554 | 1.633 |
| $\Delta R_{p2}/\Delta R_p$ | 1 | 0.936 | 0.871 | 0.802 | 0.729 | 0.653 | 0.570 | 0.478 | 0.374 | 0.248 | 0.081 |

A comparison of profiles obtained by assuming simple Gaussian and joint Gaussian distributions is given in Fig. 1. The range parameters are taken from Gibbons et al. [4].

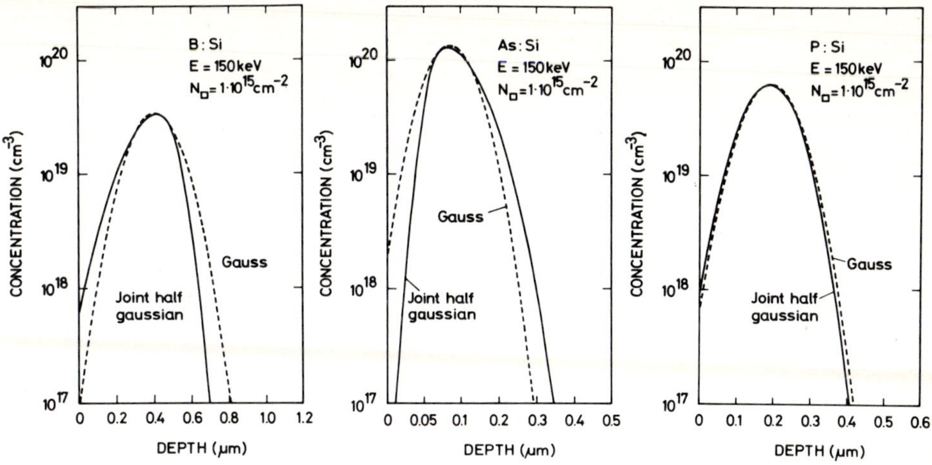


Fig. 1. Comparison of Gaussian and joint Gaussian distributions for 150 keV boron, arsenic, and phosphorus in silicon

## 3. Pearson Distributions

Many experimental investigations have shown that the simple description of implanted profiles given in the last section is not adequate for most ions in silicon and other semiconductors. It has been argued that this might be due to channeling because of the crystalline structure of the usual semiconductors. It has been found, however, that the profiles of many ions are asymmetrical, in amorphous targets as well, and higher moments have to be used to construct range distributions.

A particularly useful distribution is the Pearson type-IV function with four moments [7]. The Pearson distribution of type IV centered around the projected range $R_p$ is given by:

$$f(x) = K[b_2(x-R_p)^2 + b_1(x-R_p)+b_o]^{1/2b_2}\exp[- \frac{b_1/b_2+2a}{\sqrt{4b_2b_o-b_1^2}} \cdot \arctan \frac{2b_2(x-R_p)+b_1}{\sqrt{4b_2b_o-b_1^2}}] \qquad (5)$$

where K is obtained by the constraint:

$$\int_{-\infty}^{\infty} f(x)dx = 1 \quad . \qquad (6)$$

The four constants a, $b_o$, $b_1$, and $b_2$ are given by:

$$a = - \frac{\Delta R_p \gamma (b+3)}{A} \qquad b_o = - \frac{\Delta R_p{}^2 (4\beta-3\gamma^2)}{A} \qquad (7)$$

$$b_1 = a \qquad b_2 = - \frac{(2\beta-3\gamma^2-6)}{A}$$

where $A = 10\beta - 12\,\gamma^2 - 18$.

The parameters $R_p$, $\Delta R_p$, $\beta$, $\gamma$ in these equations are directly related to the four moments $\mu_1$, $\mu_2$, $\mu_3$, and $\mu_4$ of the distribution f(x).

The first moment $\mu_1$ is well known as the average projected range:

$$\mu_1 = R_p = \int_{-\infty}^{\infty} x\ f(x)\ dx \tag{8a}$$

The three higher moments $\mu_i$ are given by:

$$\mu_i = \int_{-\infty}^{\infty} (x-R_p)^i\ f(x)\ dx, \qquad i = 2,3,4\ . \tag{8b}$$

It is customary to use the standard deviation $\Delta R_p$, which is defined by the square root of the second moment $\mu_2$, and dimensionless expressions for the higher moments:

standard deviation $$\Delta R_p = \sqrt{\mu_2} \tag{9a}$$

skewness $$\gamma = \frac{\mu_3}{\Delta R_p^3} \tag{9b}$$

kurtosis $$\beta = \frac{\mu_4}{\Delta R_p^4}\ . \tag{9c}$$

The skewness $\gamma$ indicates the tilting of the profile, and the kurtosis $\beta$ indicates the flatness at the top of the profile.

The relation between the third and fourth moments has to be chosen to satisfy:

$$\beta > \beta_{min} = \frac{48 + 39\ \gamma^2 + 6\ (\gamma^2 + 4)^{3/2}}{32 - \gamma^2} \tag{10}$$

in order to give a Pearson distribution of type IV.

The maximum of the Pearson distribution occurs at $x = R_p + a$ unless $\gamma = 0$. For a skewness of zero, a Gaussian profile results. For a negative skewness, the peak is deeper than $R_p$ and the distribution falls off more rapidly for $x > a$ than for $x < a$. For a positive skewness, the opposite is true.

The kurtosis is 3 for a Gaussian distribution. The limit is represented by Eq. (10). An universal expression for the kurtosis is given by Gibbons [8]:

$$\beta = 2.8 + 2.4\ \gamma^2\ . \tag{11}$$

To obtain the concentration profile C(x), the Pearson distribution is multiplied by the dose:

$$C(x) = N_{\square}\ f(x) \tag{12}$$

(note that f(x) as well as K have the dimension $(\text{length})^{-1}$). In the discussion of the Pearson distribution, it has again been assumed that the profile extends from $-\infty$ to $+\infty$. To describe a more realistic, semi-infinite target, a correction of the normalizing constant K is required; K is usually computed by numerical integration.

A comparison of profiles obtained by Gaussian and Pearson distributions is given in Fig. 2. The moments are from the estimate of Gibbons [4]. Characteristic are the tails in case of Pearson distributions.

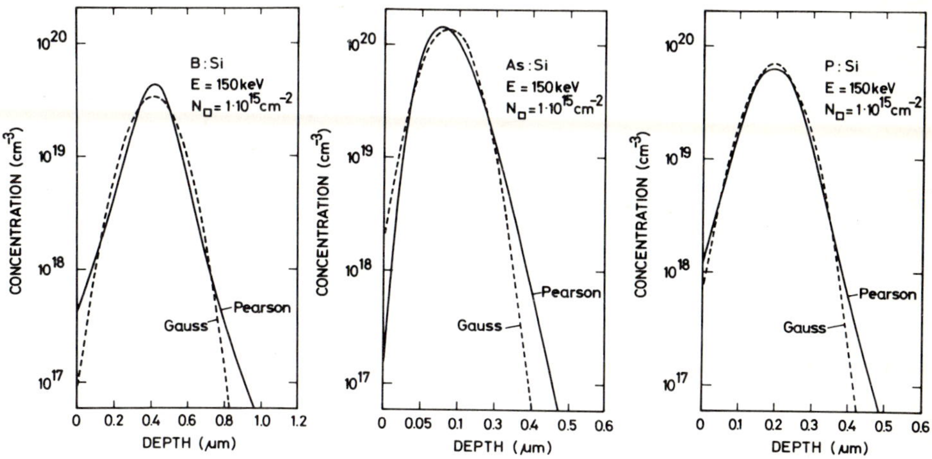


Fig. 2. Comparison of Gaussian and Pearson distributions for 150 keV boron, arsenic, and phosphorus in silicon. The moments are according to Gibbons et al. [4]

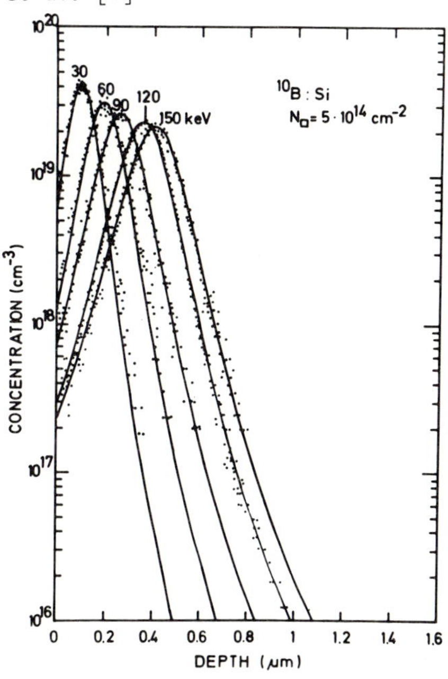


Fig.3. Experimental boron profiles in silicon for energies between 30 and 150 keV with doses of $5\times10^{14}\,cm^{-2}$ in comparison to Pearson distributions; orientation <111>

Examples for measured range distributions and the corresponding moments of boron in silicon and $SiO_2$ are given in the next figures. Figure 3 shows boron profiles in silicon for energies of 30 to 150 keV. The samples were tilted by 7° and optimally rotated to suppress the channeling effect. Nevertheless, the profile tails are probably caused by channeling; they can, however, be modeled perfectly by the Pearson distributions (drawn lines). The moments of the distributions extracted from these measured profiles are given in Fig. 4, in comparison to theoretical calculations according to Gibbons [4] and Biersack [2].

In amorphous targets, no channeling can take place. In Fig. 5, range profiles of boron in $SiO_2$ for energies between 60 and 210 keV are given. The profiles have a shallow slope towards the surface and a steep slope towards the bulk, and show no profile tails. The corresponding moments are given in Fig. 6, together with theoretical calculations.

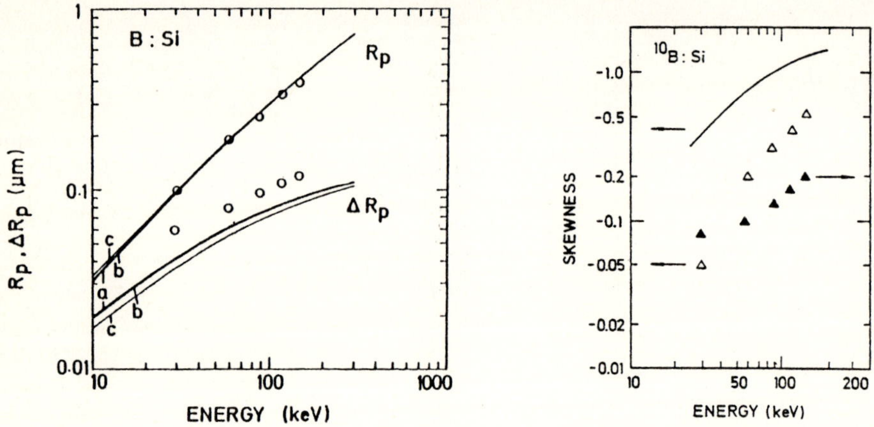


Fig. 4. Experimental range parameters of boron in silicon, in comparison to theoretical calculations [2, 4]

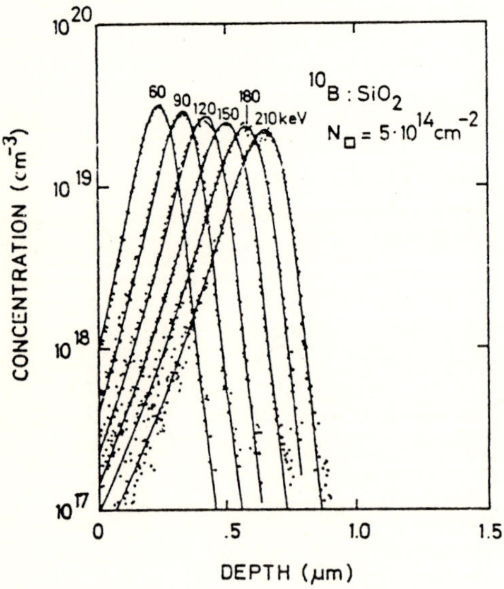


◄

Fig. 5. Boron profiles in $SiO_2$ for energies between 6o and 210 keV with doses of $5x10^{14}cm^{-2}$ in comparison to Pearson distributions

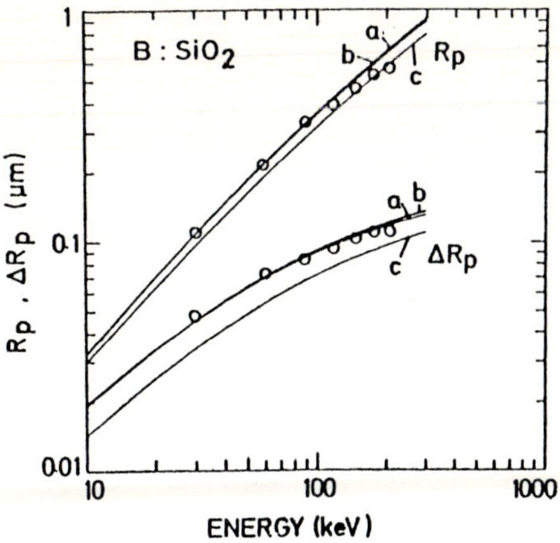


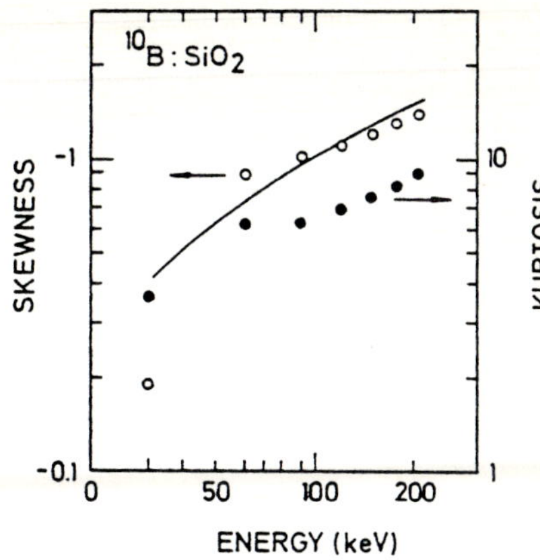


Fig. 6. Experimental range parameters of boron in $SiO_2$, in comparison to theoretical calculations [2, 4]

From the preceding figures it can be seen that Pearson distributions are well suited to describe implantation profiles of boron in silicon and $SiO_2$. Similar results were found for boron in $Si_3N_4$ and arsenic in silicon, $SiO_2$, and $Si_3N_4$ [32].

Sometimes Pearson distributions are combined with an exponential tail [7, 30] in order to reduce the long tails often found with Pearson distributions.

## 4. Other Distributions

Various other distribution functions have been proposed for the description of implantation profiles. The Edgeworth function [4] results from an expansion of the Gaussian function into Chebyshev-Hermitian polynomials, and also uses four moments. The implantation profile is given by:

$$C(x) = \frac{N_{\square}}{\sqrt{2\pi}\,\Delta R_p} \exp\left[-\frac{(x-R_p)^2}{2\,\Delta R_p^2}\right] G(x) \qquad (13)$$

$$G(x)=1 + \frac{\gamma}{6}(Z^3-3Z)+\frac{\beta-3}{24}(Z^4-6Z^2+3)+\frac{\gamma^2}{72}(Z^6-15Z^4+45Z^2-15)+\ldots$$

with $Z = (x-R_p)/2\Delta R_p^2$.

The kurtosis is calculated according to Gibbons [4] by:

$$\beta = 3 + \frac{5}{3}\gamma^2 \quad . \qquad (14)$$

In this case, for the profile, the following applies:

$$C(x) = \frac{N_{\square}}{\sqrt{2}\,\Delta R_p} \exp\left[\frac{(x-R_p)^2}{2\Delta R_p^2}\right]\cdot\left[1 + \frac{\gamma}{6}(Z^3-3Z)+\frac{5\gamma^2}{72}(Z^4-6Z^2+3)+ \right. \qquad (15)$$

$$\left. + \frac{\gamma^2}{72}(Z^6-15Z^4+45Z^2-15)\right] \quad .$$

The disadvantage of this function is that it is applicable only for small deviations from the Gaussian shape; moreover, it shows oscillations, with negative values of the function, and can therefore be used only close to the maximum of the distribution.

Other distributions such as [9]:

$$C(x) = cN_{\square}x^{m} \exp(-x/b)^{n} \qquad (16)$$

are mathematically simple, but the determination of the moments requires a non-linear search so that they appear not too well suited.

The Sonine expansion with three moments [10] gives a very good fit to Monte Carlo simulations, and probably also to experimental profiles, but is too complicated for practical applications. A further detailed discussion of different distributions can be found in the book by Winterbon [3].

A comparison between optimal Gaussian, Pearson, and Edgeworth fits to an experimental boron profile is given in Fig. 7, and shows the superior properties of the Pearson distribution.

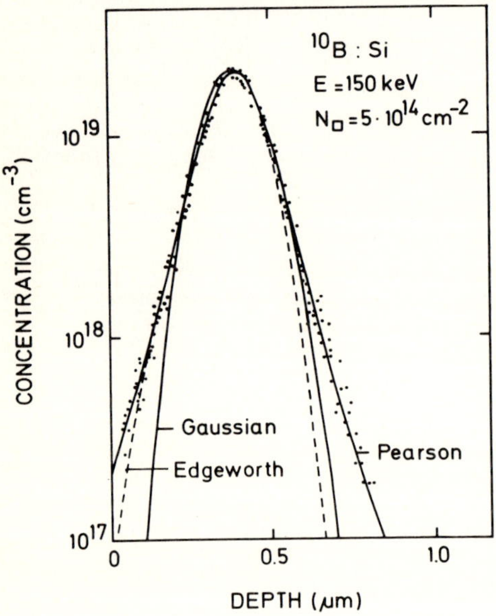


Fig. 7. Comparison of Gaussian, Pearson and Edgeworth fits to an experimental boron profile (E=150 keV, $N_{\square}=5\times10^{14}cm^{-2}$)

## 5. Two-Layer Targets

Frequently, implantations are performed in thin dielectric layers such as $SiO_2$ or $Si_3N_4$, in order to avoid a contamination of the silicon, to provide for a scattering layer, or to adjust the depth distribution of the ions. To calculate the depth distribution of such two-layer structures in an exact way, Monte Carlo simulations or the solution of Boltzmann transport equations [11] are required. This is very complicated and requires much computing time.

For practical cases in silicon technology, a simple model derived by Ishiwara and Furukawa [12] can be applied for materials whose average atomic numbers and masses are nearly equal. This is true for the most common combinations used with silicon, e.g. $SiO_2$, $Si_3N_4$, and $Al_2O_3$. Moreover, for such implants, only thin layers are used, thus further reducing the possible error. If thick layers are used, they are usually required for masking and have to stop all ions completely.

Assuming Gaussian profiles, the two parts of the profile in layer 1 and substrate 2 are given by [12]:

$$C_1(x) = \frac{N_\square}{\sqrt{2\pi}\Delta R_{p1}} \exp\left[-\frac{(R_{p1}-x)^2}{2\Delta R_{p1}^2}\right] \qquad 0 \leq x \leq d \qquad (17a)$$

$$C_2(x) = \frac{N_\square}{\sqrt{2\pi}\,\Delta R_{p2}} \exp\left[-\frac{[d+(R_{p1}-d)\Delta R_{p2}/\Delta R_{p1}-x]^2}{2\,\Delta R_{p2}^2}\right] \quad x \geq d \qquad (17b)$$

where d is the thickness of the layer. The error in using Eqs. (17 a,b) can be calculated by:

$$\eta = \frac{R_{p2} - R'_{p2}}{R_{p2}} \text{ with } R'_{p2} = R_{p1}\,\frac{\Delta R_{p2}}{\Delta R_{p1}} \quad .$$

The profile [Eqs. (17)] is derived from the profile in a thick oxide layer by a density transformation with the scaling factor $\Delta R_{p1}/\Delta R_{p2}$. This leads to a good description of the actual profile if a thick oxide is used ($d \gg R_{p1}$); in the case of a thin oxide ($d \ll R_{p1}$), however, the profile in the substrate has the range $R'_{p2} \neq R_{p2}$.

We use a different density transformation, because the profile shape in the silicon substrate is usually more critical than the profile in the oxide layer. Our model also accomodates non-Gaussian profiles, e.g. Pearson distributions, to give the two-layer profile:

$$C_1(x) = \frac{\Delta R_{p2}}{\Delta R_{p1}} \; C\left(\frac{\Delta R_{p2}}{\Delta R_{p1}}\,x\right) \qquad x \leq d \qquad (18a)$$

$$C_2(x) = C\left(x - d\left(1 - \frac{\Delta R_{p2}}{\Delta R_{p1}}\right)\right) \qquad x \geq d \qquad (18b)$$

where C(x) is the profile in bare silicon without second layer.

This model gives good results if the distribution is concentrated in the second layer. A further refinement can be made to give the exact profile in both limiting cases (thin or thick mask layers). For this purpose, the profile $C_1$ in material 1 is calculated, and the total number of atoms in this layer ($N_{\square,1}$) is obtained by integration. Then, the profile $C_2$ is calculated in the target material material 2 (assuming no masking layer), and the thickness d' is determined which contains $N_{\square,1}$ atoms. The

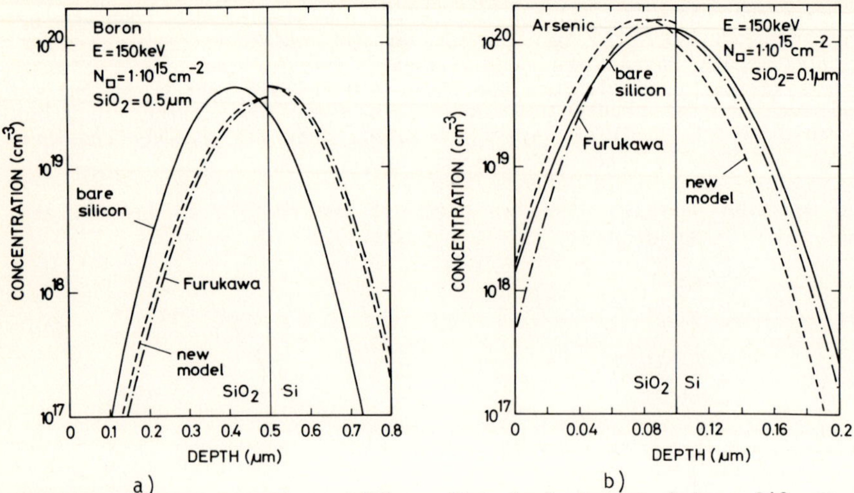


Fig. 8. Boron and arsenic profiles after implantation into a $SiO_2$-Si structure

final profile is composed of profile $C_1$ in material 1 up to d and of profile $C_2$ starting from d', thus resulting in a profile containing $N_{\square}$ ions.

In Fig. 8, the profiles resulting from boron and arsenic implantations into $SiO_2$-Si structures are shown using this model, in comparison to profiles assuming silicon for the complete structure, or performing only a density tranformation. One can clearly see the necessity of the exact profile description.

Better results, but again requiring lengthy calculations, are obtained using energy-distribution functions and approximation of range distributions by joint half-Gaussian distributions [6].

Another model published by Satya and Palanki [13] has the disadvantage of predicting arbitrary deep profiles in silicon at a given energy, if a thin mask layer with low stopping power is used.

The implantation of ions thru masking layers also results in recoil or knock-on implantation of ions. If the mass of the implanted ions is not too different from the mass of the atoms of the masking layer, a large fraction of the energy of the primary particles can be transferred to the atoms of the masking layer. These particles are then implanted into the substrate. The energy which is transferred in an impact is given by Eq. (3). Theoretical calculations of distributions of recoil implantations have been performed by Moline et al. [14], Fischer et al. [15], and recently by Sigmund [16] as well as by Hirao et al. [17]. The formalism is too complicated to deal with it here; no simple analytical profile description exists, and also no tabulated data.

However, all theories and experiments show the same phenomena: The recoiled atoms show an extremely shallow distribution, with the maximum at the surface. The primary ions come to rest much deeper in the crystal than the recoiled ones. An example of such a profile is given in Fig. 9.

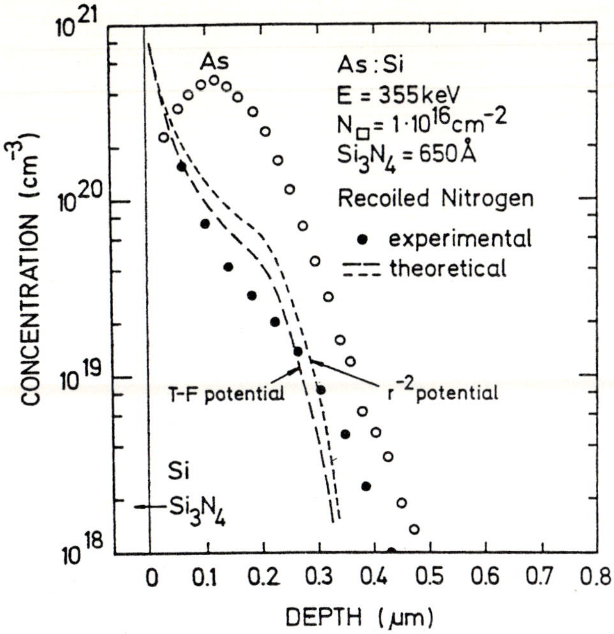


Fig. 9. Comparison of experimental and theoretical profiles for arsenic implantation at 335 keV with a dose of $1 \times 10^{16}$ cm$^{-2}$, through $Si_3N_4$ having 650 Å thickness [17].

Up to now, no detrimental effect has been found on implantation profiles, or on the diffusion of implanted species. The damage produced by the knocking ions, however, has to be considered. It may extend into the area of the p-n junction, and may degrade its properties. If, however, epitaxial layers have to be grown on such a layer, it is necessary to get rid of the damaged layer. This can be done by oxidation or etching.

Very recently, a new type of effect has been found to occur at interfaces: atomic mixing [18 - 22]. At the same time knock-on takes place, atoms from the substrate layer can also be back-sputtered into the masking layer, with the result that the interface layer is no longer well defined. This effect can become remarkable in very high dose implantations. The mixing is stronger than expected from the classical sputtering theory, and is explained by a high atomic mobility for a short time interval following the impact. At the same time, reactions between the atoms can also take place. These effects have been investigated especially in connection with silicide formation, and are of no concern in the present context.

## 6. Implantation and Sputtering

An effect usually neglected in ion implantation is sputtering. For shallow devices, however, low energies and high doses have to be used to obtain the desired junction depth and sheet resistivity. In this case, sputtering has to be considered. The most important parameter for dealing with this problem is the sputtering yield S.

According to Sigmund [23, 24], the sputtering yield for normal incidence is given by:

$$S = \frac{3}{4}\,\frac{S_n(E)\cdot\alpha(M_2/M_1)}{\pi^2 C_o U_o} \qquad (19)$$

where $C_o$ is a constant ($C_o = 1/2\ \lambda_o a^2$; $\lambda_o = 24$; $a = 0.0219$ nm), $\alpha(M_2/M_1)$ is a numerically calculable function, $U_o$ is the surface binding energy ($U_o$ = 7.81 eV for silicon [24]),and $S_n(E)$ is the nuclear energy deposition.

$\alpha(M_2/M_1)$ is given in Fig. 10 for two power potentials. Sputtering yields for several ions, depending on the energy, are given in Fig. 11. Depending on the ion mass, the yield shows a maximum in the energy range of ion implantation. For a fixed energy of 45 keV, experimental sputtering yields according to Andersen et al. [25] are given in Fig. 12, in comparison to theoretical values according to Sigmund. From these two figures, it is seen that the sputtering yield is usually between 1 and 5.

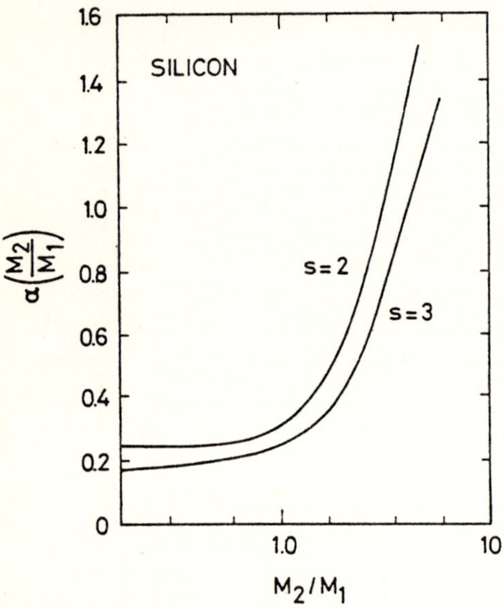


Fig. 10. Function $\alpha(M_2/M_1)$ for power potentials with s = 2 and s = 3; according to [23]

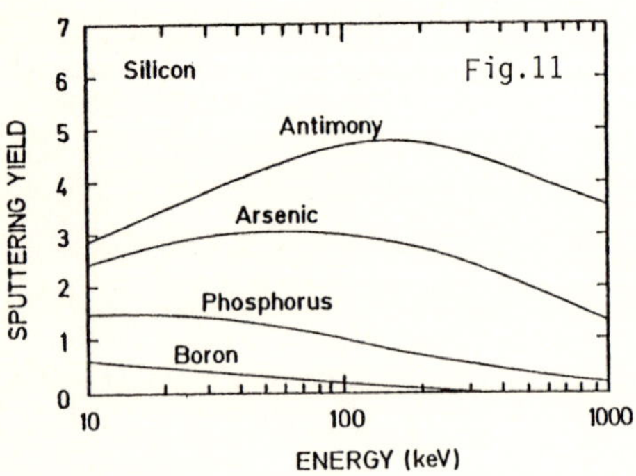


Fig. 11. Sputtering yields of antimony, arsenic, boron, and phosphorus, as a function of energy

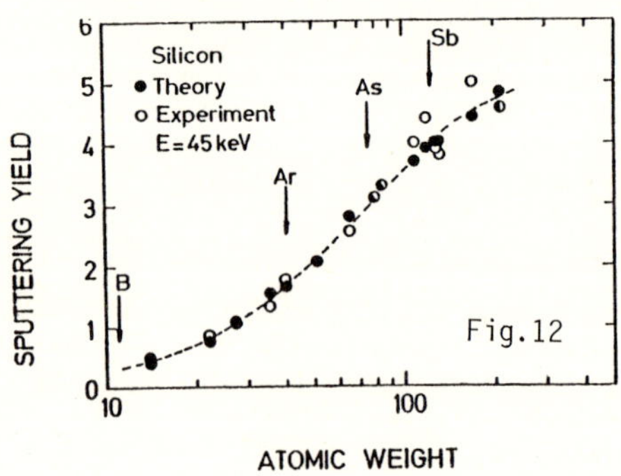


Fig. 12. Sputtering yields of ions of different mass at 45 keV, according to Andersen [25]

Table 2. Thickness of sputtered silicon layers for different ions at maximum sputtering yield. The lowest energy used for the calculations was 10 keV

| Ion | Energy (keV) | Dose ($cm^{-2}$) $10^{15}$ | $10^{16}$ | $10^{17}$ |
|---|---|---|---|---|
| B | 10 | 0.1 nm | 1 nm | 10 nm |
| P | 10 | 0.3 nm | 3 nm | 30 nm |
| As | 50 | 0.6 nm | 6 nm | 60 nm |
| Sb | 150 | 1 nm | 10 nm | 100 nm |

The sputtered silicon layer is calculated from the sputtering yield and the ion dose by:

$$d = \frac{S}{N} N_{\square} \qquad (20)$$

where N is the atomic density ($N = 5 \times 10^{22}\ cm^{-3}$ for silicon). For high impurity concentrations, however, the sputtering yield may be changed. In Table 2, the sputtered-layer thickness, after implanting different ions at the energy for maximum sputtering, is given. One can see that the effect is small for all doses below $10^{16}\ cm^{-2}$.

The modification of implantation profiles due to sputtering can be calculated using some simplifying assumptions:

a) the sputtering yield is constant,
b) no knock-on takes place,
c) the volume change due to the damage may be neglected, and
d) the profile is Gaussian.

Under these assumptions, the following is valid for the impurity profile:

$$C(x) = \frac{N}{2\,S} \left( \mathrm{erf}\, \frac{x - R_p + N_{\square} \frac{S}{N}}{\sqrt{2}\ \Delta R_p} - \mathrm{erf}\, \frac{x - R_p}{\sqrt{2} \Delta R_p} \right) . \qquad (21)$$

The saturation profile ($N_{\square} \to \infty$) is given by:

$$C(x) = \frac{N}{2\,S}\ \mathrm{erfc}\, \frac{x - R_p}{\sqrt{2} \Delta R_p} \qquad (22)$$

with the maximum concentration at the surface:

$$C_{max} = \frac{N}{2\,S}\ \mathrm{erfc}\, \frac{-R_p}{\sqrt{2} \Delta R_p} \approx \frac{N}{S} \text{ for } R_p \geq 3 \Delta R_p \quad . \qquad (23a)$$

Equations (21) to (23) were derived under the assumption that the implantation profile extends from $-\infty$ to $+\infty$ (see Eq. (1)). Assuming that all implanted ions come to rest from 0 to $+\infty$ (see Eq. (2)),

$$C_{max} = N/S \tag{23b}$$

without any restrictions. This maximum concentration is independent of the implanted dose, and depends only on the relation of atomic density to sputtering yield. In silicon, e.g. the maximum concentration for arsenic (S = 3 at 60 keV) is at $1.5 \times 10^{22}$ $cm^{-3}$, and for antimony (S = 5 at 150 keV) at $10^{22}$ $cm^{-3}$. For semiconductors with higher sputtering yields, (e.g. GaAs), this effect can be much more pronounced. In Fig. 13, a theoretical example is given for antimony in silicon. Profile changes starting from doses of $10^{16}$ $cm^{-2}$ are noticeable.

Sputtering limits in theory the maximum obtainable sheet carrier concentration. Due to solubility limits, however, this is more of theoretical interest. In Fig. 14, examples are given for antimony (theoretical) and bismuth (experimental).

In the case of non-Gaussian profiles, sputtering-modified profiles have to be calculated numerically, solving:

$$C(x,t) = \int_0^t G(x+vt')\, dt \tag{24}$$

with G being the implantation rate and v the velocity of sputtering erosion.

## 7. Lateral Spread

The lateral straggling of implanted ions has usually been neglected, since it is much lower than the range of the implanted ions. For VLSI

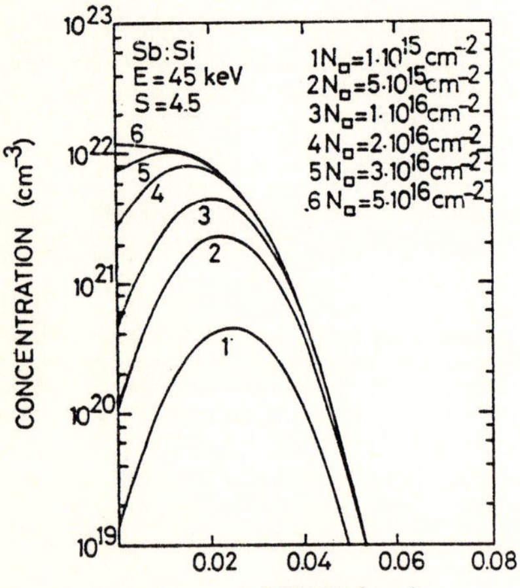


Fig. 13.

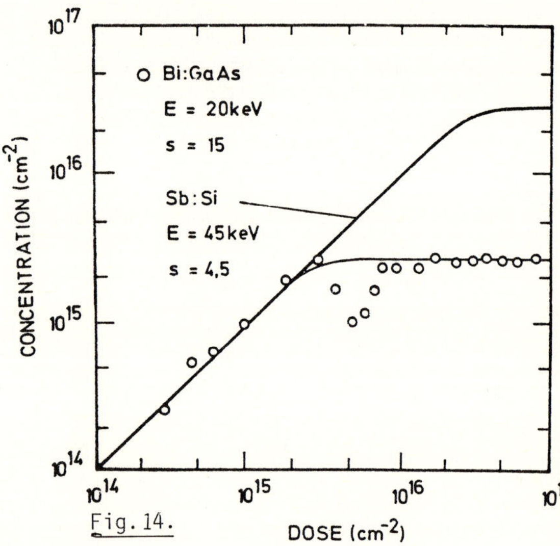


Fig. 14.

Fig. 13. Doping profiles of implanted antimony, depending on dose and sputtering yield (E = 45 keV, S = 4.5)

Fig. 14. Integral doping concentration as a function of dose and sputtering yield: a) antimony in silicon (E = 45 keV, S = 4.5); b) bismuth in GaAs (E = 20 keV, S = 15); experimental data from Tinsley [26] circuits with dimensions in the micron or submicron range, however, it is becoming more and more important. The lateral straggling is, like the range straggling, a result of the scattering of the ions; therefore, both are of the same order of magnitude.

According to the calculations of Matsumura and Furukawa [27], the two-dimensional profile in case of a Gaussian implantation profile is given by:

$$C(x,y) = \frac{N_{\square}}{\sqrt{2\pi}\ \Delta R_p} \exp\left(-\frac{(x-R_p)^2}{2\Delta R_p^2}\right) \frac{1}{2} \left[\operatorname{erfc} \frac{y-a}{\sqrt{2}\ \Delta R_{p,L}} - \operatorname{erfc} \frac{y+a}{\sqrt{2}\ \Delta R_{p,L}}\right] \qquad (25)$$

if the implantation was performed through a slit of width 2a with infinite length. In Eq. (26) $\Delta R_{p,L}$ is the lateral spread of the ions. For real masking layers with an arbitrary shape, Runge [28] found for Gaussian profiles:

$$C(x,y) = \frac{N_{\square}}{2\pi\Delta R_{p,L}\Delta R_p} \int_{-\infty}^{\infty} \left[\exp\left(-\frac{(x-\xi)^2}{2\Delta R_{p,L}^2} - \frac{(x - d_{ox}(\xi)-R_p)^2}{2\Delta R_p^2}\right)\right] d\xi \qquad (26)$$

with $d_{ox}(x)$ the thickness of the masking layer.

For non-Gaussian profiles, a multiplication with

$$\frac{1}{2} \left[\operatorname{erfc} \frac{y-a}{\sqrt{2}\ \Delta R_{p,L}} - \operatorname{erfc} \frac{y+a}{\sqrt{2}\ \Delta R_{p,L}}\right]$$

for an implantation through an ideal slit of width 2a results in the desired profile to a first approximation. The exact treatment affords a convolu-

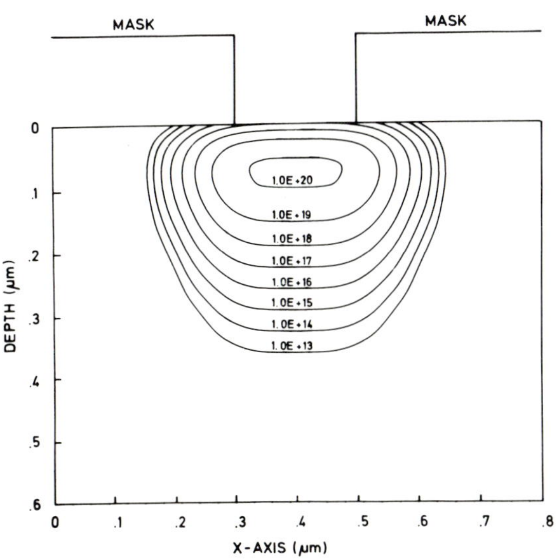


Fig. 15. Two-dimensional implantation profile through an infinite steep mask of width 0.2 µm for 150 keV arsenic with $10^{15}$ cm$^{-2}$

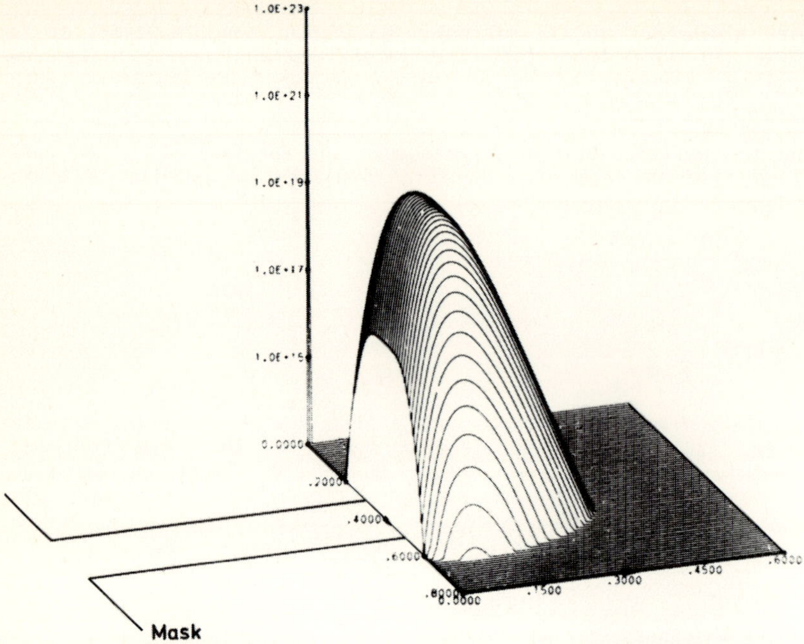


Fig. 16. Three-dimensional view of the profile shown in Fig. 15

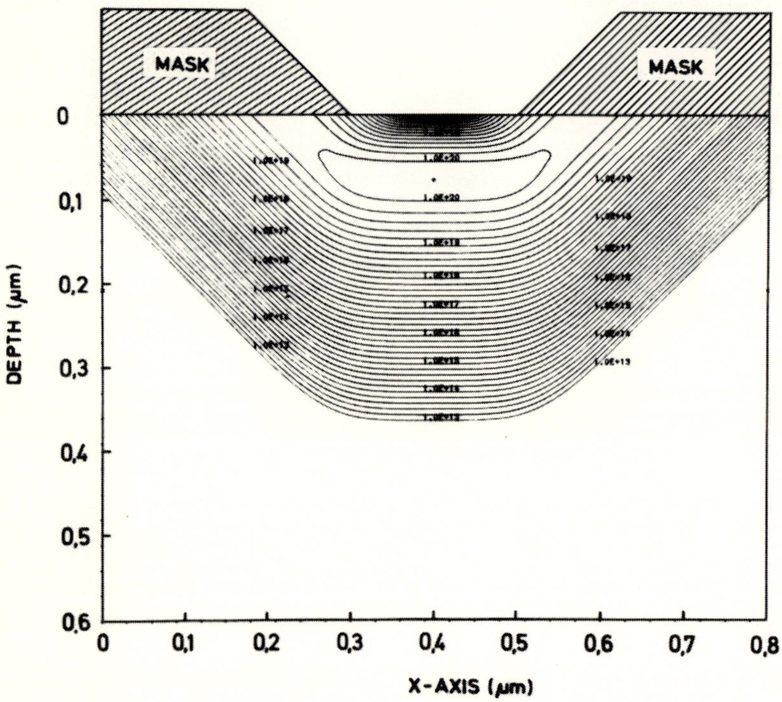


Fig. 17. Two-dimensional implantation profile through a tapered mask (45°) for 150 keV arsenic with $10^{15}\,cm^{-2}$

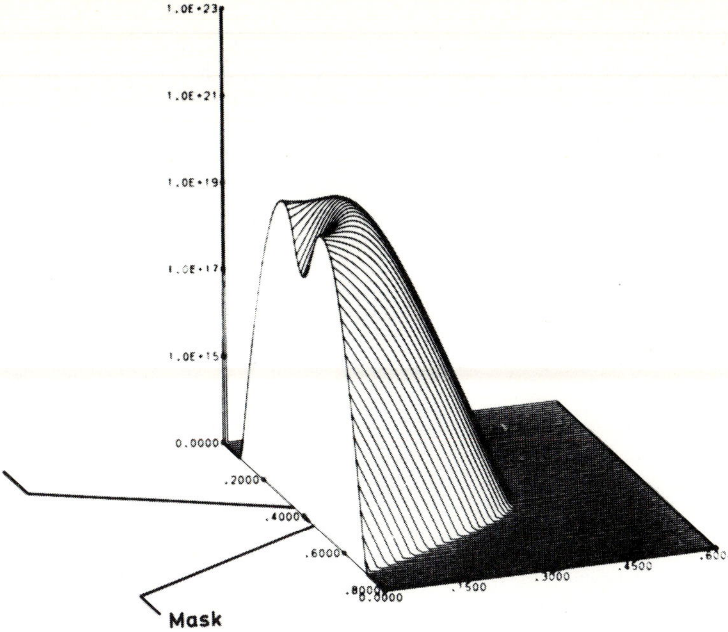


Fig. 18. Three-dimensional view of the profile shown in Fig. 17

tion of the one-dimensional profile with a Gaussian lateral distribution having a standard deviation of $\Delta R_{p,L}$.

Examples of two-dimensional profiles are given in the following figures. In Fig. 15, a profile for an implantation through an infinite steep mask is shown as an equi-concentration plot. The same profile is given in Fig. 16 in a quasi-three-dimensional plot. The same implantation through a tapered mask (45°) is shown in Figs. 17 and 18. In both cases, Pearson IV distributions were assumed in vertical direction.

## Acknowledgement

I am grateful to my co-workers K. Hoffmann and G. Prinke for many discussions.

## 8. Appendix: Range Program

In this appendix, a Fortran IV program (PEARSON) for the calculation of Pearson and Gaussian profiles is given. The program also includes profile modifications caused by sputtering.

The PEARSON program is the implantation processor of the process-modeling program ICECREM. The program is started by typing:

```
run pearson
```

The computer responds

```
ELEM    ENERGY   DOSE    RANG    STDV    GAMMA     BETA2     SPUT
<e.1>   <z.2>    <z.3>                                      [<z.8>]
<e.1>            <z.3>   <z.4>   <z.5>   [<z.6>    <z.7>]   [<z.8>]
```

<e.1> indicates which element is to be implanted (boron, phosphorus, Sb, As).

In the first form, the profile is constructed with 4 moments from the stored tables. <z.2> indicates the energy in keV, <z.3> is the dose per square centimeter.

In the second form, a Pearson distribution is produced,

```
RANGE = <range>  in µm
STDV  = <standard deviation>  in µm
GAMMA = <third moment>STDV**3
BETA2 = <fourth moment>/STDV**4
```

If <z.6> <z.7> are not specified, a Gaussian distribution is produced (GAMMA = 0, BETA2 = 3).

SPUT is the sputtering ratio, i.e.
SPUT = <number of sputtered atoms> <number of primary ions>.

```
FORTRAN IV-PLUS V3.0          14:43:44    22-Jul-82          Page 1
MAINIT.FTN;26            /TR:BLOCKS

              PROGRAM MAINIT
              REAL*8 ITEML(8)
              DIMENSION ITYPL(8),Y(130)
              INTEGER*4 ISEML(2,3)
              COMMON /IMPL/ISEM,DATAF(16)
              COMMON /PEAR/B0,B1,B2,AMX,R,S
            DATA
          2      ITEML/
          4      'ELEM    ',  'ENERGY  ',  'DOSE    ',  'RANG    ',   ! IMPLNT
          5      'STDV    ',  'GAMMA   ',  'BETA2   ',  'SPUT    '/
              DATA ITYPL/4,7*1/
          6      ISEML/ "177,7, "177,"35, "177,"175/
              CALL SYNLIN(DATAF,ISEM,ITEML,ITYPL,8,IERR)
              IF(IERR.NE.0)STOP
              DO 10 I=1,3
      10      IF((ISEM.AND.ISEML(1,I)).EQ.ISEML(2,I))GOTO 20
              STOP'SEMANTICAL ERROR'
      20      CALL IMPLIT
              DPMX=R+3*S
              WRITE(5,*)'PLOTDEPTH= ',DPMX,' ENTER REQUIRED VALUE OR /'
              READ(5,*)DPMX
              DX=DPMX/130
              H=0
              SUM=0
              VMAX=0
              DO 30 I=1,130
              Y(I)=PEARS(H)
              VMAX=MAX(Y(I),VMAX)
              SUM=SUM+Y(I)
      30      H=H+DX
    C
    C         SPUTTERING?
    C
              SPUTDP=0
              IF(ISHFT(ISEM,-7).NE.0)SPUTDP=DATAF(3)*DATAF(8)*(1E4/5E22)
    C         5E22 ATOME/CM**3,1E4 CM/MY
              NSPUT=SPUTDP/DX
              SUM=SUM*DX*1E-4
              VMAX=VMAX*DATAF(3)/(SUM*(NSPUT+1))
              IF(NSPUT.EQ.0)GOTO 55
              DO 50 I=1,130-1
              DO 50 J=I+1,MIN(I+NSPUT,130)
      50      Y(I)=Y(I)+Y(J)
      55      CONTINUE
              DO 40 I=1,130
      40      Y(I)=LOG(MAX(Y(I)*DATAF(3)/(SUM*(NSPUT+1)),VMAX*1E-4))/LOG(10.)
              CALL DRUPLO(Y,DPMX/130,7)
              CALL PLOTNDD(1)
              WRITE(1,100)DPMX
     100      FORMAT(' DEPTH TO ',1PG20.4,T130,'I')
              END
```

```
FORTRAN IV-PLUS V3.0            16:39:21      5 Jul-82            Page 2
SCR.;1                    /TR:BLOCKS

0001          SUBROUTINE IMPLIT
        C
        C     IMPLNT MODELS THE IMPLANTATION STEP
        C
        C     THE CONCENTRATION ARRAY C IS SET UP WITH A PEARSON DISTRIBUTION
        C     OR A GAUSSIAN DISTRIBUTION ACCORDING TO INPUT SPECIFICATION
        C
        C     MATERIAL ASSIGNMENT:
        C     MAT=1:BOR, MAT=2:PHOSPHOR, MAT=3:ANTIMON, MAT=4:AS, MAT=5: GA
        C
0002            COMMON /PEAR/B0,B1,B2,AMX,RR,SS,ICTL
0003            COMMON /IMPL/ISEM,DATAF(16)
0004          PARAMETER PI=3.1415926535
0005          REAL HSP(400),SGMO(46,5),
             ,DRPBO(46),DRPASO(46),DRPGAO(46),
             ,DRPSBO(46),DRPPO(46),RPGA(46),DRPGA(46),GMGA(46),
             ,RNG(46,5),SGM(46,5),GAM(46,5),RPB(46),S2(46),S3(46),S4(46),
             ,DRPB(46),S6(46),S7(46),S8(46),GMB(46),S10(46),S11(46),S12(46)
        C
        C
        C
0006          EQUIVALENCE(RNG(1,1),RPB(1)),(RNG(1,2),S2(1)),(RNG(1,3),S3(1)),
             ,(RNG(1,4),S4(1)),(SGM(1,1),DRPB(1)),(SGM(1,2),S6(1)),(SGM(1,3),
             ,S7(1)),(SGM(1,4),S8(1)),(GAM(1,1),GMB(1)),(GAM(1,2),S10(1)),
             ,(GAM(1,3),S11(1)),(GAM(1,4),S12(1)),(R,DATAF(4)),(S,DATAF(5)),
             ,(G,DATAF(6)),(B,DATAF(7)),(ENERG,DATAF(2)),(DOSE,DATAF(3)),
             ,(SGMO(1,1),DRPBO(1)),(SGMO(1,2),DRPPO(1)),(SGMO(1,3),DRPSBO(1)),
             ,(SGMO(1,4),DRPASO(1)),(SGMO(1,5),DRPGAO(1))
        C
0007          DIMENSION FM(5),FM1(5),FMO(5),FMO1(5)
0008          DIMENSION NEN(11),IENE(11)
0009          DATA NEN/0,10,20,25,30,35,37,39,41,43,45/,
             1     IENE/0,50,100,150,200,300,400,500,750,1000,1250/
0010          DATA FM /25,2*1.5,0.,1.5/,FM1/8.,2*4.5,0.,4.5/
0011          DATA S2/ 0.,
             1          0.0139, 0.0253, 0.0368, 0.0486, 0.0607, 0.0730, 0.0855,
             2          0.0981, 0.1109, 0.1238, 0.1367, 0.1497, 0.1627, 0.1757,
             3          0.1888, 0.2019, 0.2149, 0.2279, 0.2409, 0.2539, 0.2798,
             4          0.3054, 0.3309, 0.3562, 0.3812, 0.4060, 0.4306, 0.4549,
             5          0.4790, 0.5029, 0.5265, 0.5499, 0.5730, 0.5959, 0.6186,
             6          0.6744, 0.7288, 0.7819, 0.8338, 0.8846, 0.9343, 0.9829,
             7          1.0306, 1.0773, 1.1231/
0012          DATA S3/ 0.,
             1          0.0088, 0.0141, 0.0187, 0.0230, 0.0271, 0.0310, 0.0347,
             2          0.0385, 0.0421, 0.0457, 0.0493, 0.0529, 0.0564, 0.0599,
             3          0.0634, 0.0669, 0.0704, 0.0739, 0.0773, 0.0808, 0.0878,
             4          0.0947, 0.1017, 0.1086, 0.1156, 0.1227, 0.1297, 0.1368,
             5          0.1439, 0.1510, 0.1581, 0.1653, 0.1725, 0.1797, 0.1870,
             6          0.2052, 0.2236, 0.2421, 0.2608, 0.2796, 0.2985, 0.3175,
             7          0.3366, 0.3559, 0.3752/
0013          DATA S4/ 0.,
             1          0.0097, 0.0159, 0.0215, 0.0269, 0.0322, 0.0374, 0.0426,
             2          0.0478, 0.0530, 0.0582, 0.0634, 0.0686, 0.0739, 0.0791,
             3          0.0845, 0.0898, 0.0952, 0.1005, 0.1060, 0.1114, 0.1223,
             4          0.1334, 0.1445, 0.1558, 0.1671, 0.1785, 0.1900, 0.2015,
             5          0.2131, 0.2247, 0.2364, 0.2482, 0.2600, 0.2718, 0.2837,
```

```
FORTRAN IV-PLUS V3.0          16:39:21     5-Jul-82          Pase 3
SCR.;1                   /TR:BLOCKS
               6        0.3135, 0.3435, 0.3736, 0.4038, 0.4340, 0.4643, 0.4946,
               7        0.5245, 0.5552, 0.5854/
0014            DATA S6/ 0.,
               1  0.0069, 0.0119, 0.0166, 0.0212, 0.0256, 0.0298, 0.0340, 0.0380,
               2  0.0418, 0.0456, 0.0492, 0.0528, 0.0562, 0.0595, 0.0628, 0.0659,
               3  0.0689, 0.0719, 0.0747, 0.0775, 0.0829, 0.0880, 0.9228, 0.9974,
               4  0.1017, 0.1059, 0.1098, 0.1136, 0.1172, 0.1206, 0.1239, 0.1271,
               5  0.1301, 0.1330, 0.1358, 0.1424, 0.1484, 0.1539, 0.1590, 0.1636,
               6  0.1680, 0.1721, 0.1758, 0.1794, 0.1827/
0015            DATA S7/ 0.,
               1  0.0026, 0.0043, 0.0058, 0.0071, 0.0084, 0.0096, 0.0107, 0.0118,
               2  0.0130, 0.0140, 0.0151, 0.0162, 0.0172, 0.0183, 0.0193, 0.0203,
               3  0.0213, 0.0224, 0.0234, 0.0244, 0.0264, 0.0283, 0.0303, 0.0322,
               4  0.0342, 0.0361, 0.0380, 0.0400, 0.0419, 0.0438, 0.0457, 0.0475,
               5  0.0494, 0.0513, 0.0531, 0.0578, 0.0623, 0.0669, 0.0713, 0.0758,
               6  0.0802, 0.0845, 0.0888, 0.0930, 0.0972/
0016            DATA S8/ 0.,
               1  0.0036, 0.0059, 0.0080, 0.0099, 0.0118, 0.0136, 0.0154, 0.0172,
               2  0.0189, 0.0207, 0.0224, 0.0241, 0.0258, 0.0275, 0.0292, 0.0308,
               3  0.0325, 0.0341, 0.0358, 0.0374, 0.0407, 0.0439, 0.0470, 0.0502,
               4  0.0533, 0.0564, 0.0594, 0.0624, 0.0654, 0.0684, 0.0713, 0.0741,
               5  0.0770, 0.0798, 0.0826, 0.0894, 0.0960, 0.1024, 0.1087, 0.1147,
               6  0.1206, 0.1263, 0.1319, 0.1373, 0.1425/
0017            DATA S10/ 0.,
               1    0.641,   0.549,   0.459,   0.384,   0.319,   0.261,   0.209,   0.162,
               2    0.119,   0.080,   0.044,   0.011,  -0.021,  -0.049,  -0.075,  -0.100,
               3   -0.122,  -0.141,  -0.158,  -0.181,  -0.224,  -0.263,  -0.298,  -0.332,
               4   -0.365,  -0.396,  -0.424,  -0.451,  -0.475,  -0.501,  -0.523,  -0.544,
               5   -0.566,  -0.588,  -0.608,  -0.662,  -0.711,  -0.757,  -0.799,  -0.839,
               6   -0.876,  -0.911,  -0.945,  -0.976,  -1.007/
0018            DATA S11/ 0.,
               1    0.638,   0.584,   0.574,   0.570,   0.567,   0.565,   0.562,   0.560,
               2    0.558,   0.556,   0.554,   0.552,   0.550,   0.547,   0.545,   0.543,
               3    0.541,   0.538,   0.536,   0.534,   0.525,   0.517,   0.510,   0.503,
               4    0.496,   0.490,   0.483,   0.476,   0.469,   0.464,   0.459,   0.453,
               5    0.445,   0.442,   0.438,   0.421,   0.405,   0.390,   0.376,   0.362,
               6    0.350,   0.340,   0.328,   0.317,   0.307/
0019            DATA S12/ 0.,
               1    0.675,   0.647,   0.635,   0.627,   0.619,   0.611,   0.604,   0.597,
               2    0.590,   0.583,   0.576,   0.569,   0.563,   0.557,   0.551,   0.544,
               3    0.537,   0.529,   0.526,   0.522,   0.499,   0.478,   0.459,   0.443,
               4    0.427,   0.412,   0.393,   0.377,   0.361,   0.348,   0.344,   0.336,
               5    0.328,   0.320,   0.317,   0.287,   0.260,   0.236,   0.215,   0.197,
               6    0.183,   0.174,   0.163,   0.155,   0.147/
0020            DATA DRPBO/ 0.,
               * 0.0187,0.0328,0.0441,0.0534,0.0611,0.0677,0.0735,0.0784
               *,0.0829,0.0868,0.0904,0.0936,0.0965,0.0992,0.1017,0.1040
               *,0.1061,0.1081,0.1099,0.1117,0.1148,0.1176,0.1202,0.1225
               *,0.1246,0.1265,0.1283,0.1299,0.1315,0.1329,0.1342,0.1354
               *,0.1366,0.1377,0.1388,0.1412,0.1433,0.1451,0.1468,0.1484
               *,0.1498,0.1511,0.1523,0.1534,0.1544/
0021            DATA DRPASO/ 0.,
               * 0.0042,0.0068,0.0092,0.0115,0.0138,0.0160,0.0183,0.0205
               *,0.0226,0.0247,0.0268,0.0290,0.0310,0.0332,0.0352,0.0373
               *,0.0393,0.0413,0.0434,0.0454,0.0495,0.0534,0.0572,0.0611
               *,0.0649,0.0687,0.0724,0.0760,0.0797,0.0832,0.0867,0.0901
```

```
FORTRAN IV-PLUS V3.0           16:39:21      5-Jul-82           Page 4
SCR.;1                  /TR:BLOCKS
                 *,0.0936,0.0969,0.1003,0.1084,0.1163,0.1237,0.1310,0.1381
                 *,0.1448,0.1515,0.1579,0.1640,0.1699/
0022              DATA RPB/ 0.,
                 * 0.0309,0.0632,0.0952,0.1263,0.1563,0.1852,0.2131,0.2401,
                 * 0.2661,0.2914,0.3159,0.3397,0.3629,0.3854,0.4075,0.4290,
                 * 0.4500,0.4706,0.4908,0.5106,0.5490,0.5861,0.6220,0.6568,
                 * 0.6905,0.7234,0.7554,0.7867,0.8172,0.8471,0.8763,0.9049,
                 * 0.9330,0.9606,0.9877,1.0534,1.1167,1.1777,1.2367,1.2940,
                 * 1.3496,1.4038,1.4567,1.5083,1.5588/
0023              DATA DRPB/ 0.,
                 * 0.0259,0.0428,0.0555,0.0657,0.0742,0.0814,0.0876,0.0929,
                 * 0.0977,0.1019,0.1057,0.1091,0.1123,0.1152,0.1179,0.1203,
                 * 0.1224,0.1246,0.1266,0.1285,0.1319,0.1349,0.1376,0.1401,
                 * 0.1423,0.1445,0.1463,0.1481,0.1497,0.1512,0.1527,0.1540,
                 * 0.1553,0.1565,0.1577,0.1602,0.1625,0.1646,0.1663,0.1681,
                 * 0.1696,0.1710,0.1723,0.1735,0.1747/
0024              DATA GMR  / 0.,
                 *-0.0333,-0.0667,-0.1000,-0.1333,-0.1667,-0.2000,-0.2333,-0.2667,
                 *-0.3000,-0.3333,-0.3667,-0.4000,-0.4333,-0.4667,-0.5000,-0.5333,
                 *-0.5667,-0.6000,-0.6333,-0.6667,-0.7333,-0.8000,-0.8667,-0.9333,
                 *-1.0000,-1.0667,-1.1333,-1.2000,-1.2667,-1.3333,-1.4000,-1.4667,
                 *-1.5333,-1.6000,-1.6667,-1.8333,-2.0000,-2.1667,-2.3333,-2.5000,
                 *-2.6667,-2.8333,-3.0000,-3.1667,-3.3333/
0025              DATA DRPGAO/ 0.,
                 * 0.0034, 0.0056, 0.0076, 0.0096, 0.0115, 0.0134, 0.0152, 0.0170,
                 * 0.0189, 0.0207, 0.0224, 0.0242, 0.0260, 0.0277, 0.0295, 0.0312,
                 * 0.0329, 0.0346, 0.0363, 0.0379, 0.0413, 0.0445, 0.0478, 0.0509,
                 * 0.0541, 0.0571, 0.0602, 0.0632, 0.0661, 0.0690, 0.0719, 0.0747,
                 * 0.0775, 0.0802, 0.0829, 0.0894, 0.0957, 0.1017, 0.1075, 0.1130,
                 * 0.1184, 0.1236, 0.1285, 0.1333, 0.1379/
0026              DATA DRPSBO/ 0.,
                 * 0.0027, 0.0041, 0.0054, 0.0065, 0.0077, 0.0088, 0.0098, 0.0109,
                 * 0.0119, 0.0130, 0.0140, 0.0150, 0.0160, 0.0170, 0.0180, 0.0190,
                 * 0.0200, 0.0210, 0.0219, 0.0229, 0.0248, 0.0267, 0.0286, 0.0305,
                 * 0.0324, 0.0343, 0.0361, 0.0380, 0.0398, 0.0417, 0.0435, 0.0453,
                 * 0.0471, 0.0489, 0.0507, 0.0551, 0.0595, 0.0639, 0.0681, 0.0724,
                 * 0.0765, 0.0806, 0.0847, 0.0887, 0.0927/
0027              DATA DRPPO/ 0.,
                 * 0.0063, 0.0111, 0.0157, 0.0201, 0.0244, 0.0286, 0.0326, 0.0365,
                 * 0.0402, 0.0439, 0.0474, 0.0508, 0.0541, 0.0572, 0.0603, 0.0633,
                 * 0.0662, 0.0689, 0.0716, 0.0742, 0.0792, 0.0839, 0.0883, 0.0925,
                 * 0.0964, 0.1002, 0.1037, 0.1071, 0.1103, 0.1133, 0.1162, 0.1190,
                 * 0.1216, 0.1242, 0.1266, 0.1322, 0.1374, 0.1420, 0.1463, 0.1502,
                 * 0.1538, 0.1572, 0.1603, 0.1632, 0.1660/
0028              DATA RPGA/ 0.,
                 * 0.0102, 0.0169, 0.0233, 0.0295, 0.0357, 0.0418, 0.0479, 0.0540,
                 * 0.0601, 0.0662, 0.0724, 0.0786, 0.0848, 0.0910, 0.0972, 0.1035,
                 * 0.1098, 0.1161, 0.1224, 0.1287, 0.1415, 0.1543, 0.1672, 0.1802,
                 * 0.1932, 0.2063, 0.2194, 0.2326, 0.2458, 0.2590, 0.2723, 0.2856,
                 * 0.2989, 0.3123, 0.3256, 0.3591, 0.3926, 0.4260, 0.4595, 0.4929,
                 * 0.5263, 0.5595, 0.5927, 0.6257, 0.6586/
0029              DATA DRPGA/ 0.,
                 * 0.0039, 0.0064, 0.0086, 0.0108, 0.0129, 0.0150, 0.0170, 0.0191,
                 * 0.0211, 0.0230, 0.0250, 0.0270, 0.0289, 0.0308, 0.0327, 0.0346,
                 * 0.0365, 0.0384, 0.0402, 0.0421, 0.0457, 0.0493, 0.0529, 0.0564,
                 * 0.0598, 0.0632, 0.0666, 0.0699, 0.0732, 0.0764, 0.0795, 0.0827,
```

```
FORTRAN IV-PLUS V3.0              16:39:21      5-Jul-82            Pase 5
SCR.;1                      /TR:BLOCKS

         * 0.0857, 0.0888, 0.0917, 0.0990, 0.1060, 0.1127, 0.1192, 0.1254,
         * 0.1315, 0.1373, 0.1429, 0.1483, 0.1535/
          KEV(I)=NEN(1+I/100)+(NEN(2+I/100)-NEN(1+I/100))*
         *MOD(I,100)/100
          ENV(I)=IENE(1+I/5)+(IENE(2+I/5)-IENE(1+I/5))*MOD(I,5)/5
C**** START
          IEL=DATAF(1)
          IF(IAND(ISHFT(ISEM,-5),1).NE.0) GOTO 106
          B=0.
          G=0.
 106      IF(IAND(ISEM,2).EQ.0)GOTO 200
C
C         PEARSON DISTRIBUTION
C
          ENERG=MAX(0.,MIN(1000.,ENERG))
          IKEV=KEV(INT(ENERG))
          H=(ENERG-ENV(IKEV))/(ENV(IKEV+1)-ENV(IKEV))
          R=RNG(IKEV+1,IEL)+H*(RNG(IKEV+2,IEL)-RNG(IKEV+1,IEL))
          S=SGM(IKEV+1,IEL)+H*(SGM(IKEV+2,IEL)-SGM(IKEV+1,IEL))
          G=GAM(IKEV+1,IEL)+H*(GAM(IKEV+2,IEL)-GAM(IKEV+1,IEL))
 200      BMIN=6*((G**2+4.)**1.5-8.+7*G**2)/(32.-G**2)
          RR=R
          SS=S
          IF(IAND(ISEM,2).NE.0) B=FM(IEL)*BMIN+FM1(IEL)
          B=AMAX1(B,BMIN+3)
          B1=-G*S*(B+3)/(10*B-12*G*G-18)
          B0=-S*S*(4*B-3*G*G)/(10*B-12*G*G-18)
          B2=-(2*B-3*G*G-6)/(10*B-12*G*G-18)
D         TYPE*,'R,S,G,B',R,S,G,B,'S0',S0
D         TYPE*,'B0,B1,B2',B0,B1,B2
          IF(B.EQ.BMIN+3) GOTO 240
          AMX=ALOG(ABS(B0))/
         /(2*B2)-(B1/B2+2*B1)/SQRT(4*B2*B0-B1**2)*ATAN((B1)/
         /SQRT(4*B2*B0-B1**2))
          ICTL=1
          RETURN
C
C         B1**2-4*B0*B2 .EQ. 0
C
 240      ICTL=2
          AMX=ALOG(ABS(B0))/
         /(2*B2)+1/B2+2
C
C         GAUSSIAN DISTRIBUTION
C
          IF(G.EQ.0)ICTL=3
          RETURN
          END

PROGRAM SECTIONS

 Name        Size              Attributes

$CODE1   001464    410         RW,I,CON,LCL
$PDATA   000016      7         RW,D,CON,LCL
```

```
FORTRAN IV-PLUS V3.0                16:39:21      5-Jul-82              Page 6
SCR.;1                         /TR:BLOCKS

$IDATA   000004      2                  RW,D,CON,LCL
$VARS    013520   2984                  RW,D,CON,LCL
$TEMPS   000004      2                  RW,D,CON,LCL
PEAR     000032     13                  RW,D,OVR,GBL
IMPL     000102     33                  RW,D,OVR,GBL


Total Space Allocated = 015366   3451


FORTRAN IV-PLUS V3.0                16:40:03      5-Jul-82              Page 7
SCR.;1                         /TR:BLOCKS

0001                FUNCTION PEARS(X)
0002                COMMON /PEAR/B0,B1,B2,AMX,R,S,ICTL
0003              F(X)=ALOG(ABS(B0+B1*X+B2*X*X))/
             /(2*B2)-(B1/B2+2*B1)/SQRT(4*B2*B0-B1**2)*ATAN((2*B2*X+B1)/
             /SQRT(4*B2*B0-B1**2))
0004              F0(X)=ALOG(ABS(B0+B1*X+B2*X*X))/
             /(2*B2)+(B1/B2+2*B1)/(2*B2*X+B1)
0005              FE(X)=-(X/S)**2*.5
0006                PEARS=0
0007                GOTO(10,20,30)ICTL
0008     10         H=F(X-RP)-AMX
0009                GOTO 100
0010     20         IF(2*B2*(X-R)+B1.GE.0)RETURN
0011                H=F0(X-RP)-AMX
0012                GOTO 100
0013     30         H=FE(X-RP)-AMX
0014     100        IF(H.LT.-37)RETURN
0015                PEARS=EXP(H)
0016                RETURN
0017                END

PROGRAM SECTIONS

 Name       Size                        Attributes

$CODE1   000714    230                  RW,I,CON,LCL
$PDATA   000010      4                  RW,D,CON,LCL
$IDATA   000010      4                  RW,D,CON,LCL
$VARS    000010      4                  RW,D,CON,LCL
$TEMPS   000004      2                  RW,D,CON,LCL
PEAR     000032     13                  RW,D,OVR,GBL

Total Space Allocated = 001002    257
```

```
FORTRAN IV-PLUS V3.0            16:40:09      5-Jan-82            Page 8
SCR.;1                       /TR:BLOCKS

              SUBROUTINE SYNLIN(DATAF,ISEM,ITEML,ITYPL,LENG,IERR)
              BYTE INBUF(64),IHBUF1(8)
              INTEGER*4 ISEM
              REAL DATAF(8)
              REAL*8 ITEML(8)
              DIMENSION ITYPL(8)
              IERR=0
              WRITE(5,13) (ITEML(I),I=1,LENG)
              READ(5,7) INBUF
              IPOI=1
        C
        C     INTERPRETE DATA LIST
        C
              DO 104 I=1,LENG
              IANF=IANF+1
              IA=IA+1
              ITP=ITYPL(IANF)
              DO 107 J=1,8
        107   IHBUF1(J)=' '
              IPOI1=1
              DO 105 J=1,8
              IPOI=IPOI+1
              IF(INBUF(IPOI-1).EQ.9) GOTO 205
              IF(INBUF(IPOI-1).EQ.' '.OR.INBUF(IPOI-1).EQ.0) GOTO 105
              IHBUF1(IPOI1)=INBUF(IPOI-1)
              IPOI1=IPOI1+1
        105   CONTINUE
        205   NCHA=IPOI1-1
              IF(NCHA.EQ.0) GOTO 104
              ISEM=ISEM.OR.ISHFT(1,IA-1)
              GOTO (211,212,213,214,215)ITP
        C
        C     NUMERICAL DATA
        C
        211   IF((NCHA.EQ.1).AND.(IHBUF1(1).EQ.2H* )) GOTO 235
              DECODE(NCHA,2,IHBUF1,ERR=216) DATAF(IA)
              GOTO 104
        235   WRITE(5,22) I
              READ(5,*,ERR=235) DATAF(IA)
              GOTO 104
        C
        C     HOLLERITH DATA
        C
        212   DECODE(8,23,IHBUF1) DATAF(IA),DATAF(16)
              GOTO 104
        C
        C     LOGICAL DATA
        C
        213   IF(IHBUF1(1).EQ.'Y') GOTO 236
              IF(IHBUF1(1).NE.'N') GOTO 216
              DATAF(IA)=0.
              GOTO 104
        236   DATAF(IA)=1.
              GOTO 104
        C
        C     ELEMENT DESCRIPTORS
```

```
FORTRAN IV-PLUS V3.0          16:40:09      5-Jul-82             Page 9
SCR.;1                   /TR:BLOCKS

      C
      214   CALL ELMTCH(IHBUF1,J,JD)
            IF(JD.EQ.0.OR.J.LT.4)GOTO 216
      217   DATAF(IA)=J-3
            GOTO 104
      C
      C     MODEL SPECIFICATIONS
      C
      215   MDHP=0
            J=1
      110   CALL ELMTCH(IHBUF1(J),L,JD)
            IF(JD.EQ.0.AND.IHBUF1(J).NE.' ')GOTO 216
            IF(JD.EQ.0)MDHP=MDHP.OR.ISHFT(1,L)
            J=J+MAX(1,JD)
            MDHP=MDHP.OR.ISHFT(1,L)
            IF(J.LE.8)GOTO 110
            DATAF(IA)=MDHP
      104   CONTINUE
            RETURN
      216   WRITE(5,10)IA
            IERR=1
            RETURN
      2     FORMAT(E8.0)
      3     FORMAT(I1)
      7     FORMAT(64A1)
      8     FORMAT('(',I1,'X,',I1,'A2)')
      9     FORMAT(A4)
      10    FORMAT(X,'ERROR IN DATA FIELD NO. ',I1)
      13    FORMAT(X,8A8)
      15    FORMAT(X,'ILLEGAL COMBINATION OF ITEMS')
      16    FORMAT(/X,4A2,3X,6(8A2,2X))
      17    FORMAT(G15.7,X)
      18    FORMAT(A4,12X)
      19    FORMAT(12X,6(8A2,2X)/)
      20    FORMAT(12X,6(8A2,2X))
      21    FORMAT(/X,4A2,3X,36A2)
      22    FORMAT(X,'ENTER HIGH PRECISION DATA FOR FIELD ',I1,$)
      23    FORMAT(2A4)
      24    FORMAT(8A1)
            END

PROGRAM SECTIONS

 Name        Size                   Attributes

$CODE1   001632    461              RW,I,CON,LCL
$PDATA   000172     61              RW,D,CON,LCL
$IDATA   000064     26              RW,D,CON,LCL
$VARS    000136     47              RW,D,CON,LCL
$TEMPS   000004      2              RW,D,CON,LCL

Total Space Allocated = 002252    597
```

```
FORTRAN IV-PLUS V3.0                16:40:21      5-Jul-82              Page 10
SCR.#1                       /TR:BLOCKS

                SUBROUTINE DRUPLO(Y,DX,ICTL)
                PARAMETER NX=130,NY=68
                DIMENSION Y(NX)
                BYTE DRU(0:NX,0:NY)
      110       FORMAT(I3)
                IF((ICTL.AND.2).EQ.0)GOTO 11
                DO 10 I=0,NX
                DO 10 J=0,NY
      10        DRU(I,J)=' '
      11        IF((ICTL.AND.1).EQ.0)GOTO 12
                YMIN=Y(1)
                YMAX=Y(1)
                DO 20 I=2,NX
                YMIN=MIN(Y(I),YMIN)
      20        YMAX=MAX(Y(I),YMAX)
                DY=NY/(YMAX-YMIN)
                DO 30 I=INT(YMIN+.9999),INT(YMAX)
      30        ENCODE(3,110,DRU(0,(I-YMIN)*DY+.5))I
      12        IF((ICTL.AND.4).EQ.0)RETURN
                DO 40 I=0,NX
      40        DRU(I,MAX(0,MIN(INT((Y(I)-YMIN)*DY),NY)))='*'
                RETURN
                ENTRY PLOTNDD(IUNIT)
                DO 50 I=NY,0,-1
      50        WRITE(IUNIT,120)(DRU(J,I),J=0,NX)
      120       FORMAT(1X,140A1)
                END

PROGRAM SECTIONS

 Name        Size                    Attributes

$CODE1   000714    230              RW,I,CON,LCL
$PDATA   000016      7              RW,D,CON,LCL
$IDATA   000012      5              RW,D,CON,LCL
$VARS    021540   4528              RW,D,CON,LCL
$TEMPS   000002      1              RW,D,CON,LCL

Total Space Allocated = 022506   4771
```

```
FORTRAN IV-PLUS V3.0          16:40:28      5-Jul-82          Page 11
SCR.;1                    /TR:BLOCKS

0001            SUBROUTINE ELMTCH(IBUF,IEL,NCH)
         C
         C      IBUF : 2 CHARS TO BE COMPARED
         C      NCH=0: KEIN ELEMENT GEFUNDEN
         C      NCH=1: oder
         C      NCH=2: MATCH MIT 1 ODER 2 CHARS
         C      IEL=1...8 NUMMER DES ELEMENTS/MODELS
         C
0002            DIMENSION MDLST(8)
0003            BYTE IBUF(2),IMDLST(2,8)
0004            EQUIVALENCE(IMDLST(1,1),MDLST(1))
0005            DATA MDLST/'D','W','X','B','P','SB','AS','GA'/
0006            NCH=0
0007            DO 10 I=1,8
0008            IF(IBUF(1).NE.IMDLST(1,I))GOTO 10
0009            NCH=1
0010            IEL=I
0011            IF(IBUF(2).EQ.IMDLST(2,I))GOTO 20
0012     10     CONTINUE
0013            RETURN
0014     20     NCH=2
0015            RETURN
0016            END

PROGRAM SECTIONS

 Name       Size                  Attributes

$CODE1    000170    60            RW,I,CON,LCL
$IDATA    000012     5            RW,D,CON,LCL
$VARS     000022     9            RW,D,CON,LCL

Total Space Allocated = 000224    74

No FPP Instructions Generated
```

## References


1. J. Lindhard, M. Scharff, and H.E. Schiøtt, Kgl. Danske Videnskab., Selskab., Mat. Phys. Medd. 33, No. 14 (1963)
2. J.P. Biersack, Nucl. Inst. & Methods 182/183, 199 (1981)
3. K.B. Winterbon, Ion Implantation Range and Energy Deposition Distributions, Vol.2: Low Incident Ion Energies, IFI/Plenum Press (1973), New York
4. J.F. Gibbons, W.S. Johnson, and S.W. Mylroie, Projected Range Statistics in Semiconductors (Eds.: Dowden, Hutchinson, and Ross), Academic Press, Stroudsburg (1975)
5. B.Smith, Ion Implantation Range Data for Silicon and Germanium Device Technologies, Learned Information (Europe) Ltd., Oxford (1977)
6. J.F. Gibbons and S. Mylroie, Appl. Phys. Lett. 22, 568 (1973)
7. W.K. Hofker, Philips Res. Repts., Suppl. No.8 (1975)
8. J.F. Gibbons, in: Handbook on Semiconductors, Vol.3 (Ed. T.S. Moss), North Holland Publ.Comp., Amsterdam (1980)
9. E.F. Krimmel and H. Pfleiderer, Rad. Effects 19, 83 (1973)
10. Y. Yamamura and H. Inuma, Rad. Effects 38, 251 (1978)
11. L.A. Christel, J.F. Gibbons, and S. Mylroie, J. Appl.Phys. 51, 6176 (1980)
12. H. Ishiwara, S. Furukawa, J. Yamada, and M. Kawamura in: Ion Implantation in Semiconductors (Ed. S. Namba) p.423, Plenum Press, New York (1975)
13. A.V.S. Satya and H.R. Palanki in: Ion Implantation in Semiconductors (Ed. S. Namba) p.405, Plenum Press, New York (1975)
14. R.A. Moline and A.G. Cullis, Appl. Phys. Lett. 26, 551 (1975)
15. G. Fischer, G. Carter, and R. Webb, Rad. Effects 38, 41 (1978)
16. P. Sigmund, J. Appl.Phys. 50, 726 (1979)
17. T. Hirao, K. Inoue and Takayanagi, J. Appl.Phys. 50, 193 (1979)
18. W.K. Chu, M.J. Sullivan, S.M. Ku and M. Shatzkes, First Int. Conf. on Ion Beam Modification of Materials, Budapest (1978)
19. T. Ito, S. Hijya, H. Nishi, M. Shinoda, and T. Furuya, Jpn.J.Appl. Phys. 17, 201 (1978)
20. J.M. Poate and T.C. Tisone, Appl. Phys. Lett. 24, 391 (1974)
21. H. Nishi, T. Sakurai, T. Akamata, and T. Furuya, Appl.Phys. Lett. 26, 337 (1974)
22. Z.L. Liau and J.W. Mayer in "Treatise on Materials Science and Technology" (Ed. J.K. Hirvonen), Academic Press, New York (1980)
23. P. Sigmund, Phys. Rev. 184, 383 (1969)
24. P. Sigmund, Rev. Roum.Phys. 17, 823, 969, 1079 (1972)
25. H.H. Andersen and H.L. Bay, J.Appl. Phys. 46, 1919 (1975)
26. A.W. Tinsley, W.A. Grant, G. Carter, and M. Nobles in: Ion Implantation in Semiconductors (Eds: I. Ruge und J. Graul), Springer, p. 199 (1971).
27. H.Matsumura and S.Furukawa, Jap. J.Appl. Phys. 14, 1983 (1976)
28. H.Runge, Phys. Stat.Sol. (a) 39, 595 (1977)
29. J.P. Biersack in: H. Ryssel and I. Ruge, Ionenimplantation, Teubner, Stuttgart 1978
30. D.A. Antoniadis, S.E. Hansen, and R.W. Dutton, Stanford Univ. Techn. Rept. 5019-2, 1978
31. J. Bottiger, J.A. Davies, W.A. Grant, and K.B. Winterbon, Rad. Effects 11, (1971)
32. H. Ryssel, G. Prinke, K. Haberger, and K. Hoffmann, Appl. Phys. 24, 39 (1981)
33. F. Jahnel, H. Ryssel, K. Hoffmann, K. Müller, J. Biersack, and R. Henkelmann, Nucl. Inst. Meth. 182/183, 223 (1981

# Part III

# Measuring Techniques and Annealing

# Electrical Measuring Techniques

P.L.F. Hemment

Department of Electronic and Electrical Engineering
University of Surrey, Guildford, Surrey, U.K.

## 1. Introduction

The purpose of this paper is to review the electrical measuring techniques which may be used after implantation to determine the areal density of the implanted impurities and their uniformity laterally across the sample. In addition, depth profiling methods will be considered as these can yield valuable information regarding the energy and purity of the ion beam. The techniques are discussed mainly with respect to measurements on implanted silicon wafers but they may be applied more generally to any semiconductor.

It is well known that ion implantation can be used to modify any property of a material which is sensitive to atomic structure [1]. Examples include refractive index, optical absorption, electrical conductivity, carrier lifetime or even the density of the sample. It follows that ion beam induced changes in any of these properties may be used for dosimetry control, provided the relationship between the measured change and implanted dose is known. Unfortunately, the functional dependence is seldom known to any great precision and even in cases where sound theoretical models exist (e.g. conductivity), it is found that the measured changes are dependent upon experimental conditions. In essence, dose determination and control is a complex and ill-understood aspect of ion implantation - the topic should not be treated lightly.

## 2. Background

### 2.1 Why Post-Implantation Dosimetry

Ion implantation is inherently a highly controllable method of doping semiconductors and the device engineer wishes to take full advantage of this feature to implant impurities to an accurately known areal density, with good lateral uniformity and with a particular depth distribution. These parameters are required to be constant between individual wafers and between batches of wafers for the duration of a production run, which may be months or years. Unfortunately the machine operator is unable to set the parameters, dose and dose uniformity. Instead he relies upon a measurement of the ion beam current, which is integrated with respect to time to yield the total charge. If the charge state of the ions is assumed and the implanted area is known, then the incident dose, or fluence (ions/cm$^2$), can be determined. In practice it is found that changes in machine parameters can pass unnoticed and lead to gross errors in the dose and dose uniformity, which may result in devices being out of specification. Thus, post-implantation dosimetry techniques have been developed to provide quick routine methods of monitoring the performance of ion implantation systems and for cross-calibration purposes. Within a commercial implantation facility, these checks lead to improved yields of working devices and enable realistic performance norms to be defined.

## 2.2 Incident and Retained Dose

The ion beam current is an important çontrol parameter, which is used by the machine operator to determine the total ion dose (fluence) incident upon the target. On the other hand, the device engineer wishes to achieve a particular level of doping in the semiconductor. He requires to know the retained dose.

In practice the retained dose is always less than the incident dose as there is a finite probability of both ion reflection and target sputtering occurring. The discrepancy will probably be greater than 1% and may be as large as 5%. With increasing incident dose, the discrepancy becomes larger as the retained dose approaches a saturation level. This effect is illustrated in Fig. 1 which shows the variation of retained dose with incident dose (fluence) for 10, 20, and 30 keV $Kr^+$ implanted into GaAs [2]. The figure shows an extreme example which is unlikely to be encountered when implanting dopant impurities into semiconductors.

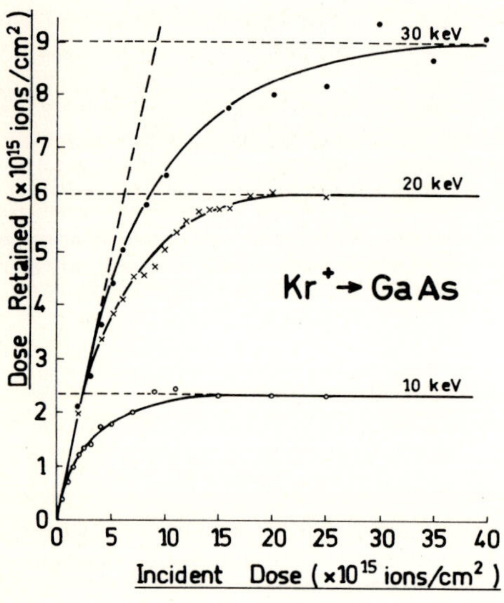


Fig.1. Ion collection data showing saturation of the retained dose, from Carter et al. [30]

Reflection. A small and approximately constant proportion of the incident ions will always be lost from the target due to ion reflection. Detailed measurements and calculations by Bøttiger and co-workers [3] have shown good agreement between theory and experiment. Their model assumes that the reflected ions are essentially those particles lying in the part of the theoretical depth distribution (assumed Gaussian) which extends in front of the surface of the sample. In practice this source of error may be ignored if $R_p \gg 2\,\Delta R_p$ where $R_p$ is the projected range and $\Delta R_p$ is the standard deviation. Reflection losses are greatest for low-energy light ions incident upon a heavy substrate, as this system has a relatively broad depth distribution. The loss is smallest for high-energy implants, as the impurity distribution is buried beneath the surface.

Sputtering. In contrast to ion reflection, the loss caused by sputtering will be highly dependent upon the dose as target erosion will occur, causing ejection of atoms previously implanted. For low doses, of say less than $10^{14}$ ions/cm$^2$, the error is small. For doses of $10^{15}$ ions/cm$^2$ - $10^{16}$ ions /cm$^2$ the error may be detectable and for higher doses the retained dose will approach a saturation value, as shown in Fig. 1. As the sputtering coefficient, defined as the number of target atoms ejected per incident ion, is dependent upon the ion energy and the masses of the ion and target atoms, the doses quoted are only indicative of the regimes.

Losses by reflection and sputtering are additive and together are the cause of the discrepancy between the incident dose (fluence) and the retained dose. Calculations by Smith [4] of the percentage error for some typical implantation systems are shown in Fig. 2. Simple Gaussian depth profiles based upon LSS range data have been assumed. The calculations show the discrepancy to be highly dependent upon the sputtering coefficient.

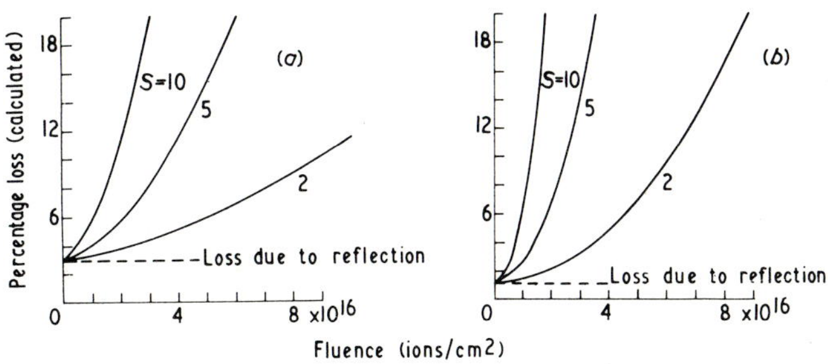


Fig. 2. Computed dose loss due to sputtering for the implantation of $P^+$ and $As^+$ in silicon (a) $P^+$ into silicon at 50 keV (b) $As^+$ into silicon at 50 keV, from Freemann [4]

## 2.3 Absolute and Relative Dose Measurement

It is, of course, always desirable to know the absolute number of impurity atoms retained within the target. But, to achieve even a modest ± 5% accuracy in the absolute dose is both difficult and very time consuming [5]. The problem is not related to instrumental errors but to an inadequate understanding of the physics of the various direct methods used for surface analysis. One of the most carefully planned international standards exercises [6] yielded discrepancies between laboratories as high as 20 % and it was only after several years and much careful experimentation that results converged towards an agreed value of the absolute dose [7].

Even so, the discrepancy between the laboratories was still about ± 2 % in contrast the indirect methods of dose determination, which must be recognised as giving relative doses, are easy to implement and require instrumentation which is readily available in most laboratories. The sheet resistance technique has been shown by many groups to have a reproducibility of better than ± 1% and a repeatability of measurements of better than

± 0.1%. For these wholly practical reasons most commercial laboratories accept and work with data based upon measurements of the relative doses.

## 2.4 Sample Contamination

Sample contamination will occur during ion implantation, with hydro-carbons, sputtered impurities and ions of different species and energy all being incident upon the surface of the sample [5]. These contaminants can lead to a degradation in the surface properties and also to dosimetry errors. But these effects can be significantly reduced by implanting through a thin protective layer (e.g. $SiO_2$) or by mechanical cleaning or chemical etching after implantation and before high-temperature processing.

## 2.5 Post-Implantation Annealing

An unavoidable feature of ion implantation is that the crystal lattice suffers damage and even amorphisation, as the energetic ions lose their kinetic energy by nuclear (elastic) scattering. If the implanted impurity is to act as a substitutional dopant then it is necessary (i) to restore good crystallinity to the implanted layer by epitaxial regrowth and (ii) to establish electrical activity of the implanted dopant through incorporation of the impurity atom onto a substitutional lattice site. This is achieved through thermal annealing at temperatures usually in excess of 2/3 of the melting point. For silicon it is customary to anneal at about 1000°C for 15-30 minutes. In Section 4 it will be shown that the thermal anneal can be the cause of large errors in the estimation of the retained dose.

In special cases where the lattice defects themselves are required as electrically active centres, as in the case with proton isolation in GaAs [9], annealing may be unnecessary [28].

## 2.6 Basic Assumptions

Determination of the retained dose calls for a detailed knowledge of the dose dependence of the measured physical property or device parameter. For example, when using any method based upon a resistance measurement it is necessary to assume that all of the implanted dopant impurities are electrically active when $n=N_d$ or $p=N_a$, where $N_d$ and $N_a$ are densities of the implanted donors and acceptors, respectively. The resistance can then be related to the carrier concentration from a knowledge of the bulk resistivity $\rho$, where

$$\rho = 1/ne\mu_e \quad \text{for donors}$$

or

$$\rho = 1/pe\mu_h \quad \text{for acceptors.}$$

n and p are the electron and hold densities and $\mu_e$ and $\mu_h$ the carrier mobilities. In Section 4 it will be shown that 100% dopant activity may not be achieved and also that the measured resistance depends upon experimental parameters such as dose rate, beam heating, anneal time and temperature and may even show a time dependence akin to an "aging" process.

In sheet resistance measurements it is assumed that the substrate makes no contribution to the measured value. This may be achieved either by implanting into a high-resistivity substrate or into material of the opposite type, when electrical isolation will be achieved by the formation of a p-n junction. Smith and Stephen [10] have observed that if the p-n junction is within about $3\Delta R_p$ of the peak of the impurity distribution then the sheet resistance will be erroneously high. In practice this error can be avoided if a lightly doped substrate is used in which $N_d/N_B \geqslant 250\ \Delta R_p$ where $N_B$ is the background doping (impurities/$cm^3$) and $\Delta R_p$ is the standard deviation of the impurity distribution [11].

As with all standards exercises, it is necessary to employ good experimental practices which include temperature control of samples during measurements, exclusion of light and mechanical vibration and cleanliness to minimise leakage currents. For resistance measurements it is necessary to form ohmic contacts, and low currents should be used to avoid Joule heating. Measurements should be made with reverse polarity, and offset voltages should be nulled.

## 3. Measurement Techniques

### 3.1 Resistance Measurements

The bulk resistance $R_B$ of a uniformly doped sample is:

$$R_B = \rho \ell / wt$$

where $\rho$ is the bulk resistivity; $\ell$, w and t are the length, width and thickness, respectively. A sample geometry more appropriate to ion implantation is that of a lamina, of thickness t, in which a square sample ($\ell$=w) has a sheet resistance $R_s$, where

$$R_s = \rho / t \ (\Omega/\square).$$

The sheet resistance has the same numerical value for a square of any size. Theoretical values of the sheet resistance of n- and p-type implanted layers in silicon have been published by Smith and Stephen [10], Fig. 3. Their results have been calculated from the relationship:

$$R_s = \int \frac{dx}{n(x)\ \mu(n(x))\ e}$$

where $\mu(n(x))$ is the carrier mobility. The calculations assume a Gaussian doping profile and that all of the impurities are electrically active. Because of the non-linear dependence of mobility upon carrier concentration [12] a family of curves has been produced for particular values of the standard deviation in the impurity depth distribution. It should be noted that $R_s$ is independent of the projected range, provided the impurity depth distribution is buried beneath the surface (see Section 2.2).

The implanted dose of ions may be calculated from the measured value of $R_s$. Several measurement geometries have been used and these include:

| | |
|---|---|
| Four-point probe | - linear and square arrays |
| Linear resistor | - two and four terminal |
| Van der Pauw | - four terminal |

Four-Point Probe. The four-point probe has proven to be a convenient tool for the determination of $R_s$, being quick to implement and requiring but a

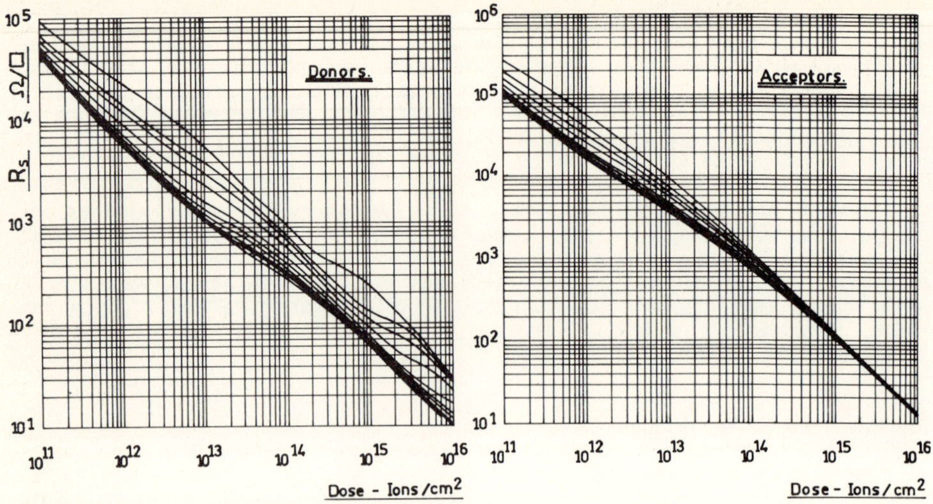


Fig. 3. Sheet resistance of ion-implanted silicon, assuming complete electrical activity of the dopant. The curves are, in descending order, for standard deviations of 0.0025, 0.0075, 0.0125, 0.025, 0.05, 0.075, 0.100, 0.125, 0.150 and 0.175µm (a) for donors and (b) for acceptors. Data taken from Ref. [10]

small number of processing steps [13 - 21]. The technique can be used over the dose range $10^{13}$ - $10^{16}$ ions/cm$^2$, corresponding to an $R_s$ range of order $10^3$ - $10^1$ Ω/□. In practice an accuracy of better than ± 1% and a repeatability of ± 0.1% can be achieved for doses of $10^{14}$ - $10^{16}$ ions/cm$^2$.

It is customary to use either a linear probe array, with equal probe spacings, or an array with the probes at the corners of a square. The method involves voltage and current measurements between tungsten carbide tips which are in contact with the implanted layer, as shown in Fig. 4. The probe spacing s is typically 0.5mm to 1.5mm, which satisfies the condition that the spacing be much greater than twice the layer thickness[13]. The sheet resistance may be calculated using:

$$R_s = \frac{k \cdot V}{I} \quad (\Omega/\square) \qquad (1)$$

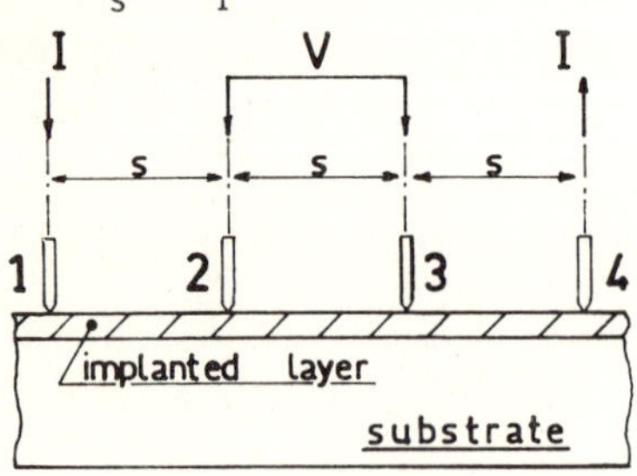


$$R_s = k . V_{23}/I_{14}$$

Fig. 4. Conventional current/voltage configuration for the linear four-point probe where k is a constant dependent upon the geometric configuration and V is the potential difference between contacts due to current I flowing through the layer. For an infinite conducting lamina, k has the values [21]

$$k = \frac{\pi}{\ell n2} = 4.5324 \quad .......... \text{ linear array}$$

or

$$k = \frac{2\pi}{\ell n2} = 9.0648 \quad .......... \text{ square array}$$

Rymaszewski[16] has calculated the geometric correction factor for the six possible combinations of potential and current contacts of the linear array (Table 1).

Table 1. Six configurations of the current and potential probes for a linear four-point probe, from Rymaszewski [16]

| Configuration | Current contacts | Potential contacts | k |
|---|---|---|---|
| a | 1,4 | 2,3 | 4.532 |
| b | 1,2 | 3,4 | 21.84 |
| c | 1,3 | 2,4 | 15.50 |
| d | 2,4 | 1,3 | 15.50 |
| e | 3,4 | 1,2 | 21.84 |
| f | 2,3 | 1,4 | 4.532 |

It is customary to use configuration (a) in Table 1, with the outer contacts carrying the current, as this leads to the largest voltage difference for a given current and equispaced probes. In practical situations, measurements will be made on finite samples. The tabulated values will only give better than 1 % accuracy in $R_s$ when these measurements are made at the centre of finite samples of linear dimensions greater than about 40 times the probe spacing. For a probe spacing of 1 mm the error will be less than 2 % over the central 50 % area of an implanted 50 mm wafer, but then increases rapidly for measurements near the edge. Values of k are listed in Table 2 for the linear array for different values of d/s, where d is the diameter of the implanted area and s is the probe spacing [13].

Table 2. Geometric correction factors k for measurements of the sheet resistance of circular samples using a linear four-point probe [13]

| d/s | 3.0 | 4.0 | 5.0 | 7.5 | 10 | 15 | 20 | 40 | ∞ |
|---|---|---|---|---|---|---|---|---|---|
| k | 2.266 | 2.929 | 3.363 | 3.927 | 4.172 | 4.362 | 4.436 | 4.508 | 4.532 |

For small samples or when the probe is near the edge of the implanted layer,as occurs during uniformity measurements, it is necessary to calculate the value of k for each of the measurement positions. The value is a maximum at the centre of the implanted area as V is a minimum. For a uniform circular layer upon which a linear probe is aligned along a radius, Logan[14] has redefined k as:

$$k = \frac{\pi}{\ell n2} \cdot \frac{1}{(1+\eta)}$$

where

$$\eta = \frac{1}{2\ell n2} \cdot \ell n \left\{ \frac{\left[1 - \left(\frac{\Delta}{r} + \frac{s}{2r}\right)\cdot\left(\frac{\Delta}{r} - \frac{3s}{2r}\right)\right]\left[1 - \left(\frac{\Delta}{r} - \frac{s}{2r}\right)\cdot\left(\frac{\Delta}{r} + \frac{3s}{2r}\right)\right]}{\left[1 - \left(\frac{\Delta}{r} - \frac{s}{2r}\right)\cdot\left(\frac{\Delta}{r} - \frac{3s}{2r}\right)\right]\left[1 - \left(\frac{\Delta}{r} + \frac{s}{2r}\right)\cdot\left(\frac{\Delta}{r} + \frac{3s}{2r}\right)\right]} \right\}$$

and r is the radius of the layer and $\Delta$ is the distance from the centre of the probe to the centre of the layer.

If the probe is orthogonal to the radius, Swartzendruber[15] has derived

$$\eta = \frac{1}{2\ell n2} \cdot \ell n \left\{\frac{\alpha_1\alpha_2}{\alpha_3\alpha_4}\right\}$$

where $\alpha_1 = (V_2-V_1)^2 + (U_1+U_2)^2$ $\qquad \alpha_2 = (V_2+V_1)^2 + (U_1+U_2)^2$

$\alpha_3 = (V_2-V_1)^2 + (U_2-U_1)^2$ $\qquad \alpha_4 = (V_2-V_1)^2 + (U_2-U_1)^2$

$U_1 = \frac{3s}{r} / D_1$ $\qquad U_2 = \frac{s}{r} / D_2$

$V_1 = \left[1 - \left(\frac{\Delta}{r}\right)^2 - \frac{9}{4}\left(\frac{s}{r}\right)^2\right] / D_1$ $\qquad D_1 = \left(1 + \frac{\Delta}{r}\right)^2 + \frac{9}{4}\left(\frac{s}{r}\right)^2$

$V_2 = \left[1 - \left(\frac{\Delta}{r}\right)^2 - \frac{1}{4}\left(\frac{s}{r}\right)^2\right] / D_2$ $\qquad D_2 = \left(1 + \frac{\Delta}{r}\right)^2 + \frac{1}{4}\left(\frac{s}{r}\right)^2$ .

Values of k have been tabulated by Swartzendruber [15] and Smith et al. [21]. In addition, Smits [13] has tabulated values for rectangular samples of various aspect ratio; see also Fig. 6.

Perloff [18] has considered the general case when the linear probe is oriented at an angle $\Theta$ to the radius of a circular wafer. He has determined values of k and finds it relatively insensitive to $\Theta$ at measurement distances greater than 4s from the edge of a 3" wafer. Simplified expressions have been obtained which are applicable when the probe separation s is $\leqslant$ 10% of the wafer radius. When $(a/s^1)^2 \gg 1$ then

$$k \rightarrow \pi/\{\ell n2 + 1/2 \cdot \ell n\ (1 + 3/4 \cdot (s^1/a)^2)\}$$

where $s^1 = s/(1-a/2r)$ and a is the distance from the edge of the wafer.

For the special geometry of a circular wafer doped on both top and bottom faces Logan [14] has shown that the sample behaves as if it is a continuous infinite sheet. The geometric correction factor has a constant value (4.532) which is independent of the position of the four-point probe.

Random variations in the contact spacing s can be significant sources of error in measurements of $R_s$. However, Rymaszewski [16] has provided a means of correcting for this error, by performing a two-configuration measurement using the values of k in Table 1, and Perloff [22] has clearly demonstrated that this procedure significantly reduces random variations. Perloff uses two configurations, namely $V_{23}/I_{14}$ and $V_{12}/I_{34}$.

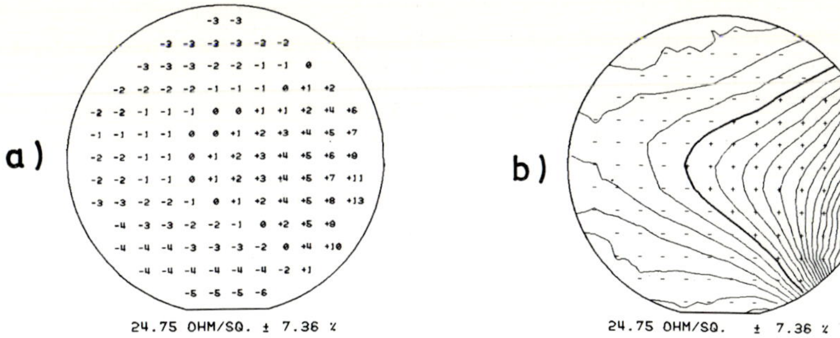


Fig. 5. Methods of displaying uniformity data: (a) deviation map and (b) iso-sheet resistance topograph, from Ref. [23]

Automatic wafer probing systems have been developed by Smith and Stephen [20] and Perloff, Wahl and Kerr [23] for measurements of wafer doping uniformity. Figure 5 shows the method developed by Perloff for displaying uniformity data (in this example, for a P-diffused p-type wafer). Measurements are made automatically at 118 test sites located on a square grid with a distance between centres of 0.222" on a 3" wafer. The data may be presented as a deviation map, a two-dimensional histogram or as an iso-sheet resistance topograph. The mean resistivity $R_s$ and the standard deviation ($\sigma$) are computed using:

$$\bar{R}_s = \frac{1}{N} \sum_{i=1}^{N} (R_s)_i \tag{2}$$

$$\sigma = \left\{ \frac{1}{N-1} \sum_{i=1}^{N} [(R_s)_i - \bar{R}_s]^2 \right\}^{1/2} . \tag{3}$$

The standard deviation may be expressed as a percentage:

$$\frac{\sigma}{R_s} \times 100\% . \tag{4}$$

The simplicity of the Four-Point Probe technique, namely the use of pointed pressure probes to make ohmic contact to the implanted layer, introduces its own problems. If the force on the probe is too large, the probe may damage the surface and pierce through the implanted layer, making a short circuit to the substrate. Smith [20] recommends that the force should be restricted to 20 gms wt., when it is possible to contact layers of 400 Å thickness (say 40 keV $As^+$). Care also should be taken to clamp rigidly both the probe assembly and wafer, to avoid relative movement. The tips of the probes must be maintained in a clean state. Smith [20] reports that reliable measurements are only possible after driving a current through all of the probes, to form good ohmic contacts.

Test Patterns. The reliability of doping uniformity measurements using the four-point probe depends critically upon a knowledge of both the wafer and probe geometry. Significant errors may be introduced if the implanted

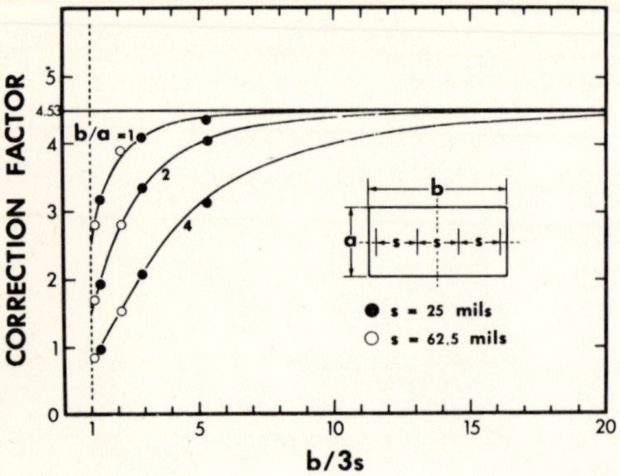


Fig. 6. Geometric correction factor k versus b/3s for a four-point probe symmetrically positioned within a rectangular test pattern of length b and width a, from Ref.[8]

layer has an irregular shape. These contraints can be overcome if photolithography is used after implantation to define a matrix of small implanted regions of fixed dimensions and adequate size to accept the probe. In this case k is constant over the whole wafer and may be determined from expressions due to Smits [13], which have been confirmed by Perloff [8] for a linear probe positioned symmetrically within rectangular test patterns, Fig.6. Unfortunately errors are still present due to variations in the probe spacing, and a new source of random error is caused by dimensional variations in the test patterns.

Linear Resistors. Some of the constraints associated with the four-point probe can be avoided if photolithography is used to form an array of 2 or 4 terminal linear resistors, as shown in Fig. 7. With this geometry the sheet resistance over the area of the resistor is:

$$R_s = \frac{W}{L}\ \frac{V}{I} \quad (\Omega/\square)$$

where W and L are the width and length, respectively.

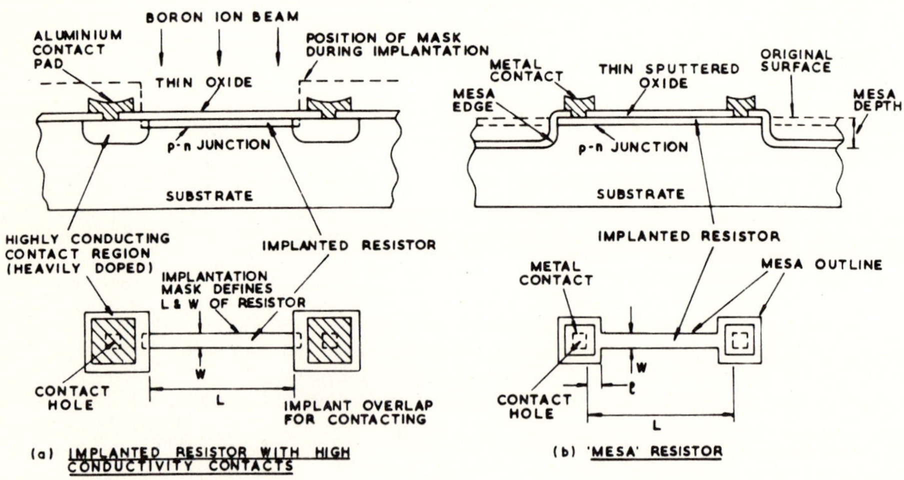


Fig. 7. Resistor structures used for sheet resistance measurements (a) resistor defined by ion implantation, (b) resistor defined by mesa etching, from Ref.[21]

A disadvantage of using resistor structures is that many more processing steps are required, including photolithography, a second implantation to form the ohmic contact pads, deposition of metal contacts, thermal processing and mesa etching. Unfortunately, each of these processes will introduce random errors, but there are advantages. The data is more easily interpreted, errors associated with the contacting probes will be minimal and reliable measurements can be made on wafers implanted with low doses or at high energies when the conducting layer will be buried.

Smith and Stephen [21] have concluded that, in general, four-point probe measurements can be as reliable and reproducible as the linear resistor and that there is little to be gained by undertaking the additional processing steps necessary to fabricate the resistors.

Van der Pauw. An alternative four-terminal resistor structure which is less sensitive than the linear resistor to dimensional variations in the photolithographic masks is that defined by Van der Pauw [24]. The geometry of a suitable structure is shown in Fig. 8. The sheet resistance is determined from the relationship

$$R_s = \frac{\pi}{\ell n2} \left(\frac{R_a + R_c}{2}\right) f(R_a/R_c) \tag{5}$$

where, with reference to Fig. 8: $R_a = V_{23}/I_{14}$ and $R_c = V_{43}/I_{12}$ .

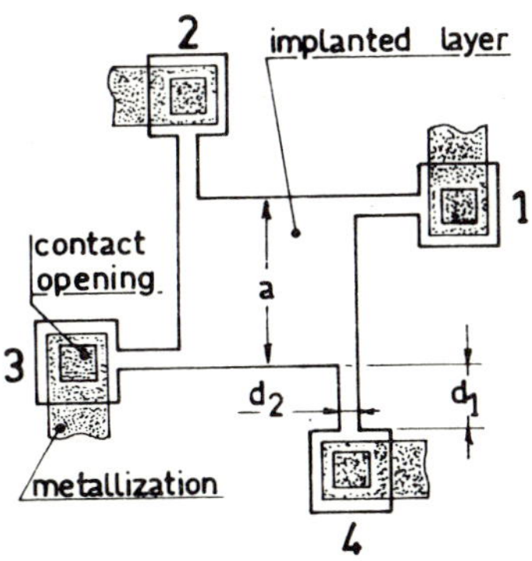


Fig. 8. Van der Pauw resistor structure

The function $f(R_a/R_c)$ satisfies a transcendental expression and has the values shown in Table 3. When the structure is symmetric $R_a = R_c$, $f(R_a/R_c) = 1$ and again

$$R_s = \frac{\pi}{\ell n2} \cdot \frac{V_{23}}{I_{14}} = 4.532 \cdot \frac{V_{23}}{I_{14}} .$$

A benefit to be gained by using the Van der Pauw geometry is that both the free-carrier concentration and mobility may be obtained, provided the thickness t of the conducting lamina is known. When

$$\rho = \frac{\pi t}{\ell n2} \frac{R_a + R_c}{2} f(R_a/R_c) \tag{6}$$

Table 3. Tabulated values of the Van der Pauw function $f(R_a/R_c)$ [24]

| $R_a/R_c$ | $f(R_a/R_c)$ |
|---|---|
| 1 | 1.0000 |
| 1.3 | 0.9941 |
| 1.8 | 0.9711 |
| 2.0 | 0.9603 |
| 3.0 | 0.9067 |
| 10.0 | 0.6993 |

and

$$R_H = \frac{t}{B} \cdot \frac{\Delta V_{24}}{I_{13}} \tag{7}$$

where $R_H$ is the Hall coefficient, and $\Delta V_{24}$ is the Hall voltage due to the applied magnetic field B, then

$$n = \frac{1}{R_H e} \tag{8}$$

and

$$\mu = \frac{R_H}{\rho} . \tag{9}$$

Equations (5), (6) and (7) are valid providing the lamina has point contacts on the periphery. For the practical case of symmetric, finite contacts on a circular sample, Versnel [25] has derived the approximate expression for k, namely:

$$k = \frac{\pi}{\ell n 2} + \frac{1}{64} \cdot \frac{\pi^3}{(\ell n 2)^2} \cdot \mu^2$$

where $\mu$ is the ratio of the length of one contact to the length of one quarter of the circumference of the circular sample. Values of k are given in Table 4 .and confirm the relative insensitivity of the Van der Pauw structure to geometric variations.

Table 4. Geometric correction factors for the Van der Pauw geometry, with contacts of finite length on the circumference of a circular sample. The data is taken from Table 1 of Ref.[25]

| $\phi$ (deg) | 0 | 2 | 5 | 10 | 15 | 20 |
|---|---|---|---|---|---|---|
| $\mu$ | 0 | 0.0444 | 0.1111 | 0.2222 | 0.3333 | 0.4444 |
| k | 4.532 | 4.534 | 4.548 | 4.548 | 4.652 | 4.758 |

Measurements by David [26] made on the symmetric structure shown in Fig.8, which is the geometry of the NBS Process Evaluation Mark Set (NBS-3), has shown better than 0.1% agreement with equation 5 as long as the condition $d_1/a \gg d_2/a$ is fulfilled. In all cases it is desirable to make measurements for two current-voltages configurations, equation 5, to eliminate geometric sources of error. In doing so, Perloff [8] has reported measurement reproducibilities of better than 0.02% for 10 µm x 10 µm resistors.

Undoubtedly, the fabrication of resistors with the Van der Pauw geometry involves many processing steps to produce the required planar structures, but the gains achieved are considerable, being excellent reproducibility, freedom to choose the size and shape of the resistor arrays, and a relative independence from errors due to photolithographic mask distortions.

Spreading Resistance. Two or four terminal measurements of the sheet resistance are usually made with a geometry much larger than the semiconductor devices incorporated in integrated circuits. A technique which permits measurements of the doping uniformity to be made on a small scale (5-10 mm), is that of spreading resistance. In this technique two contacts are required; a top sensing probe with a small radius tip, which defines the volume being measured and a large area back, or side, contact. The resistivity is then sensed in a hemispherical volume beneath the probe, typically of a radius four times that of the contact area. The tip radius can be made very small if electropolished tips are used. Mazur and Dickey [27] have reported the method and Fig. 9 shows experimental data relating the value of spreading resistance to sheet resistance of silicon wafers. These workers propose a general expression for the spreading resistance, $r_s$:

$$r_s = \rho/4a$$

where $\rho$ is the resistivity and a is the radius of the contact point.

Bicknell [28] has used the technique, with step intervals as small as 5 µm, to determine dose uniformity of $H^+$ and $He^+$ implants into highly conducting silicon. Bicknell monitored the reduction in the free-carrier concent-

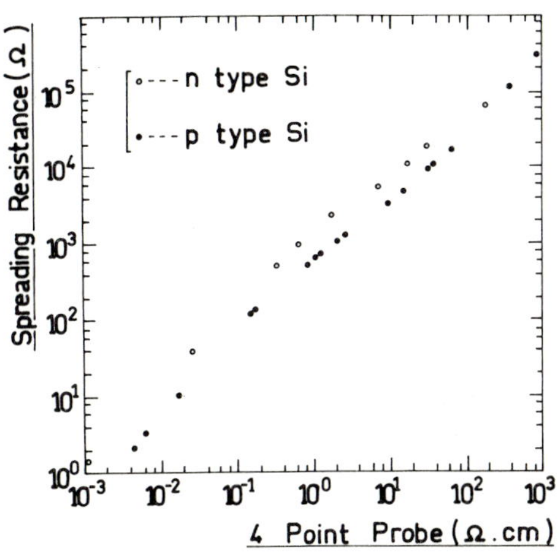


Fig. 9. Values of spreading resistance compared to four – point probe measurements for bulk silicon. Data taken from Mazur and Dickey [27]

ration due to lattice defects, and thus no thermal annealing was required. He found that the technique could be used over the dose range $10^{13}$-$10^{15}$ $H^+/cm^2$ and that the spreading resistance changes could best be represented by logarithmic relationships. For protons,

$$N = 720 \log (r_2/r_1)$$

where $r_1$ and $r_2$ are the resistances before and after irradiation and N is the dose in $\mu C/cm^2$.

Perloff [23] also reports that spreading resistance provides a convenient means of determining large local variations in doping level, and reports that the technique is effective over the dose range $5\times10^{12}$ to $5\times10^{15}$ $B^+/cm^2$, Fig. 10.

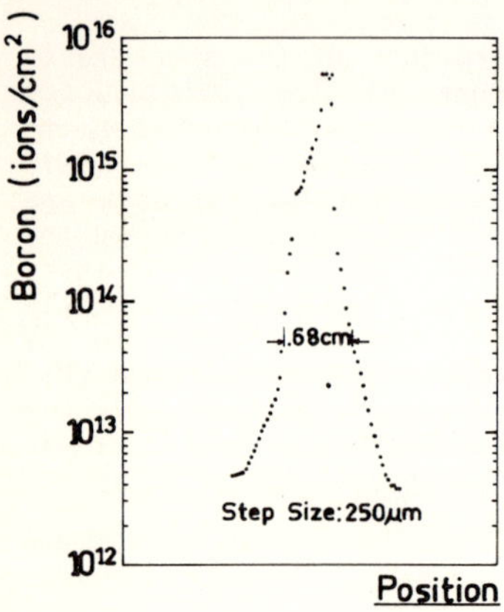


Fig. 10. Ion beam ($B^+$) profile obtained from spreading resistance measurements of a silicon wafer implanted without beam scanning [23]

3.2 Capacitance Voltage

The electrical characteristics of both the Schottky barrier and the one-sided p-n junction are similar and provide a means of evaluating doping profiles and doping uniformities, both of uniformly implanted layers and also of device structures.

The capacitance C of a $p^+$-n junction is [12]

$$C = \left\{ \frac{e\varepsilon N_d}{2(V_{bi} - V - kT/e)} \right\}^{1/2} \quad (F/cm^2) \tag{10}$$

where $N_d$ is the doping concentration, $V_{bi}$ the built-in bias and V the applied bias. Equation (10) is the basis of capacitance-voltage doping profile measurements (see Section 3.4), for which purpose it is differentiated to give:

$$N_d(W) = \frac{C^3(V)}{e\varepsilon A^2 \frac{dC(V)}{dV}} \quad \text{and} \tag{11}$$

$$W = \frac{\varepsilon A}{C(V)} \tag{12}$$

where W is the width of the depletion layer and A the area.

By applying Eqs.(11,12), Glawischnig [29] has been able to use hyperabrupt varicap (varactor) diodes for a study of small-scale variations in dose uniformity. The devices can be used over the dose range $10^{12}$-$10^{14}$ ions/cm$^2$, and the structure used for this purpose is a $p^+$-n junction in which the p-type layer is formed by implanting a dose of $1x10^{16}$ $B^+$/cm$^2$. This is followed by implantation of the ion ($As^+$ or $P^+$) whose uniformity is to be studied. The wafers are annealed at 1100°C for 10 hours to achieve the required depth profile. After selective etching, the p-n junctions are probed directly, without the addition of metallisation. The capacitors have a diameter of 80 µm, at a separation of 0.1mm, and a capacitance of about 5pF at 1 volt reverse bias. Measurements are recorded automatically and each takes approximately 0.2 seconds, making it practical to analyse an area of 7mmx7mm which will contain about 5,000 data points. Glawischnig [29] reports doping uniformity profiles with standard deviations of ± 0.5%. Some beam profiles measured by this capacitance-voltage technique are shown in Fig. 11.

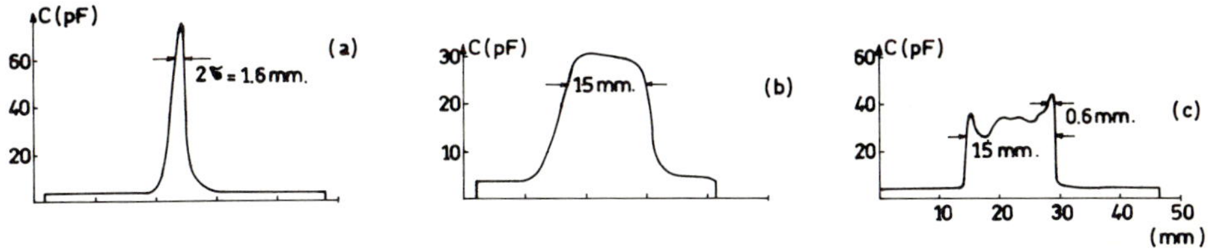


Fig. 11. Beam profiles measured by the capacitance - voltage technique. The profiles correspond to different electrostatic beam scanning conditions [29]

## 3.3 Device Parameters

The use of sheet resistance and capacitance-voltage methods may not be appropriate for all ion dose and energy combinations - in particular for low doses. In these cases, dosimetry control may be achieved by monitoring the operating parameters of partially formed or even complete devices, when it becomes possible to use commercial equipment for automatic data collection. Schmid [34] has recently reported an automatic system which will monitor up to nine MOS device parameters, including flat band voltage, threshold voltage and oxide breakdown. However, a disadvantage is the remoteness of the device parameters from the basic parameters of interest, namely the ion dose.

Threshold Voltage. For the MOSFET there is a particular gate voltage which will enable a conducting channel to form between the source and drain. The voltage at which this occurs is the threshold voltage ($V_T$) which Sze [35] has given as:

$$V_T = 2\psi_B + \frac{(2\varepsilon e N_a(2\psi_B))^{1/2}}{C_i} \tag{13}$$

where $\psi_B$ is the potential difference between the Fermi level and intrinsic level, $C_i$ the capacitance per unit area, $\varepsilon$ the semiconductor permittivity and $N_a$ the doping level of the bulk.

In a more detailed analysis additional terms must be included, and Sze has shown that the threshold voltage for a narrow-channel MOSFET is:

$$V_T = V_{FB} + 2\phi_B + \frac{(2\varepsilon e N_a(2\phi_B+V_{BS}))^{1/2}}{C_i} \cdot (1 + \frac{\pi}{2} \; \frac{W}{Z}) \qquad (14)$$

where $V_{FB}$ is the flat band voltage, $V_{BS}$ the substrate reverse bias, Z the channel length and W the depletion layer width.

Material parameters influence the threshold voltage, and the dependence upon the doping level of the bulk silicon is shown in Figure 12 for a fixed oxide charge of $5 \times 10^{11}$ charges/cm$^2$ [32].

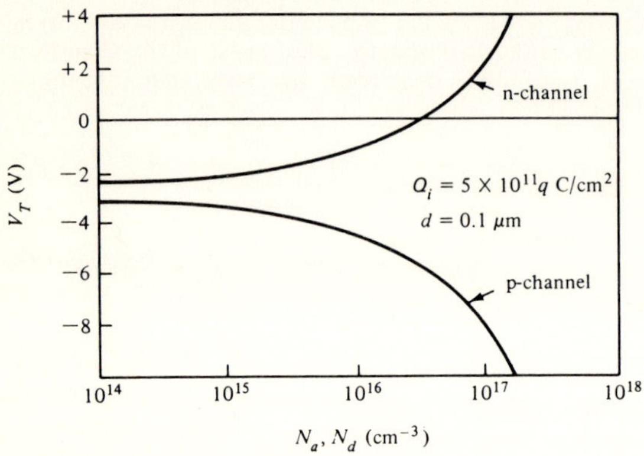


Fig. 12. Variation of threshold voltage with substrate doping for n-channel and p-channel Al-$SiO_2$-Si devices. Data taken from Streetman [32]

The threshold voltage may be determined experimentally from a linear extrapolation to zero current of the curve of source-drain current ($I_{DS}$) against gate-source voltage ($V_{GS}$) for a given small drain-source bias ($V_{DS}$) as:

$$L_{DS} = (\frac{Z}{L}) \; \mu_n \; C_i (V_{GS}-V_T) \; V_{DS} \qquad (15)$$

where L is the channel length and $\mu_n$ the electron mobility [35]. For dosimetry purposes, the change in threshold voltage ($\Delta V_T$) is determined experimentally and the implanted impurities (charge/cm$^2$) may be determined applying Eqs. (13) or (14). The overall uncertainty in the estimate of dose may be large, as it is sensitive to both variations in the device geometry and uncertainties in the electronic properties of the device materials. Figure 13 shows theoretical plots of the voltage change against implanted dose for various gate oxide thicknesses. A high sensitivity to oxide thickness is evident. If the MOS devices are to be used for dose uniformity measurements it is important to maintain tight control of this thickness parameter.

The ability to detect small doses depends upon an instrumental capability to measure small changes in $I_{DS}$ and $V_{GS}$. For n-channel devices on substrates with $N_A = 10^{14}$-$10^{15}$/cm$^3$ and an oxide thickness of 500Å it is possible to detect doses of about $5 \times 10^{10}$ ions/cm$^2$ for a $\Delta V_T$ of about 100 mV [36]. The uncertainty in the threshold voltage will be about 30 mV, whilst measurements in different laboratories show discrepancies which are typically 50-100mV [36]. Achieving these levels of reproducibility calls for careful control of all processing steps and, in particular, the inclusion of control devices on the same wafer.

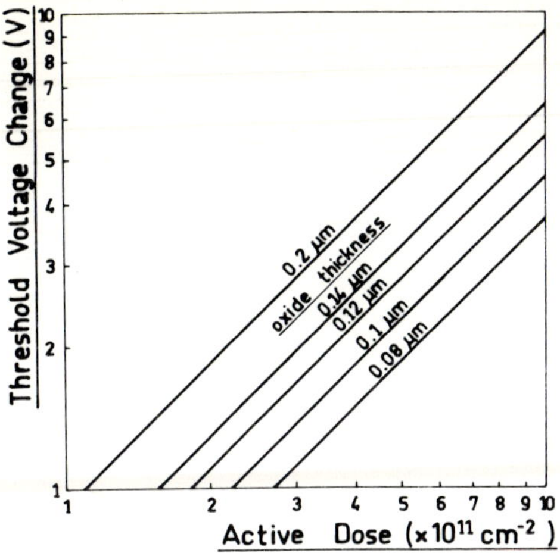


Fig. 13. Theoretical values of threshold voltage change versus electrically active dose for various oxide thicknesses (from Nicholas [59])

SLICE NO 00060/345/2 ENHT TRANSISTORS FLAT AT BOTTOM

VT TN (VSB=-2.5V)

| | | | | | | | | |
|---|---|---|---|---|---|---|---|---|
| O/C | 1.04 | O/C | 1.02 | 1.02 | 1.035 | 1.025 | 1.03 | O/C |
| -10.2 | 1.115 | 1.075 | 1.045 | 1.05 | 1.04 | 1.04 | 1.03 | S/C |
| O/C | 1.145 | O/C | 1.05 | 1.045 | 1.025 | O/C | 1.02 | .985 |
| 1.265 | 1.155 | 1.1 | 1.06 | 1.04 | 1.015 | 1.015 | 1.005 | .975 |
| 1.215 | 1.15 | 1.08 | 1.07 | 1.025 | 1.03 | 1.015 | 1.005 | .965 |
| 1.12 | 1.08 | 1.035 | 1.035 | 1 | .995 | .97 | .99 | .965 |
| 1.105 | 1.03 | 1.02 | 1.005 | .98 | 1.01 | .98 | .96 | S/C |
| O/C | 1.075 | 1.08 | 1.03 | 1 | .975 | .975 | .915 | S/C |
| O/C | O/C | 1.175 | 1.045 | 1.02 | .99 | .985 | O/C | O/C |

$\bar{V}_T = 1{\cdot}038 \quad \sigma = 6{\cdot}1\%$

Fig. 14. Measured threshold voltages for n-channel, enhancement mode, silicon gate MOS transistors implanted with a nominal dose of $2\times10^{11}$ $B^+/cm^2$ at 35 keV (from Glaccum[36])

Figure 14 shows measured values of the threshold voltage made by Glaccum [36] on n-channel, enhancement mode, silicon gate MOS transistors. These devices form a 9x9 matrix on a 2" silicon wafer, which was implanted with $2\times10^{11}$ $B^+/cm^2$ at 35 keV. A complex processing schedule then followed which included poly-silicon deposition at 600°C (15 minutes), photolithography and implantation of source and drain contacts using $6\times10^{15}$ $P+/cm^2$ at 100 keV. This was followed by thermal treatments at 550°C (1 hour), 875°C (50 minutes), phosphorus oxide deposition at 400°C and finally a drive-in anneal at 1050°C (40 minutes), to form a junction at a depth of 1 µm. The threshold was determined by using Eq. (15) and extrapolating the curve to zero current. Full depletion of the implanted layer was achieved by applying a substrate bias of -2.5 volts. The result of this elaborate processing was a shift in threshold voltage of about 0.8 volts. The actual value of threshold voltage measured on each device is shown in Fig. 14. The mean value is 1.038 volts with a standard deviation of 6.1%. This large spread is believed to be typical of this type of measurement and arises from variations of device geometry across the wafer. Independent experiments show that measurements on a single device can be repeated with a spread of less than ± 1%.

Saturation Current and Pinch-Off Voltage. The uniformity of selenium-implanted gallium arsenide wafers, over the dose range $10^{12}$-$10^{13}$ ions/$cm^2$, has been determined by Livingstone et al. [37], who monitored the saturation current $I_{DS}$ between source and drain of MESFET structures where:

$$I_{DS} = Nev_{sat}Wd \qquad (16)$$

with N the implanted dose (100% activity), $v_{sat}$ the saturated drift velocity, W the device width and d the thickness of the active layer. These workers report a spread of 4% in $I_{DS}$ over the central area of the wafer. Geometric errors were kept to a minimum as $I_{DS}$ could be measured before deposition of the gate electrode. Reassessment of the doping uniformity from variations in the pinch-off voltage, measured after deposition of the gate, showed an apparent increase to a value of 10%. This clearly demonstrates that each processing step introduces new sources of error, which will be reflected in an apparent degradation of the dose uniformity.

## 3.4 Depth Profiles

The ion beam which is incident upon the sample does not consist only of a single monoenergetic species of particle. Included in the primary beam will be neutral, molecular and multi-charged particles [5, 31], which through charge exchange and disassociation processes have been able to pass through the electrostatic and magnetic filters in the beam transport system. If reproducible doping is to be achieved it is necessary to monitor implanted wafers to confirm that there are but few of these unwanted particles and that the electrical doping profile is not distorted. Whilst surface analysis techniques, such as SIMS[38] and RBS[39], are most suited for detailed studies, the use of electrical depth profiling techniques can provide a relatively quick assessment of the beam purity, using equipment to be found in most electrical measurement laboratories.

The problem of beam purity is most severe when implanting multicharged ions or when using implanters with a post-analysis acceleration stage. An

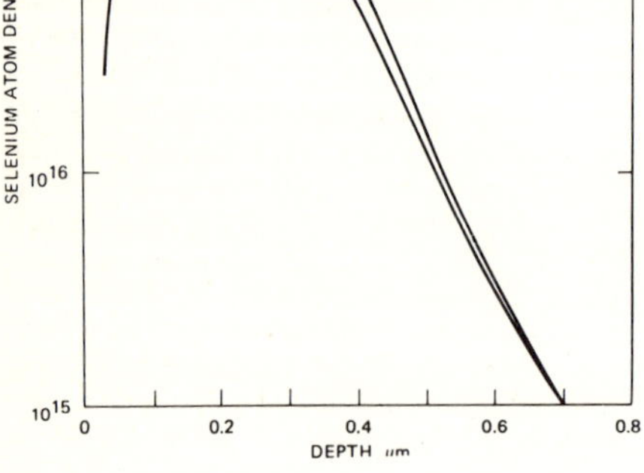


Fig. 15. Comparison of doubly ionized selenium ($Se^{++}$, 600 keV) implantations made in different implanters (from Jamba [40])

example has been discussed by Jamba [40], who used SIMS to show the difference in the depth profiles of nominal 600 keV $Se^{++}$ implants carried out in two different machines (see Fig. 15). The distorted depth distribution is assumed to be due to disassociation of 300 keV $Se_2^+$ to give 150 keV $Se^+$.

Two techniques commonly used to determine depth profiles are capacitance-voltage methods and differential sheet conductance. The former is non-destructive whilst the latter requires sample thinning using a layer removal technique. Alternative, but less precise methods include sample beveling combined with spreading resistance [27] or contact resistance [41] measurements.

Differential Sheet Conductance. The free-carrier and mobility depth profiles may be obtained using [42][43]

$$\mu_i = \frac{\left(\frac{R_H}{\rho_s^2}\right)_i - \left(\frac{R_H}{\rho_s^2}\right)_{i+1}}{\left(\frac{1}{\rho_s^2}\right)_i - \left(\frac{1}{\rho_s^2}\right)_{i+1}} \tag{17}$$

and

$$n_i = \frac{\left(\frac{1}{\rho_s}\right)_i - \left(\frac{1}{\rho_s}\right)_{i+1}}{e t_i \mu_i} \tag{18}$$

where $R_H$ and $\rho_s$ are the Hall coefficient and sheet resistivity, respectively, which have been defined in Section 3.1. These parameters are measured before (i) and after (i+1) removal of the i-th layer of thickness $t_i$. The implanted dose ($cm^{-2}$) may be obtained from $\sum n_i t_i$, assuming 100% electrical activity.

The main source of error in the profile will be uncertainty in the depth scale. For silicon this error may be minimised by using one of a number of highly controllable anodic oxidation methods in which an oxide of known thickness is formed and then chemically dissolved. Barber [44] has characterised a system using ethylene glycol with 0.04N $KNO_3$, 2.5% water and 1-2 g/litre of $Al(NO_3)_3 \cdot 9H_2O$. It is reported that the system is relatively insensitive to current density and temperature and that at room temperature silicon is consumed at a rate of 2.20 Å/V, up to 280 volts. For removal of thin layers of compound semiconductors it is customary to use either anodic oxidation or dilute etchants [45][46].

Capacitance-Voltage. Depth profiling by capacitance-voltage methods [47-50] is based upon Eqs. (11) and (12), which enable the doping concentration to be determined at the edge of the depletion region. The technique calls for a small a-c voltage to be superimposed upon the d-c reverse bias applied across a p-n junction or Schottky diode. Commercial instruments are available to process the signals and automatically plot the doping profile in a time interval of order tens of seconds over a depth extending, generally, from the zero bias depletion width to the depletion depth at which reverse breakdown occurs. These constraints generally restrict the use of the technique to samples implanted with low to medium doses. Integration of the doping profile then permits the areal density of the implanted impurities to be determined.

Interpretation of capacitance-voltage is made difficult when there are trapping levels present within the forbidden gap [51]. These levels will empty and fill with characteristic time constants, and this process can give a frequency dependence to the measured capacitance which will lead to distortion of the depth profile. Great care should be exercised in selecting instrumentation which operates at frequencies which avoid this cause of error. When monitoring samples implanted with the usual dopant species (shallow levels) it is desirable to operate at high frequencies.

Gordon [52] has extended the depth which may be probed by using a MOS structure which permits doping profiles to be measured up to the insulator-semiconductor interface for the dose range $10^{11}$-$10^{12}$ ions/cm$^2$. The measurement accuracy is quoted as 10% although a difference as large as 23% was recorded between the calculated and actual doses in control samples. Pulsed capacitance measurements [34] on MOS structures enable greater depths to be probed than under d-c biassing. Doping profiles to depths greater than the depletion width also have been achieved in GaAs by Ambridge and Faktor[53] using an electrochemical technique which permits controlled dissolution of the layer whilst continuously monitoring the capacitance.

## 4. Limitations of Electrical Measurements

The assumption is made that good experimental practices have been followed (see Section 2.6), and that experimental conditions have been chosen to avoid introducing errors due to carrier freeze out, degeneracy and non-linear dependences upon dose. Despite these precautions large discrepancies in the estimated dose and dose uniformity may still occur due to impurity and defect interactions thermally activated during the anneal schedule. In this section examples will be given which show the magnitude of these discrepancies and the conditions under which they may occur.

Beam heating of the sample during implantation and the use of multi-temperature anneal schedules may both be the cause of discrepancies in the measured value of the sheet resistance of ion-implanted layers. Smith et al. [54] have shown that even after a high-temperature anneal the sheet resistance may not have a unique value, depending only upon ion dose and energy. Figure 16 shows the dependence of the sheet resistance upon anneal

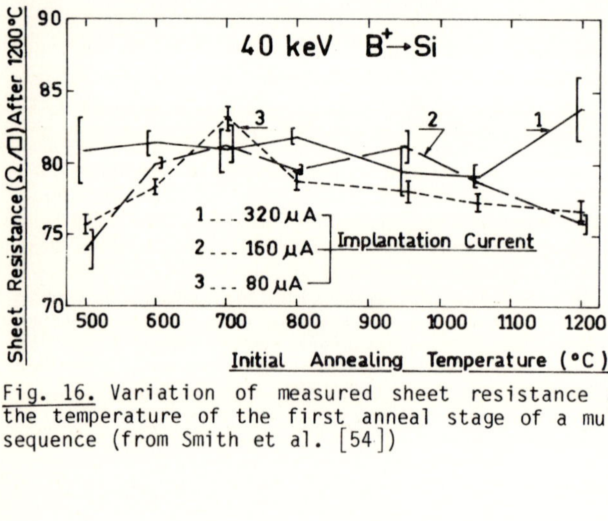


Fig. 16. Variation of measured sheet resistance after 1200°C anneal with the temperature of the first anneal stage of a multi-temperature annealing sequence (from Smith et al. [54])

temperature for 21 similar (111) silicon wafers, each implanted with $1\times10^{15}$ $B^+/cm^2$ at 40 keV, in a machine using mechanical scanning. The wafers were implanted in three batches, each batch of seven wafers was implanted with the same beam flux (irradiance, in units of $W/m^2$), and hence each wafer in a batch experienced the same, but unknown, temperature. The wafers were all annealed for 30 minutes, but within each batch, the wafers had a different thermal cycle. The first wafer was annealed at 500°C for 30 minutes, cooled and then heated to 60C°C and cooled again. It followed this cycle up to 1200°C. The second wafer started the cycle at 600°C, the third at 700°C and the remaining wafers at 800°C, 950°C, 1050°C and 1200°C. In the figure, the final value of sheet resistance, measured after the 1200°C anneal, is plotted against the temperature of the first anneal.

Each wafer was measured with a Four Point Probe [20] at 81 different positions and sheet resistance maps were plotted. In Figure 16, the variance for each wafer is shown as a vertical bar. These bars do not represent experimental errors, but show that the sheet resistance and the variance depend upon the implantation irradiance and the anneal history. Variations of five per cent and larger are evident. For these samples an anneal at 1050°C appears to give the best uniformity and a value of sheet resistance which is only weakly dependent upon the beam irradiance. Other measurements by Smith [54] on $As^+$, $P^+$ and $Sb^+$ implanted wafers, show even larger variations and confirm that annealing to 1200°C does not guarantee a unique value for the measured sheet resistance.

Discrepancies in the measured sheet resistance may also be introduced by the use of shorter time and lower-temperature anneals, which appear to be necessary to achieve device geometries suitable for VLSI. Recent measurements by Scovell and co-workers[55] have highlighted this problem. Table 5 shows the sheet resistance of silicon implanted with $6\times10^{15}$ $As^+/cm^2$ at 150 keV and annealed at 700°C for various times.

Table 5. Dependence of sheet resistance upon anneal conditions for silicon implanted with $6\times10^{15}$ $As^+/cm^2$ at 150 keV (Scovell et al. [55])

| Anneal conditions | $R_s$ (Ω/□) | % activity |
|---|---|---|
| Theoretical | 26 | 100 |
| 700°C,pulse | 31.7 | ⩾ 85 |
| 700°C,30 min | 45.7 | ⩾ 58 |
| 700°C,1000 min | 52.3 | ⩾ 48 |

Other experiments by Scovell et al. [56] have confirmed that beam heating during implantation can be a cause of gross variations in sheet resistance, even after annealing. Silicon wafers were implanted with $As^+$ at 150 keV, using electrostatic beam scanning to achieve a uniform dose of $6\times10^{15}$ $As^+/cm^2$. The wafers were implanted at different beam currents, as shown in Table 6, and then were annealed in flowing nitrogen for 30 minutes at 600°C. It will be seen (wafers a→d) that the measured values differ between wafers, with the lowest resistance being measured on the

Table 6. Measured sheet resistance of wafers annealed at 600°C for 30 minutes after implantation at different beam current densities with $6x10^{15}$ $As^+/cm^2$ at 150 keV (from Scovell et al. [56])

| Sample | R (Ω/□) | Beam current(μA/cm²) | Beam power (mW/cm²) |
|---|---|---|---|
| a | 118 | 1.25 | 188 |
| b | 64 | 6.25 | 936 |
| c | 60 | 0.65 | 98 |
| d | 38 | 0.05 | 7.5 |
| e | 37 | 0.5 | 75 |

wafer (d) implanted at the lowest beam current. Wafer (e) was a control which was mounted on an efficient heat sink.

As part of an exercise to establish the limitations of four-point probe measurements for the evaluation of doping uniformity, Perloff and Markert [57] have studied the long-term stability of the sheet resistance of wafers implanted with $5x10^{14}$ $B^+/cm^2$ at 150 keV and annealed at 950°C for 30 minutes. They find a time dependence both in the measured sheet resistance and the uniformity of the sheet resistance across the wafer. After about 150 days a saturation value is approached, akin to an "aging" process, but upon heat treatment at 450°C for 30 minutes in nitrogen, the sheet resistance returns to its original value. Their results, plotted over a period of 350 days, are shown in Fig. 17.

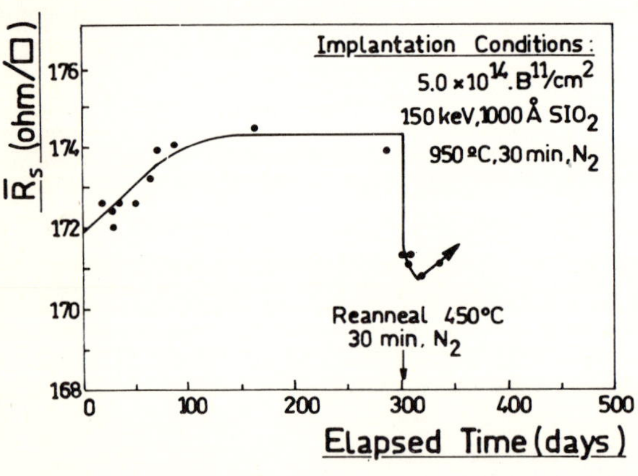


Fig. 17. Variation of the mean sheet resistance as a function of time for a single silicon wafer implanted with $5.0x10^{14}$ $B^+/cm^2$ at 150 keV and annealed at 950°C for 30 min in nitrogen (from Perloff and Markert [57])

The results just quoted have been obtained from sheet resistance measurements. However, all electrical techniques may show changes due to impurities and residual lattice defects being thermally activated. In practice, therefore, the choice of measurement technique will be based upon the level of sensitivity required to detect the implanted dose. The techniques which have been reported are listed in Table 7.

Table 7. Electrical measuring techniques

| Dose range($cm^{-2}$) | Measurement technique | Comments |
|---|---|---|
| High doses $10^{14}$-$10^{16}$ | Sheet resistance: Four-point probe | Highly developed. Good reproducibility, can be better than 0.5%. Errors associated with geometry of probe and wafer. |
| $10^{13}$-$10^{16}$ | Sheet resistance: Van der Pauw | Wafer processing required. Geometric errors small. |
| Wide dose range | Spreading resistance | Relatively unproven, errors large. Simple with very good spatial resolution. |
| Medium dose $10^{12}$-$10^{14}$ | Capacitance: p-n junction, Schottky diode | Wafer processing required. Good spatial resolution. |
| Low doses $5x10^{10}$-$10^{13}$ | Device parameters: Threshold voltage, saturated current, pinch-off voltage | Data interpretation difficult. Extensive wafer processing. Good spatial resolution. Very sensitive to geometric tolerances. |

## 5. Standards Exercises

It is, of course, normal practice for laboratories to run their own standards exercises but, unfortunately, the results are seldom published. However, a detailed and carefully controlled international standards exercise was carried out during April to June 1980 and the results were reported by Gan and Perloff [58] at the Third International Conference on Ion Implantation Equipment and Techniques held in Kingston, Ontario in July 1980. The stated objectives were the determination of (i) a typical value for the sheet resistance of silicon implanted with $5x10^{14}$ $B^+/cm^2$ at 150 keV, (ii) dose and dose uniformity variations between implanters and (iii) differences in the performance of the implanters due to the scanning system. The authors conducted a survey involving 85 implanters at 44 companies in the USA and Europe. The wafers used in this exercise were supplied, annealed and measured in the co-ordinators, own laboratory. Maintaining this level of control over wafer processing ensured that the differences in the measured sheet resistance could only be caused by variations in the performance of the implanters.

Gan and Perloff conclude that (i) dose variations between implanters may be much greater than 1%, (ii) implanters employing mechanical scanning generally show smaller variations between machines than those employing electrostatic or hybrid scanning, (iii) mechanical and hybrid systems generally give a higher dose, (iv) sheet resistance non-uniformity ($\pm 1\sigma$) values of better than ± 1% can be achieved irrespective of scanning method and (v) a sheet resistance value of 169.7 Ω/□ may be associated with a dose of $5x10^{14}$ $B^+/cm^2$ implanted at 150 keV. Values of sheet resistance range from 150.6 Ω/□ to 192.6 Ω/□ whilst the non-uniformity ranges from ± 0.37% to ± 4.77%, with a median at ± 0.72%. More details of this investigation are given in the chapter,by Current, Perloff and Gutai.

As a result of this survey and other, unpublished work, it is clear that ion implanters do drift out of specification, causing dose errors not just of a few percent but as large as tens of percent. Standards exercises serve to improve our understanding of ion beam systems and also enable checks to be made of the performance of the equipment. For these reasons, standards exercises and surveys are to be encouraged. In the future, there would be value in running an exercise involving $As^+$ or $Sb^+$ implantation into silicon when the dose could be estimated from indirect electrical measurements ($R_S$) and compared with absolute measurements by Rutherford backscattering [39].

Acknowledgements

It is a pleasure to acknowledge the time so freely given and the great value of discussions with B.J. Smith, S. Thompson, P.A. Leigh, K. Nicholas, G. McGinty, P. Scovell and A.C. Glaccum. The author wishes to thank B.J. Smith for data received before publication, L. Huaume for art work and Mrs. A. North for typing the manuscript.

References

1. G. Dearnaley, J.H. Freeman, R.S. Nelson, and J. Stephens, Ion Implantation (North Holland, Amsterdam 1973)
2. G. Carter, J.N. Baruah, and W.A. Grant, Rad. Effects 16, 107 (1972) and Rad.Effects 16, 101 (1972)
3. J. Bøttiger, J.A. Davies, W.A. Grant, and K.B. Winterbon, Rad. Effects 11, 61, 69 (1971)
   J. Bøttiger, H. Wølder Jongensen, and K.B. Winterbon, Rad. Effects 11, 133 (1971)
4. J.H. Freeman, in: Applications of Ion Beams to Materials, Warwick, 1975. Inst.of Phys. Conf. Ser. No.28.
5. P.L.F. Hemment, Proc. of Second Inter. Conf. on Low-Energy Ion Beams, Bath, 1980. Inst. of Phys. Conf. Ser. No. 54.
6. J.E.E. Baglin, in: Proc. Conf. on Ion Beam Surface Layer Analysis (Plenum Press, New York) Vol. 1, p.313, 1976.
7. J. L'Ecuyer, J.A. Davies, and N. Matsunami, Nucl. Instr. and Meth. 160, 337 (1979)
8. D.S.Perloff, F.E.Wall, and J. Conragon, J.Electrochem. Soc.124, 582 (1977)
9. P.L.F. Hemment, in: Applications of Ion Beams to Materials, Warwick, 1975, Inst. of Phys. Conf. Ser. No. 28.
10. B.J. Smith and J. Stephen, Theoretical Calculations of Resistance of n- and p-Type Implantations in Silicon, AERE-R7097.
11. J.F. Gibbons, W.S. Johnson, and S.W. Mylroie, Projected Range Statistics (2nd Edition, Halsted Press, Wiley, NY 1975)
    B. Smith, Ion-Implanted Range Data for Si and Ge Device Technology, Research Studies 1978.
12. S.M. Sze, Physics of Semiconductor Devices (2nd Edition, Wiley-Interscience 1981)
13. F.M. Smits, Bell Syst. Tech. J. 37, 711 (1958)
    Burger and Donovan, Fundamentals of Silicon Integrated Device Technology, Vol. 1, p.319 Prentice Hall 1967.

W.M. Bullis, Standard Measurements of the Resistivity of Silicon by the Four-Probe Method, NBSIR 74-496, August 1974.
Measuring Resistivity of Silicon Slices with a Colinear Four-Probe Array, ASTM F84-73.
14. M.A. Logan, Bell Syst. Tech. J. 40, 885 (1961)
15. L.J. Swartzendruber, NBS, Tech. Note 199 (1964)
16. R. Rymaszewski, J. Phys. E, 2, 2, 170 (1969)
17. L.B. Valdes, Proc. Inst. Radio Engrs. 42, 420 (1954)
18. D.J. Perloff, J. Electrochem. Soc. 123, 1745 (1976)
19. F.E. Wahl, and D.J. Perloff, Proc. 8th Int. Conf. on Electron & Ion Beam Science & Technology, Seattle, WA., May 1978.
20. B.J. Smith, and J. Stephen, Revue de Phys. Appl. 12, 493 (1977)
21. B.J. Smith, J. Stephen, and G.W. Hinder, Measurement of Doping Uniformity in Semiconductor Wafers, AERE-R7085.
22. as Ref. 19.
23. D.S. Perloff, F.E. Wahl, and J.T. Kerr, Proc. Seventh Int. Conf. Electron and Ion Beam Science and Technology, 1976.
D.S. Perloff, F.E. Wahl, and J.D. Reimer, Solid State Technol. 20, 31 (Feb. 1977)
24. L.J. Van der Pauw, Philips Res. Repts. 13, 1 (1958)
25. W. Versnel, Solid-State Electr. 21, 1261 (1978)
26. J.M. David, Nat. Bur. Stand. Spec. Publ. 400-19, 44 (April 1976)
27. R. G. Mazur and D. H. Dickey, J. Electrochem. Soc. 113, 255 (1966)
J. Kudoh, J. Electrochem. Soc. 123, 1751 (1976)
Proc. NBS Spreading Resistance Sympos. NBS Spec. Publ. 400-10 (1974)
28. R. W. Bicknell, J. Phys. D 9, 1953 (1976)
29. H. Glawischnig, K. Hoerschelmann, W. Holtschmidt, and W. Wenzig, Nucl. Inst.& Meth. 189, 291 (1981)
30. As Ref. 2.
31. As Ref. 4.
32. B.G. Streetman, Solid State Electronic Devices, Prentice-Hall, New Jersey 1980.
33. P.D. Scovell, private communication.
34. G.E.Schmid, Nucl. Inst.& Meth. 189, 219 (1981)
35. S.M. Sze, Physics of Semiconductor Devices (2nd Edition, John Wiley 1981), p.477
36. A.C. Glaccum, private communication.
37. A.W. Livingstone, P.A. Leigh, N. McIntyre, I.P. Hall, J. A. Bowie, and P.J. Smith, Solid-State Electronics, to be published.
38. C.W. White, and W.H.Christie, Sol. Stat. Tech. 23, 109 (Sept. 1980)
39. W.K.Chu, J. W. Mayer, and M.A. Nicolet, Backscattering Spectrometry (Academic Press, New York 1978)
40. D.A. Jamba, Nucl. Inst. & Meth. 189, 253 (1981)
41. R.C. Goodfellow, A. C. Carter, R. Davis, and C. Hill. Electr. Lett. 14, 328 (1978)
42. R.L. Petritz, Phys. Rev. 110, 1254 (1958)
43. N.G.E. Johansson, J.W. Mayer, and O.J. March, Solid-State Electr. 13, 317 (1970)
44. H.D. Barber, H.B. Lo, and J.E. Jones, J. Electrochem. Soc. 123, 1404 (1976)
45. A. Yamamoto and C. Memura, Electr. Lett. 18, 63 (1982)
46. A. Smith, Electr. Lett. 4, 332 (1968)
47. J. Hilibrand and R.D. Gold, RCA Rev. 21, 245 (1960)
48. J. A. Copeland, IEEE Trans. Electron Devices ED-16, 445 (1969)
49. G.L. Miller, IEEE. Trans. Electron Devices, ED-19, 1103 (1972)
50. C.P. Wu, E.C. Douglas, and C.W. Mueller, IEEE Trans. Electron. Devices ED-22, 319 (1975)
51. E.H. Rhoderick, Metal-Semiconductor Contacts (Oxford University Press 1978)

52. B.J. Gordon, IEEE Trans. Electron Devices ED-27, 2268 (1980)
53. T. Ambridge and M.M. Faktor, Electr. Lett. 10, 10, May (1974)
54. B.J. Smith, J. Stephen, D. Chivers, and M. Fisher, Annealing of Ion Implanted Silicon, AERE-R-9868.
55. P.D. Scovell, private communication.
56. P.D. Scovell, ESSDERC 81, Toulouse 1981.
57. D.S. Perloff and M.J. Markert, Microelectronic Measurement Technology Seminar, San Jose, CA, March 1981.
58. J.N. Gan and D.S. Perloff, Nucl. Inst. and Meth. 189, 265 (1981)
59. K.H. Nicholas, Acta Electronica 19, 95 (1976)

# Wafer Mapping Techniques for Characterization of Ion Implantation Processing

M.I. Current, D.S. Perloff, and L.S. Gutai

Philips Research Laboratories Sunnyvale, Signetics Corporation
Sunnyvale, CA 94086, USA

## Abstract

This paper discusses the use of sheet resistance, capacitance-voltage and device parameter mapping techniques to characterize the dose accuracy and doping non-uniformity of ion-implantation equipment. Examples are provided which illustrate the use of these techniques for establishing performance norms for production implantation equipment, for diagnosing equipment malfunctions and for characterizing phenomona such as wafer heating effects in high current implanters.

## 1. Introduction

Reliable exploitation of the advantages of ion implantation for IC device fabrication, or for some of the more recent metallurgical applications, requires careful control of ion species, energy and dose level. Particularly in the case of IC fabrication, additional controls must be exercized on such conditions as wafer temperature during implantation and the uniformity of the implanted dose on larger diameter wafers. Many of the characterization tools which have been used to develop the technology of ion implantation processing, such as SIMS, RBS, TEM and spreading resistance-bevel profiling, involve destructive testing of a small portion of the implanted material. These "sampling" methods are not generally useful for efficient monitoring of either implant uniformity over a single wafer or wafer-to-wafer variations in the large number of wafers which are typical in a modern IC production line.

The need for efficient, accurate tools which can provide information on the spatial variation of implantation process parameters on a large volume of test wafers has led to the adaptation of several surface conductivity measurement techniques as full wafer mapping tools. Of particular importance are sheet resistance measurements, using either a four-point probe or van der Pauw patterns, and various capacitance-voltage techniques such as impurity profiling. Applications of these tools include evaluation of ion implantation equipment, process development, on-line monitoring of production process conditions and implanter calibration. In addition, these tools have proved to be sensitive to new process conditions, such as wafer heating during high dose implantation and the temperature homogeneity of advanced annealing equipment.

## 2. High-Dose Characterization: Sheet Resistance Measurements

Of the post-implant techniques commonly used for characterizing dose accuracy, sheet resistance measurements are perhaps the most easily implemented. Although this method gives an indirect measure of the total dose, no other technique now available offers the repeatability, sensitivity and convenience required to detect and display routinely small spatial variations in doping.

The sheet resistance $R_s$ of an impurity layer of junction depth $z_j$ is defined by:

$$R_s^{-1} = \int_0^{z_j} n(z)e\mu(n)dz \qquad (1)$$

where n(z) represents the concentration of activated impurities at a depth z, e is the electronic charge and $\mu(n)$ represents the mobility which, in general, is a function of impurity concentration. According to Eq. 1, $R_s$ will be a complicated function of dose, energy and annealing conditions. This is because both n(z) and $\mu(n)$ depend on the degree of electrical activity and the extent to which damage is removed by annealing. For fixed energy and annealing conditions, however, $R_s$ and dose are found to be related by a simple empirical expression which enables one to deduce dose values from sheet resistance measurements in a straightforward manner.

Two vehicles are used for sheet resistance measurements of junction-isolated implanted impurity layers: the four-point probe array (Fig. 1) and the van der Pauw sheet resistor (Fig. 2). The four-point probe is used directly on the implanted layer, without recourse to the lithographic or metallization steps employed with van der Pauw resistors. Its applicability, however, is generally restricted to dose levels $\geq 10^{13}$ cm$^{-2}$, since reproducible ohmic contacts between the probe tips and the silicon wafer are difficult to achieve at lower dose values. In contrast, the van der Pauw resistor may be used at dose values down to about $10^{12}$ cm$^{-2}$ by employing heavily doped contact regions to carry current between the test probes and the implanted layer.

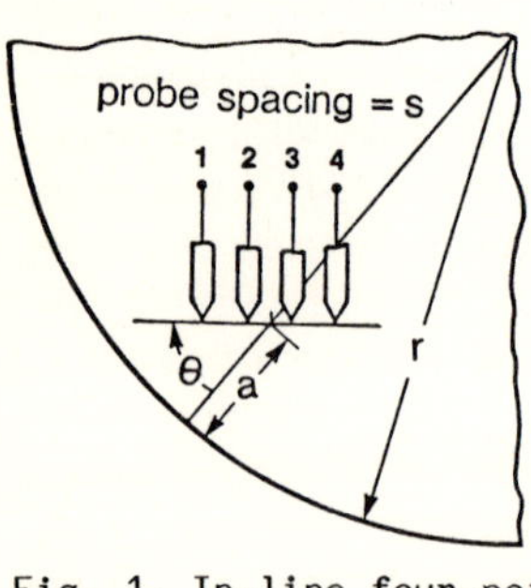


Fig. 1. In-line four-point probe

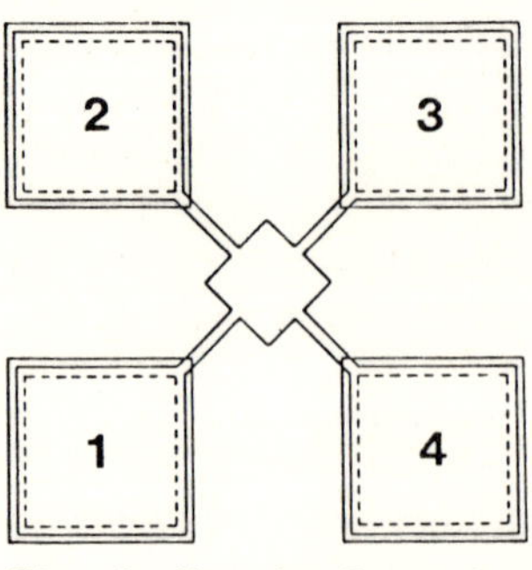


Fig. 2. Van der Pauw sheet resistor

## 2.1 The van der Pauw Resistor [1]

Van der Pauw sheet resistors of the type represented in Fig. 2 are readily fabricated on silicon wafers using standard photolithographic methods. A thick stopping layer of silicon dioxide is ordinarily used to define the resistor geometry. For a nominally symmetric van der Pauw structure, such as that shown in Fig. 2, the sheet resistance is given by:

$$R_s = \frac{\pi}{\ell n2}\left(\frac{R_a + R_c}{2}\right) \tag{2}$$

where $R_a = V_{23}/I_{14}$ and $R_c = V_{34}/I_{12}$ are obtained by employing the test configurations shown in Fig. 3. Use of two independent test configurations in conjunction with Eq. 2 eliminates geometric sources of error associated with photolithographic and etching variability.

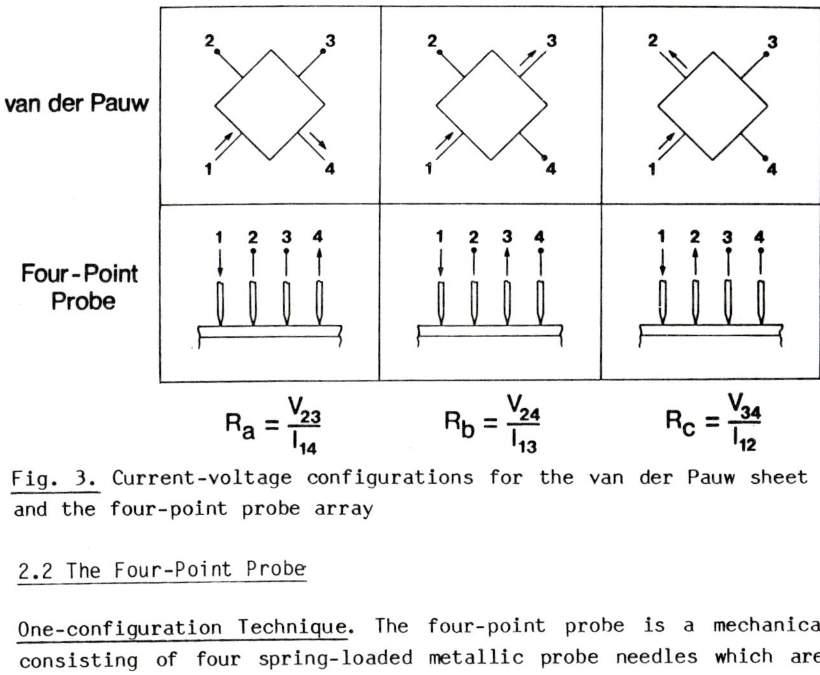


Fig. 3. Current-voltage configurations for the van der Pauw sheet resistor and the four-point probe array

## 2.2 The Four-Point Probe

One-configuration Technique. The four-point probe is a mechanical device consisting of four spring-loaded metallic probe needles which are brought into contact with the surface of the silicon wafer. Ordinarily, a constant current is introduced through the outer probes and the voltage drop measured across the two inner probes of an in-line four-point probe, as shown schematically in configuration "a" of Fig. 3. If the average probe spacing is at least twice the thickness of the impurity layer, the sheet resistance will be proportional to $R_a = V_{23}/I_{14}$. One may then write:

$$R_s = k_a R_a \tag{3}$$

where, for a circular wafer, the geometrical correction factor $k_a$ will be a function of the parameters r/s, a/s, and $\theta$, as shown in Fig. 1 [2].

A mechanical four-point probe assembly will almost always exhibit small random variations with respect to its nominal probe spacings. Thus, even if the quantities r, s, θ and a are accurately known, variations in $k_a$ will occur from measurement to measurement which will have a proportionate effect on the values computed for $R_s$. This behavior becomes particularly important when evaluating highly uniformly doped layers, since the variability in $R_s$ associated with the implantation step may be comparable to that induced by probe spacing variations. In addition, mechanical devices such as clips and clamping rings may mask the edge of the wafer during implantation, thereby invalidating the assumption of a circular wafer. These sources of error make it difficult to distinguish actual implant non-uniformity from measurement-related contributions. They can, however, be virtually eliminated by employing the technique which will now be described.

Two-configuration Technique. Figure 3 illustrates two other useful permutations of the four-point probe current and voltage leads. Use of either $R_b$ or $R_c$ in conjunction with $R_a$ makes it possible to determine $k_a$ "electrically" when the probe tips are oriented along the radius of a circular wafer. For this case [2,3]:

$$k_a = \frac{\pi}{\ln 2}\left(1 + \frac{1}{\xi}\right) f(\xi) \tag{4}$$

where

$$\xi = \frac{R_a}{R_c} = \frac{R_a/R_b}{R_a/R_b - 1} \tag{5}$$

and $f(\xi)$ is obtained by solving the transcendental equation:

$$\left(\frac{\xi - 1}{\xi + 1}\right)\frac{\ln 2}{f} = \cosh^{-1}\left(\frac{1}{2} \exp \frac{\ln 2}{f}\right) . \tag{6}$$

For measurement convenience it is useful to choose configurations "a" and "b", rather than "a" and "c", to determine the parameter ξ. This is because the resistance values $R_a$ and $R_b$ are comparable in magnitude and significantly larger than $R_c$. To an accuracy of at least 0.05%, we can represent $k_a$ vs. $R_a/R_b$ as a second order polynominal:

$$k_a = -14.696 + 25.173(R_a/R_b) - 7.872(R_a/R_b)^2 \tag{7}$$

where $R_a/R_b$ is in the range 1.20 to 1.32.

The requirement that the probe array must be parallel to the radius is readily satisfied in most manual wafer probing systems. For detailed uniformity mapping, however, it is useful to employ an automated X-Y stepping table. The angle θ, therefore, will depend on the location of the probe array on the wafer. Fortunately, it is possible to use the two-configuration measurement technique without appreciable error, even when the probe array is not oriented parallel to the wafer radius (θ = 0). For example, at a distance of 5 probe spacings from the wafer edge, the error will be $\leq$ 0.1% if one employs Eq. 7 to compute the correction factor $k_a$.

Comparison of the Two Techniques. Figure 4 provides experimental data which illustrate the advantages of the two-configuration measurement technique over the more traditional one-configuration method. The periodic variation in $R_S$ along the diameter of an electrostatically scanned, boron-implanted wafer, is clearly seen in the two-configuration data of Fig. 4 [4]. In contrast, the one-configuration data include the effects of variable probe spacings.

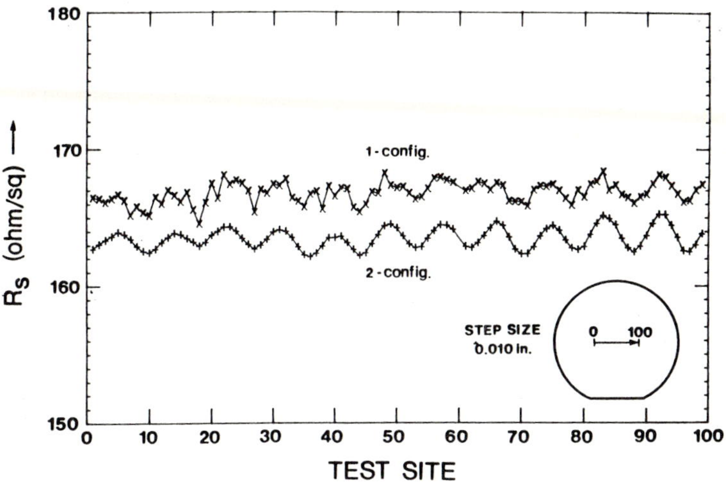


Fig. 4. Variation in sheet resistance along the diameter of a boron-implanted wafer illustrating the improved precision associated with the two-configuration measurement technique

## 3. Sheet Resistance Wafer Mapping

### 3.1 Performance Norms for Dose Accuracy and Uniformity

Implanter Survey. A set of performance norms for dose accuracy and uniformity can be extremely valuable as a basis for cross-calibrating and monitoring the performance of ion implantation equipment. One method of developing such norms is to perform a survey in which silicon wafers are implanted under nominally identical conditions on a cross-section of production implanters. Such a survey, involving eighty-five implanters at forty-four companies in the United States and Europe, was conducted during the months of April, May and June, 1980 [5]. Its purpose was to obtain the following information:

i) "Typical" values for sheet resistance and non-uniformity for a benchmark set of implantation conditions.

ii) Implanter-to-implanter dose and non-uniformity variations.

iii) Possible performance differences between types of scanning systems.

Experimental Conditions. A blank 76 mm diameter n-type silicon wafer with a 1050 Å ± 50 Å thick layer of thermally-grown silicon dioxide was supplied for each implanter tested. After implantation with boron at a nominal dose of $5.0 \times 10^{14}$ cm$^{-2}$, energy of 150 keV and tilt angle of 7°, the wafers were annealed for 30 minutes at 950°C in a dry nitrogen ambient.

Following removal of the oxide layer in hydrofluoric acid, the implanted wafers were measured at 118 uniformly distributed test sites using a precision mechanical four-point probe array in conjunction with a computer-controlled d.c. test system. As described below, an equi-value contour map was generated for each wafer, and the mean $\bar{R}_s$ and standard deviation $\sigma(R_s)$ computed.

To assure that sheet resistance values obtained from the four-point probe technique were accurate and repeatable, the following precautions were taken:

i) The two-configuration technique described above was used to compute the correction factor at each point on the wafer, effectively removing geometrical sources of error associated with probe-spacing variations and the circular wafer boundary.

ii) Testing was performed on groups of 8-12 wafers. A "control" specimen was included with each group for which the repeatability of $\bar{R}_s$, $\sigma(R_s)$ and the appearance of the contour map were carefully monitored.

iii) Four-point probe sheet resistance data from a blank test wafer were shown to agree to within 1% with the data obtained from a wafer containing van der Pauw resistors.

Survey Results. The frequency distributions for the mean wafer sheet resistance $\bar{R}_s$ and non-uniformity $\sigma(R_s)/\bar{R}_s$ are shown in Figs. 5 and 6, respectively. Values of $\bar{R}_s$, which range from 150.6 to 192.2 ohms/sq, are distributed about a grand mean of 169.7 ohms/sq with a standard deviation of $\sigma(\bar{R}_s)$ = 8.1 ohms/sq (Fig. 5). According to Fig. 6, values for wafer

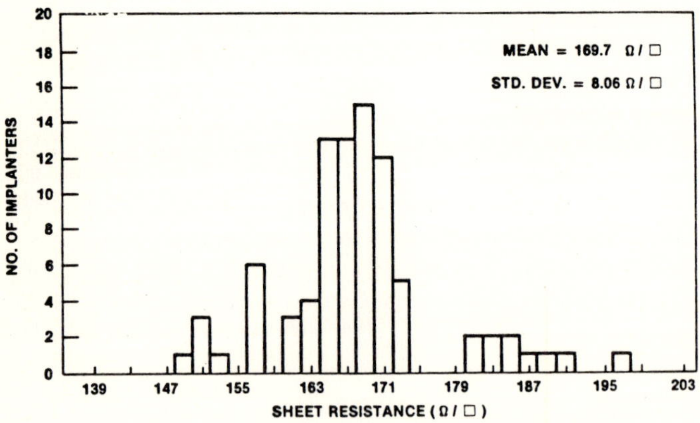


Fig. 5. Frequency distribution of the 85 implanters surveyed showing the number of implanters vs. mean sheet resistance $\bar{R}_s$

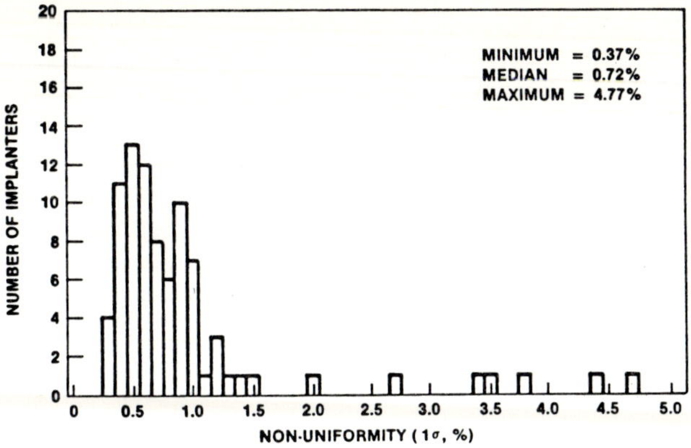


Fig. 6. Frequency distribution of the 85 implanters surveyed showing the number of implanters vs. the percentage non-uniformity $\sigma(R_s)/\overline{R}_s$

non-uniformity range from 0.37% to 4.77%, with the median at 0.72%. Of the implanters evaluated, only eight exhibited what might be considered excessive non-uniformity, i.e., $\sigma(R_s)/\overline{R}_s \geq 1.5\%$.

## 3.2 Implanter Diagnostics

Maps of sheet resistance variations across a single wafer have proved to be highly effective tools for evaluation of the scan uniformity of an ion implantation machine. These maps can be used not only to verify that a particular implanter unit is functioning properly, but to diagnose a wide variety of malfunctions as well.

In Figs. 7 through 14, four-point probe sheet resistance contour maps are presented which demonstrate the sensitivity and diagnostic capabilities of the contour mapping technique. In each case the heavy contours represent the wafer mean, $\overline{R}_s$, whereas the lighter contours differ from the mean in increments of 1%. The symbols +, -, and * represent, respectively, test sites for which $R_s \geq \overline{R}_s$, those for which $R_s < \overline{R}_s$ and those for which $R_s$ failed to satisfy a sorting criterion designed to detect anomalous individual readings.

Figure 7 is obtained from a modern electrostatic-scanning system, for which $\overline{R}_s$ = 172.1 ohm/sq and $\sigma(R_s)/\overline{R}_s$ = 0.46%. The small variation in $R_s$ is due to the wafer having been tilted by 7°, with the top oriented away from the scanning plates. The gradient in sheet resistance is caused by more ions per unit area reaching the bottom of the wafer than the top.

Another map that illustrates the high sensitivity of the technique is given in Fig. 8. The pattern shown is for a mechanical-scanning system in which speed variations have caused the sheet resistance to increase from the left side of the wafer to the right side. In this case $\overline{R}_s$ = 169.2 ohm/sq and $\sigma(R_s)/\overline{R}_s$ = 0.66%.

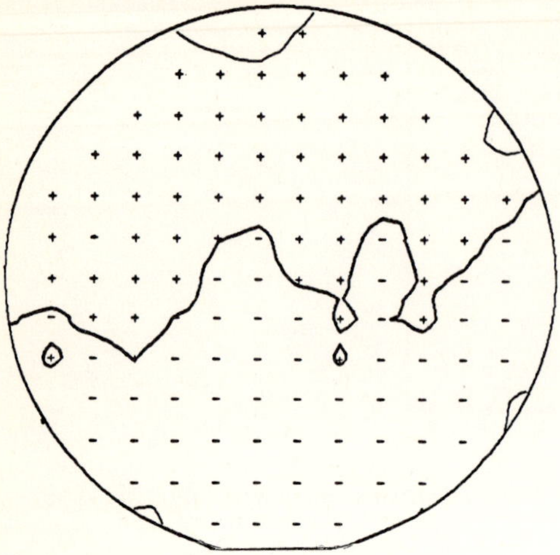

Fig. 7. Sheet resistance contour map showing the effects of a 7° tilt angle in an electrostatic-scanning system ($\overline{R}_S$ = 172.1 ohm/sq, $\sigma(R_S)/\overline{R}_S$ = 0.46%)

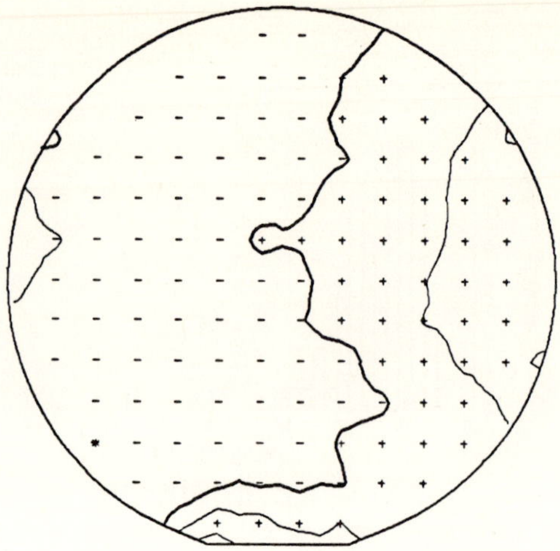

Fig. 8. Sheet resistance contour map showing the effects of speed variation in a mechanical-scanning system ($\overline{R}_S$ = 169.2 ohm/sq, $\sigma(R_S)/\overline{R}_S$ = 0.66%)

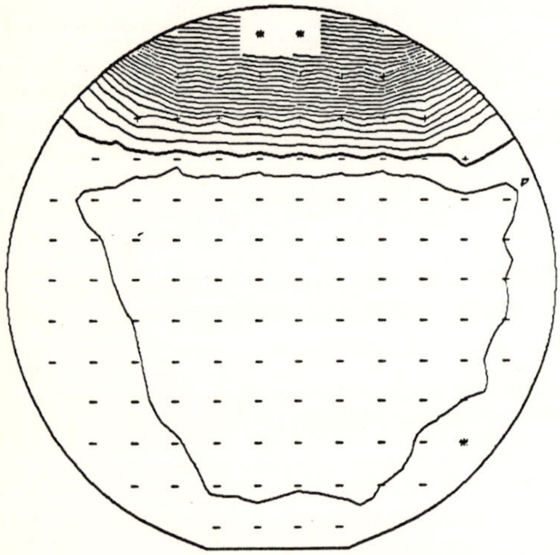

Fig. 9. Sheet resistance contour map showing the effects of underscanning in a hybrid-scanning system ($\overline{R}_S$ = 166.8 ohm/sq, $\sigma(R_S)/\overline{R}_S$ = 3.52%)

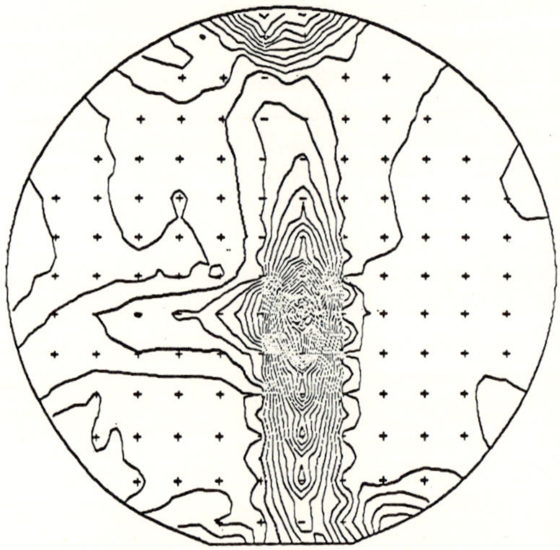

Fig. 10. Sheet resistance contour map showing the effects of neutrals in an electrostatic-scanning system ($\overline{R}_S$ = 154.8 ohm/sq, $\sigma(R_S)/\overline{R}_S$ = 4.44%)

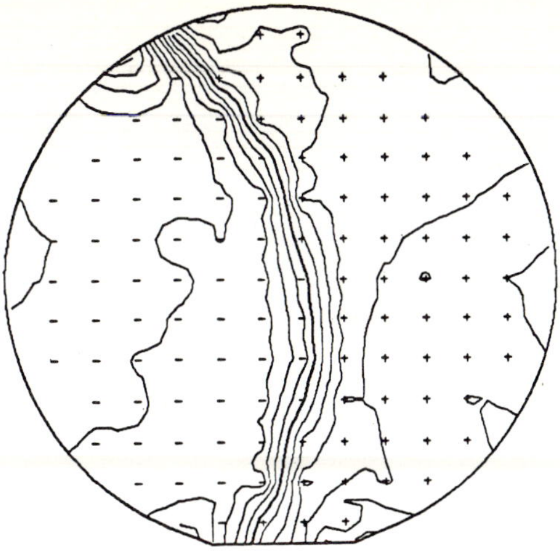

Fig. 11. Sheet resistance contour map showing the effects of failure in the quadrupole lens power supply of an electrostatic-scanning system ($\overline{R}_S$ = 170.6 ohm/sq, $\sigma(R_S)/\overline{R}_S$ = 3.86%)

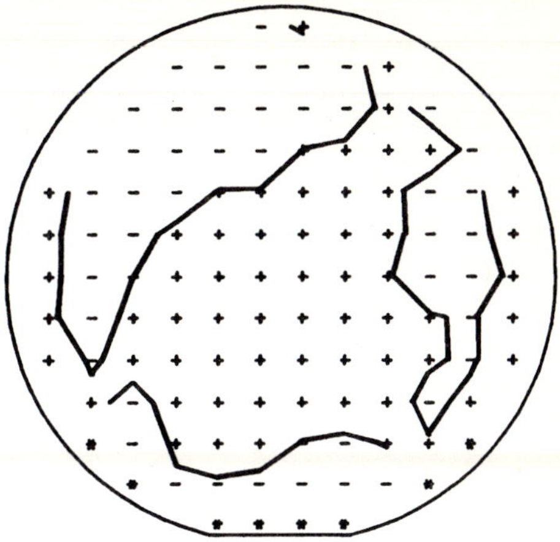

Fig. 12. Representative sheet resistance contour map for the Nova-80 high-current implanter ($\overline{R}_S$ = 36.98 ohm/sq, $\sigma(R_S)/\overline{R}_S$ = 0.29%)

Excessive non-uniformities which occur because of equipment problems are shown in Figs. 9 through 11. Figure 9 is obtained from a hybrid system with electrostatic scan in one direction and rotational scan in the other direction. Here the value of $\overline{R}_S$ = 166.8 ohm/sq and $\sigma(R_S)/\overline{R}_S$ = 3.52%. The stripes parallel to the flat are due to the fast rotational scan which gives good uniformity in that direction. The top of the wafer shows higher sheet resistance values due to underscanning by the electrostatic plates.

An electrostatic-scanning system in which the neutral trap deflection plates are placed before the X and Y scanning plates may produce a pattern such as that shown in Fig. 10. Here the value of $\overline{R}_S$ = 154.8 ohm/sq and $\sigma(R_S)/\overline{R}_S$ = 4.44%. This behavior is due to the formation of neutral atoms within the deflection and Y-scanning plates under poor vacuum conditions.

Figure 11 exhibits a pattern due to the failure of the quadrupole lens power supply in an electrostatic-scanning system, with $\overline{R}_S$ = 170.6 ohm/sq and $\sigma(R_S)/\overline{R}_S$ = 3.86%. The center stripes and large non-uniformity may have been caused by changes in beam shape.

Figures 12 through 14 show representative sheet resistance contour maps for 10 mA arsenic implants obtained from several of the modern high-current ion implanters. These implants were all made through a screen oxide of 300 Å at an energy of 80 keV and a dose of $4 \times 10^{15}$ ions/cm$^2$. The anneal used was 1000°C for 30 min. in a $N_2$ ambient.

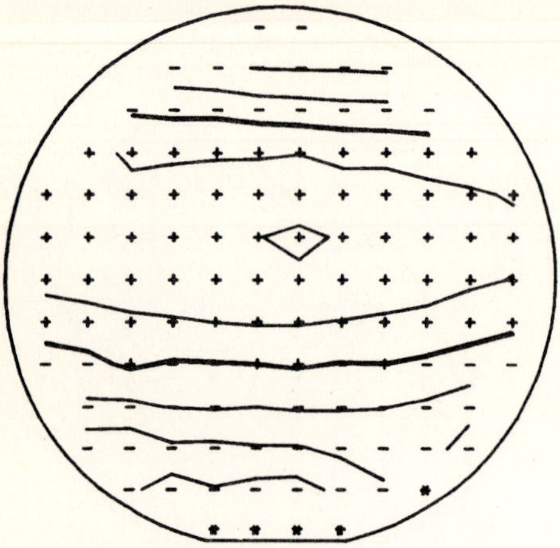

Fig. 13. Representative sheet resistance contour map for the Extrion 80-10 high-current implanter ($\bar{R}_S$ = 40.73 ohm/sq, $\sigma(R_S)/\bar{R}_S$ = 1.52%)

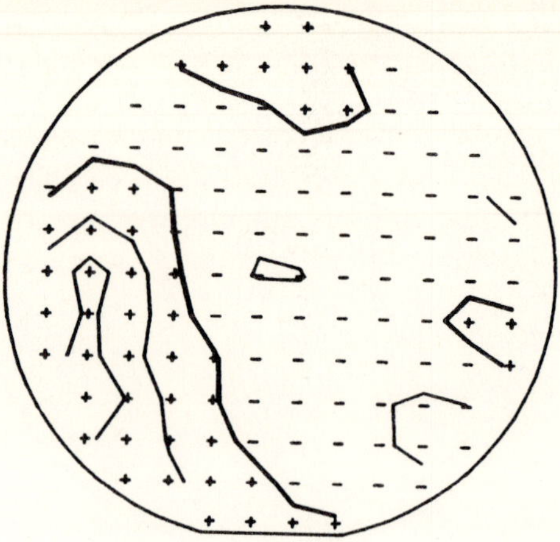

Fig. 14. Representative sheet resistance contour map for the Applied Implant Technology AIT-IIIX high-current implanter ($\bar{R}_S$ = 43.42 ohm/sq, $\sigma(R_S)/\bar{R}_S$ = 0.87%)

The Nova-80 implanter employs a system in which wafers mounted on a rapidly spinning disk are translated across the beam location. The map in Fig. 12 demonstrates a high degree of sheet resistance uniformity, with an average $\bar{R}_S$ = 36.98 ohms/sq and a non-uniformity of $\sigma(R_S)/\bar{R}_S$ = 0.29%. Sheet resistance non-uniformities of less than 1.0% were often observed for the Nova system at beam currents of 10 mA.

The Extrion 80-10 machine also employs an end station in which wafers are mounted on a rapidly spinning disk. However, in contrast to the Nova system, the ion beam is scanned magnetically across the disk radius. As shown in Fig. 13, slight irregularities in the scan waveform of the magnet field strengths produce a sheet resistance variation along the radial direction. For this implant, $\bar{R}_S$ = 40.73 ohms/sq while $\sigma(R_S)/\bar{R}_S$ = 1.52%. Non-uniformities for implants at 10 mA ranged from 1.1 to 2.0% for the Extrion 80-10 machine.

Equipment-related dose variations for high-current implants are also seen in wafer maps obtained from the Applied Implant Technology Model AIT-IIIX implanter, as shown in Fig. 14. $\bar{R}_S$ for the AIT test was 43.42 ohms/sq while $\sigma(R_S)/\bar{R}_S$ = 0.87%. The sheet resistance non-uniformity performance of the AIT-IIIX machines was found to vary between 0.75 to 1.25% for beam currents of 10 mA. The principal scan uniformity patterns for the AIT systems were related to the velocity variations of the carousel assembly in the fast scan direction (left to right in Figs. 8 and 14).

The average sheet resistance values $\bar{R}_S$ for these three implants varied by 6.4 ohms/sq with a typical value of 40 ohms/sq. This amounts to a 15% variation in the total arsenic dose. This variation in dose accuracy for the high-current implanters is considerably larger than the distribution observed for the medium-current implanters in the survey described in the previous section.

### 3.3 Effects of Annealing Conditions

The measured sheet resistance after annealing is determined by the carrier concentration and mobility profile in the implanted layer (See Eq. 1). For anneals in inert or slightly oxidizing ambients, increases in both the anneal time and temperature will produce reductions in the final sheet resistance through increases in the junction depth and dopant activation level. The specific choice of anneal conditions will have a significant

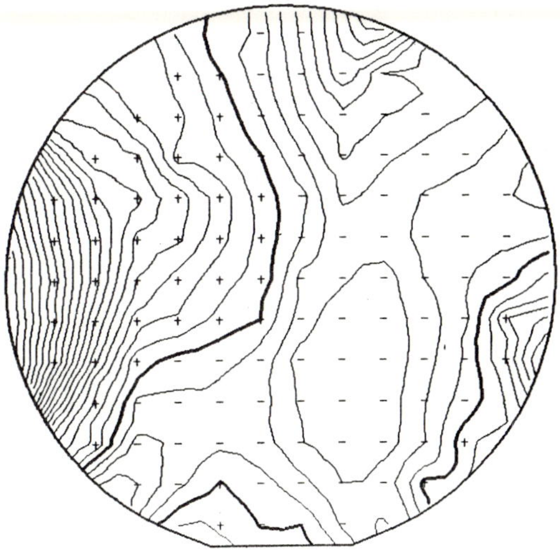

Fig. 15a. Sheet resistance contour map for an arsenic-implanted wafer annealed at 950°C ($\overline{R}_S$ = 105.6 ohm/sq, $\sigma(R_S)/\overline{R}_S$ = 4.0%)

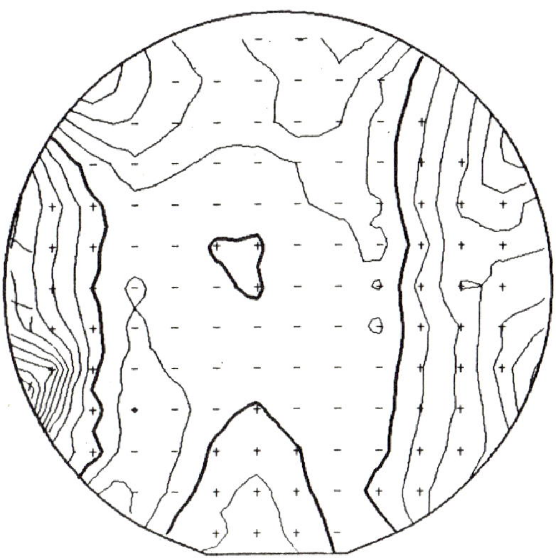

Fig. 15b. Sheet resistance contour map after annealing the same wafer at 1150°C ($\overline{R}_S$ = 84.5 ohm/sq, $\sigma(R_S)/\overline{R}_S$ = 1.7%)

effect on the measured sheet resistance uniformity over a wafer. This is illustrated in Fig. 15, where results are shown for a single arsenic-implanted wafer sequentially annealed at temperatures between 950 and 1250°C. As the anneal temperature was increased from 950 to 1250°C, the average sheet resistance decreased from 105.6 to 60.1 ohms/sq. The non-uniformity in sheet resistance across the wafer also decreased markedly: from 4.0% for the 950°C anneal to 1.1% for the final map after the 1250°C anneal.

Results such as those shown in Fig. 15 emphasize the need to use standardized anneal conditions when comparing the dose level and uniformity performance for a number of different implanters or when the objective is the control of an ongoing production facility.

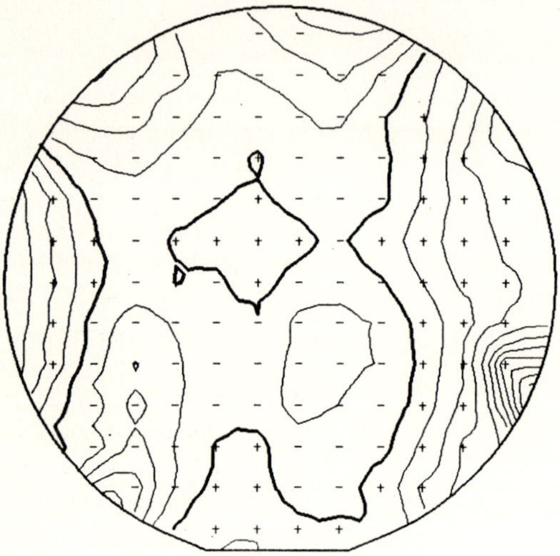

Fig. 15c. Sheet resistance contour map after annealing the same wafer at 1200°C ($\overline{R}_S$ = 74.3 ohm/sq, $\sigma(R_S)/\overline{R}_S$ = 1.5%)

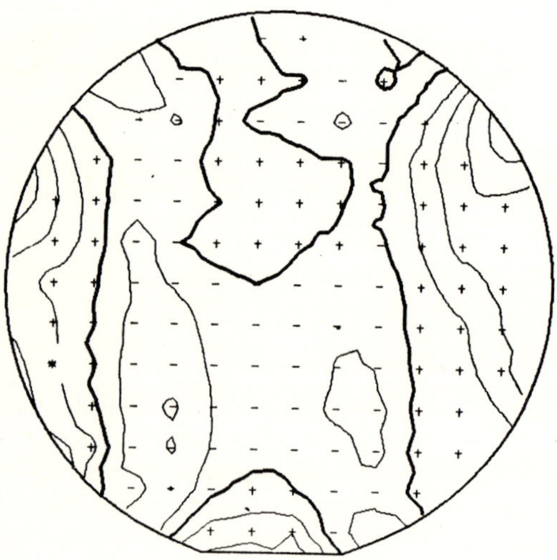

Fig. 15d. Sheet resistance contour map after annealing the same wafer at 1250°C ($\overline{R}_S$ = 60.1 ohm/sq, $\sigma(R_S)/\overline{R}_S$ = 1.1%)

### 3.4 Wafer Heating During Implantation

The challenges which must be overcome in order to use beam currents of 10 mA and higher for IC device manufacture include:

i) absolute dose accuracy and uniformity (discussed in the previous section);

ii) contamination of the wafer surface with metals and dopants from previous ion beam species;

iii) damage to dielectric layers due to high rates of charge buildup during implantation; and

iv) deleterious effects in silicon and photoresist materials from wafer heating.

The effects of wafer heating during implant are the most pressing issues, since beam powers on the wafer now extend up to nearly 2 kW. Implants in silicon at temperatures near 400°C have resulted in a variety of radiation-enhanced annealing effects leading to partial recrystallization of the implanted layer. These efects can be seen in the formation of optical reflection "color bands" [6] which have been studied with TEM and RBS techniques [7]. Moreover, measurements of sheet resistance values for high dose phosphorus implants have shown significant effects for implantation temperatures between 25 and 180°C [8].

Figure 16 shows a sheet resistance contour map for a high dose boron implant using the relatively low beam current of 2 mA. The measured non-uniformity of the sheet resistance is quite good, $\sigma(R_S)/\overline{R}_S$ = 0.42%, but the "bull's-eye" pattern is the result of a radial temperature variation from the center of the wafer to the clamped edges. The implanter used in this example employs a mechanical-scanning, carousel assembly which was very unlikely to produce an actual doping variation of the pattern shown in Fig. 16.

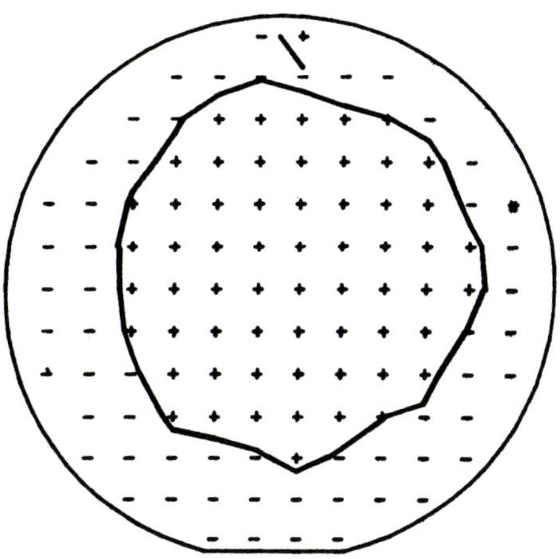

Fig. 16. Sheet resistance contour map for a high dose implant showing the effect of wafer heating ($\sigma(R_S)/\overline{R}_S$ = 0.42%)

Another example of the effects of poor heat transfer through the back of a wafer is shown in Fig. 17 for a high dose arsenic implant at 6.25 mA, where a direct comparison was made between wafers mounted on a carousel plate with a "heat sink" pad beneath the wafer (Fig. 17a) and with a bare metal back (Fig. 17b). Since the two wafers shown in Fig. 17 were mounted on adjacent locations on the same carousel plate, the scan uniformities would be expected to be quite similar. The effects of local temperature gradients over the wafer without good thermal contact to the mounting plate (Fig. 17b) resulted in a sheet resistance non-uniformity of $\sigma(R_S)/\overline{R}_S$ = 8.43%, while the wafer with adequate thermal contact (Fig. 17a) showed a non-uniformity of 0.62%. Sheet resistance variations related to temperature gradients across single wafers have also been reported for high dose phosphorus implants [6].

The ability of sheet resistance mapping to reveal more subtle local heating effects on a single wafer is shown in Fig. 18. This wafer was implanted with arsenic at a beam current of 15 mA with a spinning disk system which used two metal clips at the locations marked with the asterisks. The sheet resistance contours show a pattern which is perpendicular to the fast scan direction. This suggests that the additional heat transfer from the wafer to the disk in the vicinity of the metal clips resulted in local temperature variations over the wafer during implantation at a beam power of 1.2 kW.

These results show that dopant activation is highly sensitive to the conditions existing during implantation for the case of high-dose implants. The degree of ion beam induced damage and annealing has been linked to the wafer temperature and is perhaps also affected by the dose rate. Although this relationship is not well understood at this time, sheet resistance contour mapping is perhaps the most sensitive tool available at

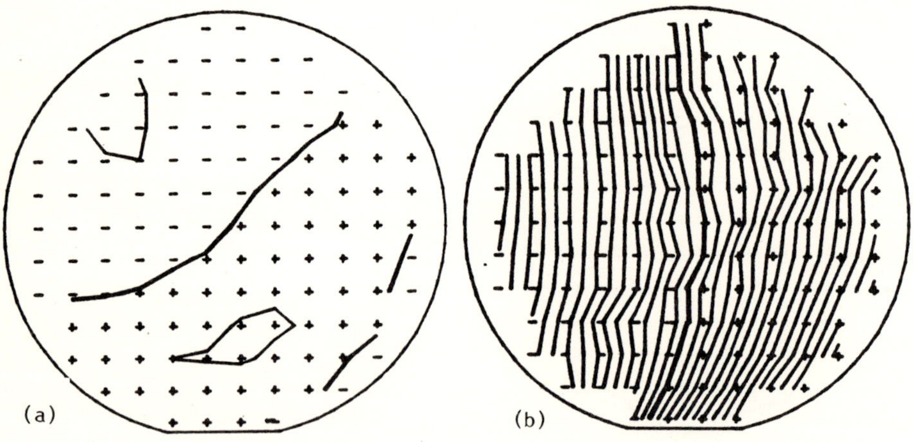


Fig. 17. Comparison of sheet resistance uniformity between two wafers implanted in a high-current implanter (a) with and (b) without a heat sink pad beneath the wafer

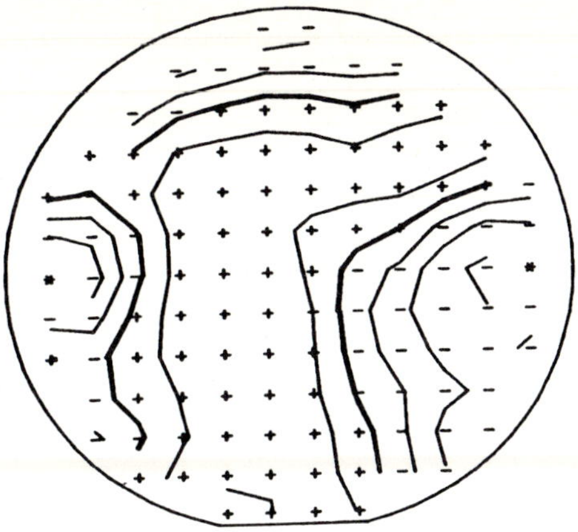

Fig. 18. Sheet resistance contour map for a high-dose implant showing heat transfer from the wafer to the mounting plate at the two clip locations (denoted by *)

this time for rapid-turnaround diagnosis of the onset of wafer heating effects in high-current implantation.

## 4. Low-Dose Characterization

The sheet resistance measurement technique described in Sections 2 and 3 is generally limited to doses > $10^{12}$ cm$^{-2}$ for the van der Pauw resistor (with prediffused pad areas) and > $10^{13}$ cm$^{-2}$ for the four-point probe [5]. Below these dose levels, the technique becomes susceptible to surface effects, junction leakage, excessive measurement current and ambient temperature variations. Moreover, because the measurement requires junction isolation, one becomes increasingly sensitive to variations in background doping as the net dose decreases.

The primary application of implant doses in the range of $10^{10}$-$10^{11}$ cm$^{-2}$ is to altering the threshold voltage of MOS transistors. The following sub-sections will discuss the use of device parameter and capacitance-voltage methods for characterizing implant non-uniformity at these dose levels.

### 4.1 Device Parameters

The threshold voltage of an MOS transistor can be derived from the well-known parallel plate condenser formula:

$$V_T = V_{To} + \alpha(qD_I/C_{ox}) \tag{8}$$

where $V_{To}$ is the threshold voltage of the unimplanted MOS transistor, q is the electron charge, $C_{ox}$ is the oxide capacitance per unit area, and $D_I$ is the net dose that penetrates the oxide layer. The correction fact-

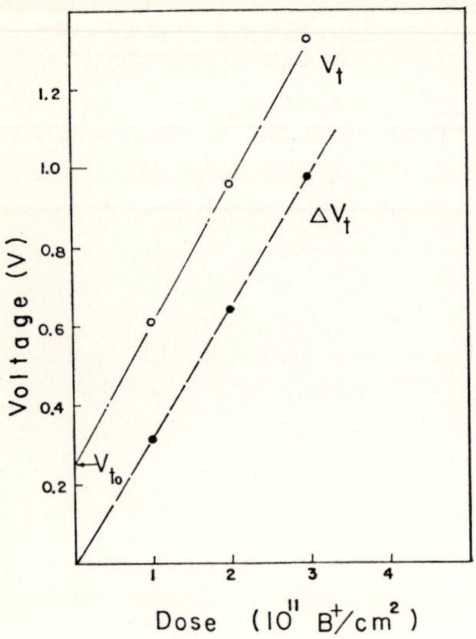


Fig. 19. Threshold voltage and threshold voltage shift as a function of implant dose

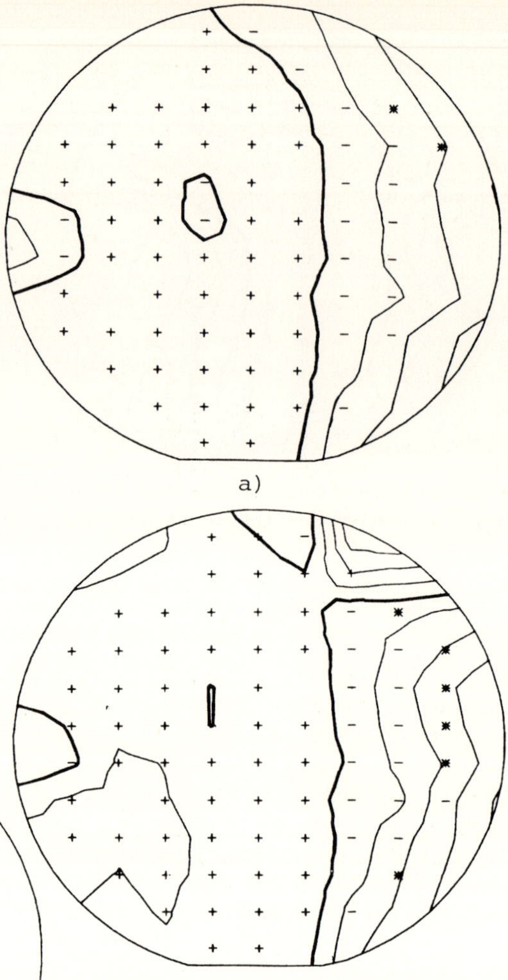


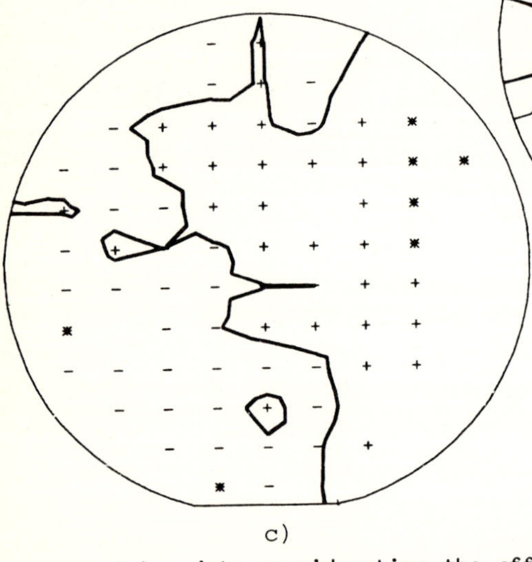


Fig. 20. Contour maps of (a) unimplanted transistor threshold voltage, $\bar{V}_{To}$= 0.14 volts; (b) implanted transistor threshold voltage, $\bar{V}_T$ = 0.93 volts; (c) threshold voltage shift, $\bar{V}_T - \bar{V}_{To}$= 0.81 volts. The contour interval is 0.10 volts

or α takes into consideration the effect of the finite depth and spread of the implant for shallow layers [9]. For example, α = 0.95 for the case of a boron implant with an energy of 45 KeV through an 800 Å thick oxide layer into a silicon substrate having a background doping concentration of $10^{15}$ $cm^{-3}$.

The threshold voltage $V_T$ and the threshold voltage shift $\Delta V_T = V_T - V_{To}$ are plotted in Fig. 19 as a function of the machine dose.

(Each data point represents an average of measurements from 2 - 6 wafers with about 100 test transistors on each wafer.) Over this dose range both the average threshold voltage and the average threshold voltage shift increase linearly with implanted dose. Contour maps of the threshold voltages $V_T$ and $V_{To}$ are shown in Figs. 20a and 20b, respectively. The strong similarity between the two maps suggests that the main factors causing these variations are related to the oxide properties rather than to the implanted dose non-uniformities.

Using the threshold voltage shift rather than the threshold voltage itself allows one to eliminate the effect of parameters not related to the implantation step. This is shown in Fig. 20c, in which the contour map of $\Delta V_T$ shows a pattern which would be expected for implant non-uniformities characteristic of a mechanically-scanned system (see Fig. 8). The variations in $\Delta V_T$ in Fig. 20c have a percent standard deviation of 2.3%.

## 4.2 High-Frequency Capacitance Technique

Range of Applications. It is well known that a capacitance-voltage (C-V) measurement on a p-n junction, Schottky diode or MOS capacitor is capable of revealing the carrier concentration distribution beneath the device [10]. The high-frequency capacitance measurement is a well developed and convenient technique. Using commercial apparatus, the concentration range can easily be extended to implanted doses of $10^{10}$ ions/cm$^2$. The major advantage of the C-V technique over the sheet resistance and the threshold shift methods is that it gives a direct measure of the implanted depth profile.

A simple relationship between the high-frequency capacitance (as a function of the applied voltage on the gate of an MOS capacitor) and the impurity concentration (as a function of the depth beneath the oxide/semiconductor interface) can be derived [11] based on the following assumptions:

i) sharp depletion layer edge and complete ionization of impurities in the depletion region ("depletion approximation"), and

ii) no mobile oxide and/or interface charges:

$$N(W) = -(1/q\varepsilon_s)C^3(dC/dV)^{-1} \tag{9a}$$

$$W = \varepsilon_s(1/C - 1/C_{ox}) \tag{9b}$$

where q is the electron charge; C is the series equivalent capacitance of the oxide capacitance $C_{ox}$ and the semiconductor space charge capacitance; N(W) is the dopant concentration at the depletion layer edge W; and $\varepsilon_s = 1.037 \times 10^{-12}$ farad/cm$^2$ is the dielectric constant of silicon. The assumptions used to derive Eqs. 9 limit their use to deep implants with relatively high doses. However, their usefulness can be extended with suitable correction factors to the evaluation of shallow, low-dose implants [12,13,14].

Test Structures. Measurements of low-dose implant profiles by C-V techniques generally involve the fabrication of an MOS capacitor stucture, although Schottky barrier diodes have also proved useful for this purpose [10]. The simplest MOS capacitor structure is made by using a mercury ball as a contact electrode over a thin oxide layer, as shown in Fig. 21a. Since this method requires no furthur processing after anneal, it is widely used for in-line process control. However, a significant disadvantage of this method is the lack of tight control on the capacitor area.

Evaporation of aluminum dots onto the surface of the implanted and annealed wafer provides better control over the metal contact area, as shown in Fig. 21b. However, the dimensions of the depletion region under the contact, represented by a dotted line in the figure, is not restricted to the aluminum dot area. This lack of precision in the dimensions of the MOS capacitor makes the simple aluminum dot and the mercury probe methods unsuitable for the high-precision, automated measurements required for wafer mapping applications.

Fabrication of a guarded ring structure, as shown in Fig. 21c, provides tighter control on the depletion region dimensions and prevents surface inversion of the capacitor for deep depletion conditions. However, this structure requires a fast capacitance measurement apparatus (< 1 msec.) and high carrier lifetimes for accurate measurements at low doses.

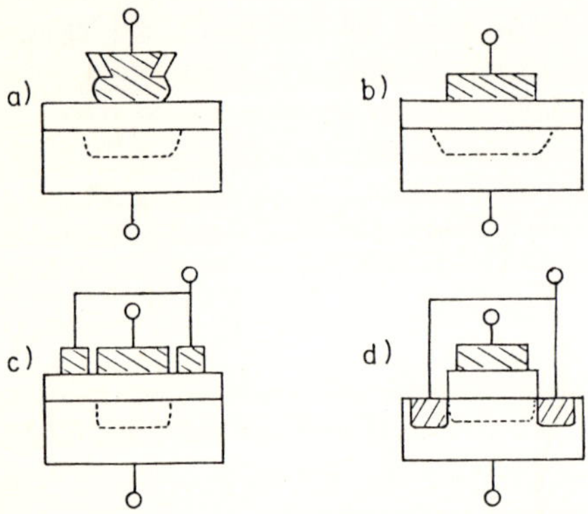


Fig. 21. MOS capacitor structures useful for low-dose profiling: (a) mercury ball; (b) aluminum dot; (c) guarded ring; and (d) guarded junction

The highest performance structure for low-dose, automated testing is a junction guarded capacitor, shown in Fig. 21d. The reverse biased guarded junction sweeps the minority carriers from the oxide interface under deep depletion conditions. This structure allows measurements of net implant dose with a resolution of 0.2%. Junction guarded capacitors are particularly useful when they are fabricated along with other process-control test structures and used for automated characterization of IC production wafers.

Contour Mapping. C-V measurements can be used to directly characterize dose variations across a wafer. In the example shown in Fig. 22, net dose measurements were made using a guarded capacitor structure with an area of $1.5 \times 10^{-3}$ $cm^2$. In this case, isolation was provided by a thick oxide layer augmented by a boron-implanted layer. The capacitor was implanted with 45 keV boron ions through an oxide layer of 750 Å. High-frequency C-V measurements were made on approximately 80 test sites over the wafer. The measured mean dose was $1.9 \times 10^{12}$ $cm^{-2}$ with a standard deviation of $2.8 \times 10^{11}$ $cm^{-2}$. Those sites which gave results differing from the mean value by more than $\pm$ 1.7 standard deviations are marked with an asterisk. Sites which gave inconsistant individual measurements or showed high leakage current levels were discarded from the sample and account for the blank regions in the regular test pattern.

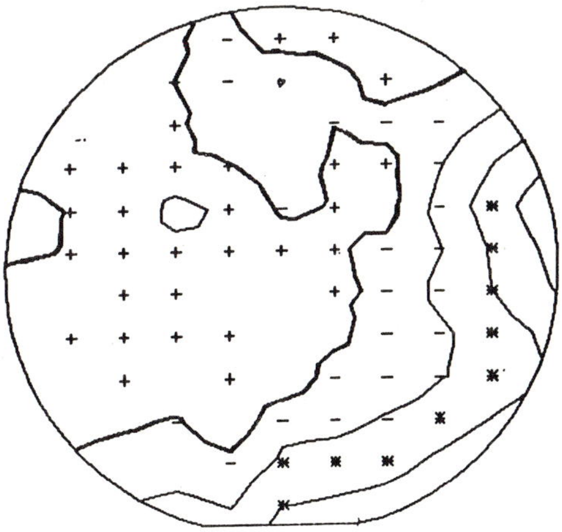

Fig. 22. C-V dose contour map (mean dose $1.9 \times 10^{11}$ $cm^{-2}$, contour interval $5 \times 10^{10}$ $cm^{-2}$)

## 5. Summary

Wafer mapping tools have been described in this paper which provide accurate and efficient evaluation of ion implantation procedures and make it possible to diagnose common implanter malfunctions. Automated testing requires that the resolution of the measurement technique in such applications be greater than the non-uniformity characteristic of "normal" implanter operation. For most practical cases, this condition can be satisfied as long as the measurement repeatability is 0.5% or better.

In the high-dose range (greater than $10^{13}$ ions/$cm^2$), automated sheet resistance measurements with a mechanical four-point probe have proved to be well suited for the control of ion implantation processes. The use of a two-configuration measurement technique and test-site interpolation allows for accurate, reproducible measurements and easily interpreted presentation of the data. The sensitivity of sheet resistance measurements to wafer heating effects during implantation has proved to be a particularly valuable tool for evaluating the high-current (10 mA) ion beams used for high-dose implantation.

In the range of doses used for MOS threshold voltage adjustments ($10^{10}$ -$10^{12}$ ions/$cm^{-2}$), MOS device parameters and C-V measurements have been utilized for generating contour maps of implant dose non-uniformity. Contour maps of threshold voltage shift variations over a wafer show implant non-uniformity patterns similar to those observed at higher doses with sheet resistance mapping techniques. Considerable work remains to be done on measurement procedures for the low-dose range. In particular, it is essential to determine the influence of non-implant processing parameters in order to develop these tools to the same level of utility and confidence as has been achieved with sheet resistance measurements at higher doses.

Acknowledgements

We wish to thank the staff of the Signetics Corporate Quality Control Lab for their continued support in the development and implementation of wafer mapping procedures for characterization of ion implantation. We also acknowledge the cooperation of A. Wittkower of Nova/Eaton, P. Hanley of Extrion/Varian and B. Raicu of Applied Implant Technology in our studies of the effects of high beam currents and T. Hazendonk for helpful suggestions in the theoretical analysis of MOS device parameters.

References

1. L. J. van der Pauw, Philips Res. Rep. 13, 1 (1958)
2. D. S. Perloff, J. Electrochem. Soc. 123, 1745 (1976)
3. R. Rymaszewski, J. Sci. Inst. 2, 170 (1969)
4. The wafer was implanted in an electrostatic scanning system in which the X- and Y-scan plate frequencies had values which resulted in a Lissajous-type scan pattern.
5. J. N. Gan and D. S. Perloff, Nuc. Inst. Meth. 189, 265 (1981)
6. D. G. Beanland and D. J. Chivers, J. Vac. Sci. Tech. 15, 1536 (1978)
7. D. K. Sadana, M. Strathman, J. Washburn, and G. R. Booker, J. Appl. Phys. 51, 5718 (1980)
8. D. G. Beanland, W. Temple, and D. J. Chivers, Solid-State Electronics 21, 357 (1978)
9. J. R. Brews, Physics of the MOS Transistor, Applied Solid State Science, Supplement 2A, D. Khang, Ed., Academic Press (1981)
10. R. G. Wilson and D. M. Jamba, NBS Special Pub. 400-71 (1982)
11. J. Hilibrand and R. D. Gold, RCA Rev. 21, 245 (1960)
12. K. Ziegler, E. Klausman, and S. Kar, Solid-State Electronics 18, 189 (1975)
13. J. Verjans and R. J. Overstraeten, Solid-State Electronics 18, 911 (1975)
14. G. Baccarani, M. Rudan, G. Spadini, H. Maes, W. Vandervorst, and R. van Overstraeten, Solid-State Electronics 23, 65 (1980)

# Non-Electrical Measuring Techniques

P. Eichinger and H. Ryssel

Fraunhofer-Institut für Festkörpertechnologie, Paul-Gerhardt-Allee 42
D-8000 München 60, Fed. Rep. of Germany

## Abstract

The goal of non-electrical measurements in the field of ion implantation is to determine quantitative depth profiles of the total concentration of the implanted species, generally together with an evaluation of the damage produced by the ion bombardment. Thermal or beam annealing of implanted samples results in a rearrangement of the impurity and host-material configuration via diffusion and lattice reconstruction. Following a short review of the analytical problems encountered in implantation work and the various measuring techniques, it is shown that secondary ion mass spectroscopy (SIMS) offers the most favorable solution for implantation-related studies if combined with complementary nuclear measurement techniques, especially neutron and high-energy ion beam analysis, which are of particular usefulness in the high-concentration range and in cases where interfaces are to be included in the analysis. Making use of the channeling effect, high-energy ions can also provide valuable information about lattice damage and incorporation of impurities on lattice sites, while for the classification of extended defects, X-ray topography and transmission electron microscopy (TEM) are indispensable tools.

## 1. Introduction

In the classical application of ion implantation - non-thermal doping of semiconductor layers - the electrically active doping profile and the total atomic concentration profile of the dopant species are essentially identical for concentrations significantly below the solubility limit. In this case, electrical measurements of parameters such as sheet resistivity or Hall mobility may sufficiently describe the effects of ion implantation. These measurements are generally easier to perform, and give the ultimately relevant information in terms of device performance, despite some implications which have been reviewed in the contribution by Hemment.

However, there are several important applications of ion implantation which are hardly, or not at all, accessible by electrical methods:

- In the high-concentration range ($10^{19}$ impurities per $cm^3$ or more, e.g. in bipolar process technology), prominent deviations exist between the electrically active and the total impurity profile. A knowledge of the latter is important because it acts as a diffusion source, and is essential for an understanding of the post-anneal profiles, e.g., for the purpose of process modelling, eventually combined with high-concentration and multiple-element diffusion effects.

- In semiconductor device technology, implantation into insulating or polycrystalline layers, or implantation of electrically inactive species (for damage gettering, radiation-enhanced diffusion, etc.), becomes increasingly interesting, as well as the formation of compound insulators such as silicon nitride or silicon oxide by high-energy implantation (buried layers) or very low energy implantation.
- The field of lattice damage by direct or secondary (knock-on) implantation and its annealing behavior are hardly accessible to electrical measurements.
- The technique of ion implantation is steadily expanding its use, also in fields other than semiconductor applications. The general term "ion beam modification of materials" includes hardening, wear reduction or corrosion protection of metal surfaces.

These few examples may illustrate the importance of non-electrical measurements in implantation work; very often they complement electrical data and vice versa. Generally, non-electrical measurement techniques require a considerable effort of time and equipment, and are only feasible for basic process development rather than for routine process control. In order to distinguish the type of non-electrical measurements to be discussed here from the more classical methods of analytical chemistry, this field is often called "instrumental (or physical) analytics". In this sense, the term analytics will be used throughout the following sections.

## 2. Structure of Implantation-Related Analytical Problems

Non-electrical measurements of ion-implanted layers, in general, have two goals: to determine the total impurity distribution and/or to investigate the crystalline quality of the layer, especially in connection with specific annealing procedures. Sometimes ion-implantation-induced stress or volume change, or chemical effects, are also of interest; for the latter problems, reference is made to the special literature. Usually, the implanted species is well defined, as is the substrate which consists either of a homogeneous material - of high purity in semiconductor work - or of a two-layered structure (e.g., $SiO_2$ on polycrystalline Si or crystalline Si). Therefore, there is generally no need for methods identifying complex elemental compositions as, e.g., in geological or environmental work, unless contamination has to be included in the analysis. Furthermore, the composition of the outermost surface layer (adsorption layer) is of little significance, since the active structures are buried a few tenths (or less) of a micron below the surface - a situation very much different from the analytical task, e.g., in catalyst research. Finally, the samples to be analyzed are usually not rare and unique specimens in the sense that only very limited amounts of material are available for analysis.

Lateral resolution is certainly of interest, especially in the connection with lateral spread of implanted ions, as the structures in microelectronic circuits shrink to dimensions comparable to the projected range of the implanted ions. The required lateral resolution is then of the order of less than a tenth of a micron, which is presently only achievable with electron beam techniques (an example will be given later). If the resolution of an analysis technique is worse than that, it is mostly of marginal importance in ion implantation studies.

Having thus narrowed the field of the most commonly encountered analytical problems in ion implantation work by excluding multielement and/or surface (i.e. submonolayer) analysis for our discussion, as well as spurious sample identification and micrometer imaging, let us now further assess the kind of information which is generally essential. In this context, work with semiconductors will be our main objective (although occasionally, reference will be made to more general applications in materials technology). For this purpose, a representative step in silicon process technology will be selected and analyzed from the viewpoint of the analytical requirement, which can be generalized further:

As an example, a silicon wafer, covered by a 50 nm thick layer of thermal oxide is implanted with different doses ($10^{15}$ to $10^{16}$ $cm^{-2}$) of As (implantation energy 150 keV), and annealed in different ambients at different temperatures. This basic process sequence is, e.g., of interest for emitter formation or VLSI MOS source/drain implantation. The following information should be obtainable by analytical methods:

- depth distribution of As in the oxide with depth resolution of ca. 10 nm and sensitivity of ca. $10^{17}$ $cm^{-3}$
- segregation coefficient at the oxide-silicon interface for As
- depth distribution of As in the substrate with depth resolution of ca. 10 nm, sensitivity range $10^{20}$ -$10^{16}$ $cm^{-3}$
- oxygen distribution in the substrate originating from knock-on implantation
- primary and residual damage profile from the oxidation, As implantation and secondary O implantation
- fraction of the total impurities occupying regular lattice sites as a function of concentration (depth)
- because of the many process parameters (dopant, energy, annealing technique, oxide thickness, crystal orientation), a large set of samples has to be analyzed in a reasonable time.

These tasks can be translated into analytics in the following terms:

- probing depth in the range of one micrometer
- sensitivity in the higher ppb range for selected impurities
- dynamic range up to seven orders of magnitude
- depth resolution of the order of ten atomic layers
- stability with matrix changes (interfaces)
- classification and quantification of lattice disorder with depth
- economical aspects (equipment, sample preparation and measurement time).

It is quite clear that no single analytical tool can fulfil all these requirements at once, especially if further requirements originating in other tasks are included which are not intrinsic to the given example. These may deal with the exact composition of implantation-generated compound surface or buried layers (e.g. $Si_3N_4$), or gettering of trace amounts of metals in ion-implanted damaged layers. Rather than to try to present a universal solution to the whole spectrum of analytical problems originating from ion implantation work, the following sections will give a survey of the most commonly used methods. Emphasis will be laid on the complementary information obtainable from different techniques. In this sense, special attention will be given to secondary ion mass spectroscopy (SIMS) because of its almost unmatched sensitivity and dynamic range, and to nuclear methods because of their insensitivity to matrix effects and ability to roughly evaluate lattice disorder and substitutional fractions of impurities. However, for damage structures other than random primary damage, e.g. for dislocation networks, stacking faults, etc., the reader's attention will be directed to the methods of X-ray topography and transmission electron microscopy, the latter being a subject of the following contribution by Mader. The relevance of the above selection of measurement techniques in the field of ion implantation may be illustrated by an evaluation of the papers presented at the 1982 International Conference on Ion Beam Modification of Materials, Albany [1]: Out of a total number of 127 contributions, 50 refer to nuclear measurement techniques using high-energy ion beams or neutrons, 13 quote SIMS measurements, and 4 show TEM pictures. Other analytical techniques, e.g., optical and electron-spectroscopic methods, X-ray techniques or Mössbauer spectroscopy, are scattered over 20 papers.

## 3. Review of Measurement Techniques: Impurity Profiling

When scanning the abundant number of monographs or review articles and books dealing with surface or near-surface analytical methods, one easily encounters almost a hundred different techniques, all of them characterized by a combination of capital letters: SIMS, SNMS, SIPS, RBS, AES, LEIS, XPS, ESCA, PIXE, NAA, CPAA, etc., to name only a few of them. A rather confusing picture is the consequence, for the reader looking for the most suitable solution for his analytical problem, in our case resulting from ion implantation processing. The reason for this variety of acronyms is the sometimes puristic or schematic viewpoint of the reviews, which subdivides the methods according to the kind of excitation of the sample (photons, electrons, ions, neutrons), and the excitation products used for analysis, which are again quanta or particles. Since nearly all kinds of excitations and excitation products have some special analytical capabilities and may further be subdivided in specific energy ranges and equipment configurations, the possible combinations are almost innumerable. In order to avoid a copying of already existing excellent reviews, and making use of the formerly outlined characteristics of ion implantation analytics, a different scheme will be used for this survey of analytical methods. This scheme will be based on the kind of information obtainable: e.g., if highest sensitivity in depth profiling is the goal, controlled layers have to be removed continuously or stepwise, analyzed "alone" and differentially compared, thereby getting rid of all background from the substrate. On the other hand, if absolute concentrations are essential with undisturbed substrate portions serving for calibration, non-destructive methods are preferable. This holds especially true if damage is to be investigated because, of course, damage information can hardly be preserved by layer removal. As already mentioned, the non-destructiveness of a analytical method is by itself of minor importance in the context of the problems considered here, because generally, there is no lack in sample size or even numbers of ion-implantation-processed specimens, as long as process development is concerned. Nevertheless, if we translate "non-destructiveness" into "simultaneously available depth information with quantitative reference to the unimplanted substrate", these methods may be of considerable interest. Furthermore, a great need for non-destructive methods results from the fact that layer removal techniques like sputter etching inevitably alter the subsurface composition, giving rise to analytical "artefacts", especially in alloys, at interfaces and in regions of high concentrations.

### 3.1 Abrasive Techniques

#### 3.1.1 Differential Analysis of the Removed Material

A variety of techniques is available which serve to remove in a controllable way material from solid surfaces, either continuously or stepwise.

Among these methods, sputtering with keV ions, spark source or plasma erosion, or laser beam evaporation are used to continuously "dissolve" the surface layers into a highly excited gaseous state, composed of atoms and molecules of the abraded material (as neutrals or in different ionization states), together with recoiling primary particles and, eventually, new reaction products induced by the catalytic action of the surface. A simplified scheme of these complex interactions is shown in Fig. 1, for the most widely used case of sputtering. Included are some effects of the bombarding ions on the remaining surface layer, such as damage, primary and recoil implantation. Not included in this figure are electron or quantum emission from the surface layer or from the ejected particles as a consequence of the primary excitation, which also have important analytical applications.

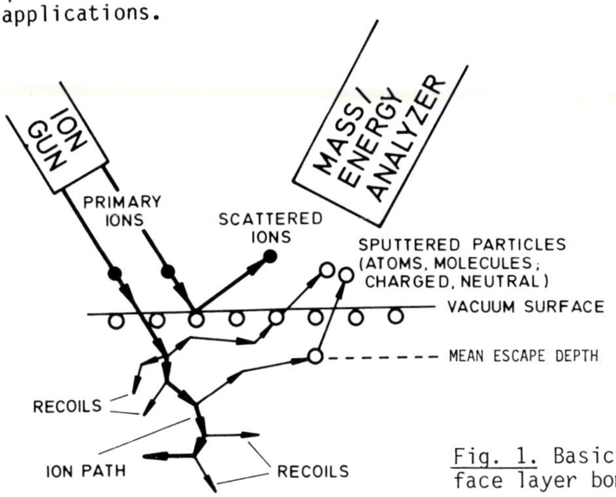


Fig. 1. Basic processes induced by surface layer bombardment with keV ions

Layer removal by sputtering will be discussed in somewhat more detail in Chapter 4 in the context of secondary ion mass spectroscopy (SIMS), where the mass distribution of the secondary ions, ejected in a limited solid angle and energy range, is analyzed. For the present survey, it is sufficient to recall that the sputtering yield, i.e. the number of removed particles per incident ion, is about one in the most common configurations, e.g. for 10 keV $Ar^+$ ions normally incident on silicon. Thus, for the profiling of 1 µm of silicon with an atomic density of $5 \cdot 10^{22}$ cm$^{-3}$, the analysis time is in the order of one hour, with current densities in the order of 100 µA per cm$^2$. The total area to be sputtered off is usually 100 µm in diameter, and time-to-depth calibration is performed by measuring the sputtered crater by a electromechanical surface probe after the analysis. For the main types of modern SIMS equipment, as well as for a discussion of the most relevant experimental methods and parameters of this technique, the reader is again referred to the next chapter. Two examples may demonstrate the excellent capabilities of the SIMS profiling technique for impurity profiling in semiconductors. Fig. 2 shows an as-implanted depth profile of boron in silicon extending over a concentration range of six orders of magnitude with a depth-profiling sensitivity of $10^{14}$ - $10^{15}$ cm$^{-3}$ [2]; an arsenic profile in silicon with a sensitivity in the order of $10^{15}$cm$^{-3}$ is shown in Fig. 3 [3]. Both measurements were made with commercially available instruments. Also for bulk-impurity analysis, SIMS can offer excellent detection sensitivities for many elements, as is shown in the following table for gallium arsenide [4]:

Table 1. Detection limits in GaAs bulk analysis.

| Element | ppma | $cm^{-3}$ |
|---|---|---|
| B | 0.005 | $2x10^{14}$ |
| Be | 0.001 | $4x10^{13}$ |
| Mg | 0.001 | $4x10^{13}$ |
| Al | 0.01 | $4x10^{14}$ |
| Si | 0.01 | $4x10^{14}$ |
| Cr | 0.0005 | $2x10^{13}$ |
| Mn | 0.0005 | $2x10^{13}$ |
| Cu | 0.02 | $2x10^{15}$ |
| Zn | 0.01 | $1x10^{15}$ |
| Ge | 0.2 | $1x10^{16}$ |
| Sn | 0.2 | $1x10^{16}$ |
| Fe | 0.001 | $6x10^{13}$ |

For a silicon matrix, detection limits for important impurities in the following orders of magnitude can be obtained: $10^{13}cm^{-3}$ (B), $10^{15}cm^{-3}$ (As, C, P), $10^{17}$ $cm^{-3}$ (O,N), $5\cdot10^{17}$ $cm^{-3}$(H).

Although this presentation is focussed on near-surface depth profiling rather than on bulk impurity analysis, it should be mentioned that spark source mass spectroscopy (SSMS [5]) offers detection limits in the ppba

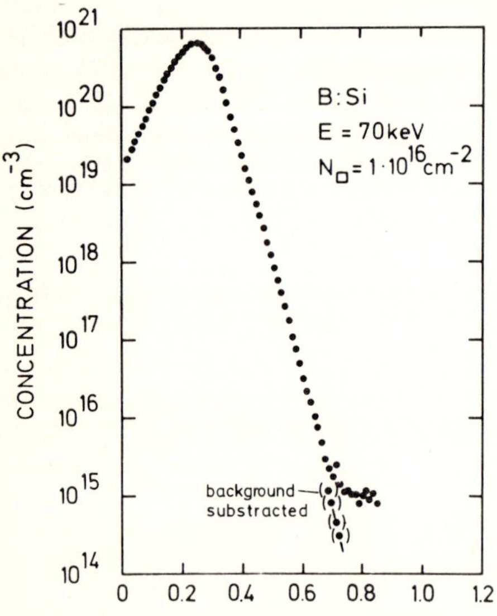


Fig. 2. Concentration-depth profile of boron implanted into silicon, as measured using the SIMS technique [2]

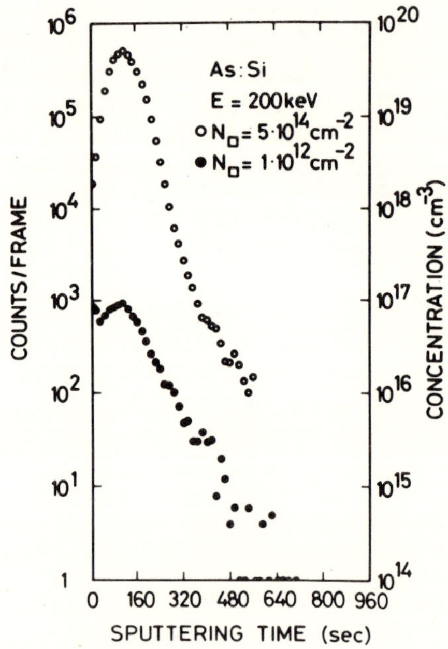


Fig. 3. SIMS profile of arsenic implanted into silicon [3]

range (3.12). In Fig. 4, an example of such a spectrometer is given. It contains a spark ion source where solid bars of the semiconductor being investigated are vaporized and ionized; the resulting ions are then accelerated. An electrostatic analyzer is used for the energy, a magnet for the mass selection. The separated ions are registered using a photographic plate, an array of semiconductor detectors or a multiplier (by sequential detection). For profiling, this technique is (in its present state) inadequate because of the poor lateral control of the eroded material. Similar arguments hold for plasma or glow discharge mass spectroscopy (GDMS [6]), where a rf planar diode glow or a dc discharge [7] is used both to sputter-etch the sample and to ionize a representative fraction of the neutral sputtered atoms via the interaction with metastable plasma particles and electrons, prior to their mass-spectrometric detection. The use of a narrow, collimated beam of keV ions together with electrostatic areal scanning, as in SIMS and its modifications, is presently the superior tool for continuous in situ material removal, despite some inherent difficulties (like atomic mixing or preferential sputtering) to be discussed below.

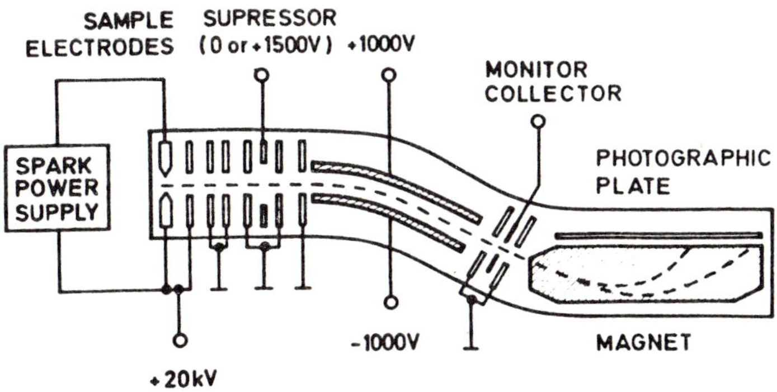


Fig. 4. Principle of Spark Source Mass Spectroscopy

The most critical implication of SIMS in general originates in the sensitivity of the fraction of the ionized secondary products - generally small as compared to the neutrals - to the matrix composition, to absorbed surface layers or even to built-in potentials in the specimen. In the above-given examples with relatively low impurity concentrations in homogeneous semiconductor materials, this implication is of negligible influence if a sufficient sputtering rate is preserved together with ultra-high vacuum conditions. It becomes crucial, however, in implantation studies as well, as soon as impurity concentrations are to be profiled extending over multilayer structures such as oxide-semiconductor, polycrystalline-monocrystalline semiconductors and their respective interfaces. Other examples are compound semiconductors with varying compositions like GaAlAs or built-in p-n junctions [8]. Furthermore, in the domain of very high concentration of the implanted species (forming of dielectric layers by implantation, modification of mechanical surface properties in metals), the ionization yield may become concentration-dependent.

It is known that the emitted neutral particles, contrary to the ionized species, do not exhibit this matrix sensitivity. Therefore, a technique has been developed which collects the neutrals and uses post-ionization by the electron component of a special low-pressure argon plasma, prior to mass analysis with a quadrupole mass spectrometer. This technique is

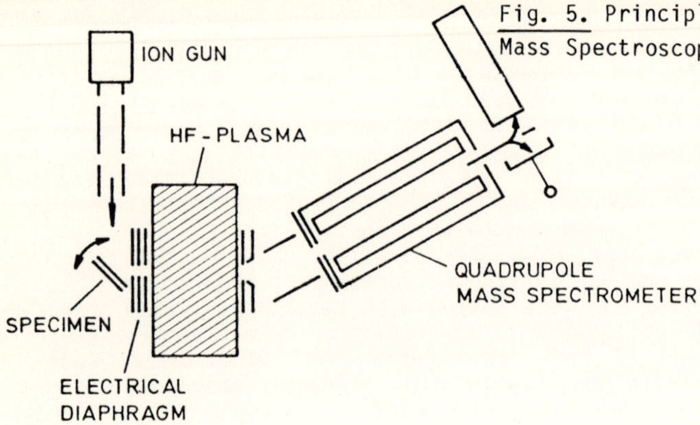


Fig. 5. Principle of Sputtered Neutral Mass Spectroscopy [9]

called Sputtered Neutral Mass Spectroscopy (SNMS), and is shown schematically in Fig. 5 [9]. Although the post-ionization process efficiency is below $10^{-2}$, it may be balanced by the large amount of ejected neutrals, so that only little loss of sensitivity for a given application may result, with the advantage of a reduced sensitivity against matrix effects. Up to now, this method has mainly been used for surface analysis of metals; its depth-profiling capabilities, especially in the context of ion implantation, are not yet evaluated. A similar situation may apply for Sputter Induced Photon Spectroscopy (SIPS), where the spectral optical emission of the particles is evaluated, rather than their mass. Apparently, similar implications (chemical effects) hold for photoemission and ionization probabilities [10]. By monitoring the light emitted by sputtering in an argon dc glow discharge (Glow Discharge Optical Spectroscopy, GDOS [11]), good depth-profiling characteristics have been reported in GaAs, with dynamic detection limits of approx. $10^{18}$ - $10^{19}$ $cm^{-3}$ for Ge, B and Mg, using a comparatively inexpensive and simple arrangement.

Let us now turn to a completely different approach to near-surface impurity depth profiling, based nevertheless on the same analytical principle: the dissolution of a controlled amount of material by chemical methods (stripping) with subsequent instrumental analysis of the liquid solvent. In work with ultra-pure semiconductors, this analytical concept suffers, generally, from the fact that available solvents, mainly acids, contain impurity levels comparable to those to be monitored if multiplied by the dissolution factor (e.g., volume of solvent divided by volume of dissolved material). Therefore, e.g., optical methods such as emission or absorption spectroscopy with their excellent detection sensitivities for many elements give only very poor information in this context. By making use of a simplified radiotracer method, however, this difficulty can be overcome: If the impurity to be analyzed is converted fractionally into a radioactive isotope by neutron activation of the specimen before analysis, and the solution is evaluated by a suitable nuclear radiation measurement technique, no additional background signal will evidently be introduced by the solvent. On the other hand, proceeding this way, the matrix constituents may be activated as well, and therefore, this method is only feasible for very limited impurity/matrix combinations without involving "hot chemistry", i.e. chemical separation of highly radioactive compounds, or true radiotracer techniques, i.e. implantation of radioactive material in nonactive, natural substrates. Nevertheless, for implantation studies in silicon and its oxides or nitrides, this method has given useful results due to the fact that silicon has a low cross section for neutron activation and the activity decays with a half-life of 2 1/2 hours. Important dopants such as As, P or Sb are much more easily activated and their activities decay considerably more slowly (of the order of days). Neutron activation service is offered by all research reactor installations, and with silicon samples irradiated for 24 h after a two-day decay period at the reactor, negligible radioactivity in the sense of radioprotection will result from the implanted species and contaminations. The samples can be processed and measured at the implantation lab with relatively inexpensive equipment. Fig. 6 [12] shows an arsenic depth profile after implantation and wet oxidation of the sample recorded with this technique. Stripping has been performed in steps of 23 nm by buffered HF in the oxide, and the silicon surface has been oxidized anodically before stripping. The gamma radiation resulting from the activation of As has been monitored using a commercial gamma spectrometer. From the analytical viewpoint, it is remarkable that the behavior of the dopant can be quantitatively traced over the interface region, a problem not easily accessible, e.g., with SIMS due to the matrix effects on secondary ion yield and sputtering. Other dopant elements in silicon which can be measured with the same technique are Al, Ga, In or Sb. Neutron activation of P results in beta radiation, to be monitored using beta counting equipment, e.g., a Cerenkov counter [13]. Some definitions of useful parameters describing nuclear reactions will be given in section 3.2.1.

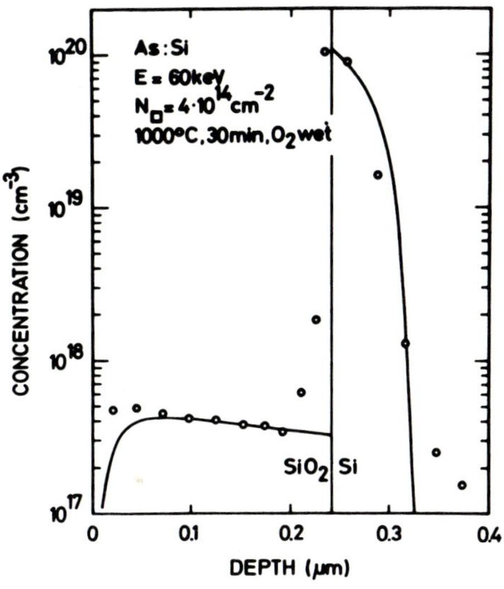


Fig. 6. Arsenic profile in $SiO_2$/Si, measured by neutron activation analysis [12]

### 3.1.2 Analysis of the Exposed Surface Layer

If material is removed, e.g. by sputter etching in a vacuum system, the freshly exposed surface layer of the specimen can also be the source of information for impurity depth profiling. Among the various surface analysis techniques in this context, electron-spectroscopic methods are to be mentioned, with the electron emission originating either from high energy photon irradiation (Electron Spectroscopy for Chemical Analysis, ESCA, or X-ray Photon Spectroscopy, XPS), or from transitions brougth about by high energy electron irradiation (Auger Electron Spectroscopy, AES). Furthermore, the energy distribution of the reflected primary ions can be analyzed (Low Energy Ion Scattering, LEIS), and photon emission from the excited surface can be studied. Most of these methods give information about the composition, chemical nature, electronic structure or atomic configuration of the outermost surface, and have found little application in impurity depth profiling in the field of ion implantation as yet. An exception is AES, because it is complementary to SIMS in the sense that, especially in the high impurity-concentration domain, it is virtually free from matrix effects (with the trade-off of a considerably poorer sensitivity of approx. 0.1 atomic percent). The lateral resolution can be as good as 30 nm [14]. AES can be incorporated into a SIMS apparatus with relatively little impact on the SIMS performance, or into a scanning electron microscope.

An AES arrangement consists of an electron gun producing a highly focused monochromatic beam with energies in the keV range, which can remove an inner shell electron from a target atom. This vacancy will be filled by an outer shell electron. The energy freed in this act will be released either radiatively in the form of an X-ray quantum, as in the Electron Microprobe (EMP), or transferred non-radiatively as kinetic energy to another electron from the outer shell in the so-called Auger process which dominates for the excitation of light elements. Like the energy of the X-ray quanta, the energy of the Auger electrons is characteristic for the emitting element. For ESCA measurements, a monochromatic X-ray source is incorporated. The mean free path of Auger electrons and photoelectrons is only of the order of 1 nm, which makes the method extremely surface sensitive as compared to characteristic X-ray detection. (With electron excitation, the

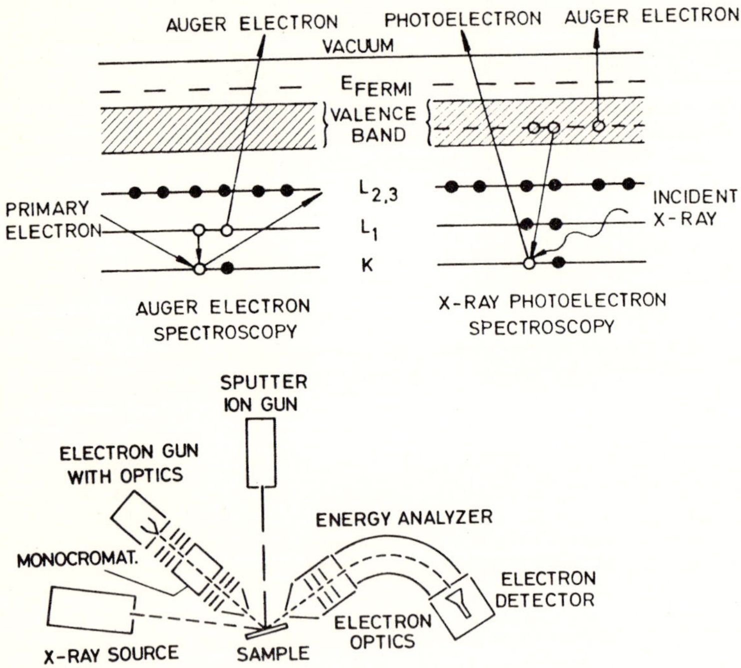


Fig. 7. Principle of Auger Electron and Photoelectron Spectroscopy [15]

X-rays originate from an excitation volume having a diameter of the order of 1 μm.) By measuring the spectral intensity of the emitted electrons with an electron spectrometer - usually a Cylindrical Mirror Analyzer (CMA) - the atomic composition of the surface layer can be established and chemical information can be obtained (chemical energy shift). Fig. 7 shows the basic process of Auger electron and photoelectron emission [15] and the schematic experimental arrangement. A typical AES measurement is given in Fig. 8, showing oxygen and sulfur surface enrichment on a zone-refined Fe foil [16]. In order to discriminate the Auger lines against the slowly varying background in the energy spectrum N(E), an electronic differentiation of the energy distribution is usually performed and dN/dE is displayed. It has to be pointed out that in-depth profiling using sputtering AES measures the actual surface composition, while SIMS measures the composition of the sputtered fraction. These may be different due to the sputtering process (see section 4). AES is quantitative because inner-shell excitation takes place, contrary to SIMS, where the ionization is strongly influenced by the outer electrons, i.e., the chemical nature of the bonds.

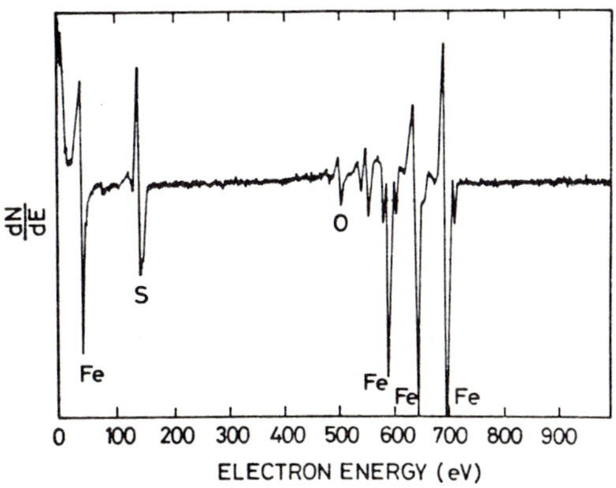


Fig. 8. AES analysis of a steel surface [16]

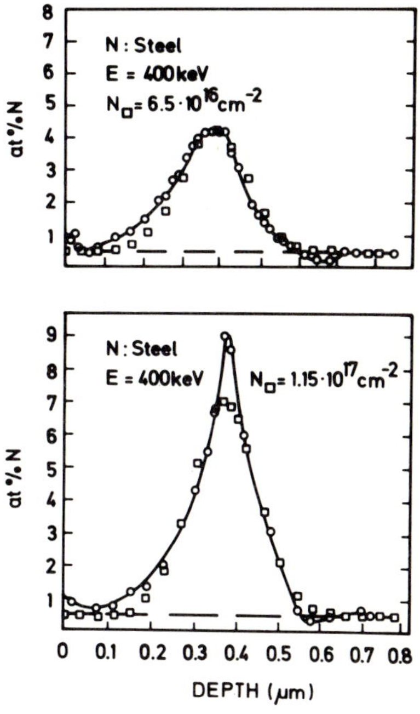


Fig. 9. Nitrogen depth profiles in ► steel, measured with AES [17]

A depth profile is shown in Fig. 9 [17] for high-dose nitrogen implantation into steel. Additional information about the atomic configuration of the incorporated nitrogen is derived from the shape of the Auger lines in this technique.

## 3.2 Non-Destructive Methods

If the specimen is to remain essentially unchanged during the analysis, depth profiles of impurities can be obtained in two ways:

- the sample is excited by a suitable radiation homogeneously (or slowly varying with penetration depth), down to a depth below the surface considerably greater than the extension of the profile to be monitored. Depth information must then be conveyed by the excitation products leaving the sample surface to be detected.

- the depth information is connected with the incident radiation; i.e., by choosing specific parameters of this radiation, the solid is only excited in a given depth (resonance techniques).

Both types of interaction mechanisms are of considerable importance for analytical purposes.

### 3.2.1 Homogeneous Excitation

Neutrons or high-energy ions are feasible types of radiation, penetrating deep enough into the solid to probe a relatively thick surface layer via nuclear interactions. These interactions result in the formation of a compound nucleus which decays, giving rise to high-energy quantum and particle radiation with energies characteristic for the specific nucleus (Nuclear Reaction Analysis, NRA). This is similar to the characteristic X-ray (or Auger electron) production where quanta or electrons interact with inner-shell electrons to produce quanta (or Auger electrons) in the keV range which serve to identify by their energy the chemical element. The difference is that with nuclear reactions, energies are in the MeV range because of the much stronger nuclear forces in comparison to the electrostatic forces responsible for the energetic arrangement of the electrons.

Of particular interest are reactions giving rise to the emission of heavy particles such as protons or alpha particles (in contrast to light particles like electrons), because (1) they have enough energy to travel several μm in the solid without being deflected, and (2) they lose a

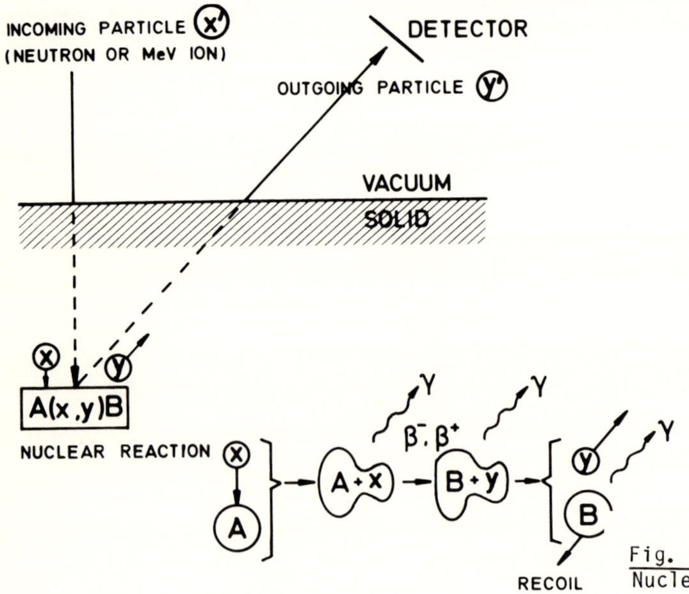


Fig. 10. Principle of Nuclear Reaction Analysis

sufficient amount of energy on their way to the surface through electronic stopping so that their depth of origin can be identified via this energy loss. This situation is schematically depicted in Fig. 10. A nuclear reaction is usually written in the form A(x,y)B, where A represents the nucleus hit by the particle x, which leads to the formation of a new nucleus B and the emission of y. In Fig. 10, we find x' and x together with y' and y for the incoming and outgoing particles in vacuum, resp., at the interaction site, to indicate the energy difference due to electronic stopping. Furthermore, the formation, transformation and decay of the compound nucleus are schematically shown. These processes can be accompanied by the emission of characteristic gamma radiation and beta or positron emission. As in chemical reactions, a nuclear reaction involves an amount of energy. A quantity Q is defined as the net energy gain in the reaction, so that positive Q values generally result in reaction products carrying more energy than the incoming particle (exothermic reactions). The Q values are in the MeV range. Finally, to describe a nuclear reaction fully, we need the cross section $\sigma$ as a function of the energy of the particle initiating the reaction (x): Imagine a homogeneous particle flow of $j_x$ particles per cm² and s impinging on a homogeneous areal distribution of $N_A$ target nuclei per cm². The number of reactions r which will take place per unit time and area is then clearly proportional to $j_x$ and $N_A$, the factor of proportionality being the cross section with the dimension of cm²: $r = \sigma j_x N_A$. The cross section for nuclear interactions is of the order of $10^{-24}$ cm$^2$, a unit called 1 barn, and sometimes referred to as the "area of a nucleus" A for a given experiment. Nuclear interactions, in general, are not isotropic, i.e., the outgoing particle is emitted with different probabilities into different angular directions with respect to the incoming beam. Therefore, a differential cross section $\sigma(\Omega)$ is defined, with $\sigma(\Omega)d\Omega$ being the cross section for secondary particle emission into a solid angle interval $d\Omega$.

As an example, the reaction $^{16}O(d,\alpha)^{14}N$ will be considered in more detail: A deuterium ion beam with an energy of 900 keV hits an oxygen-implanted solid surface. Oxygen nuclei of the mass number 16 convert into nitrogen nuclei (m.n.14) by emitting alpha particles. The reaction is exothermic with a Q value of 3 MeV; the resulting alpha particles then

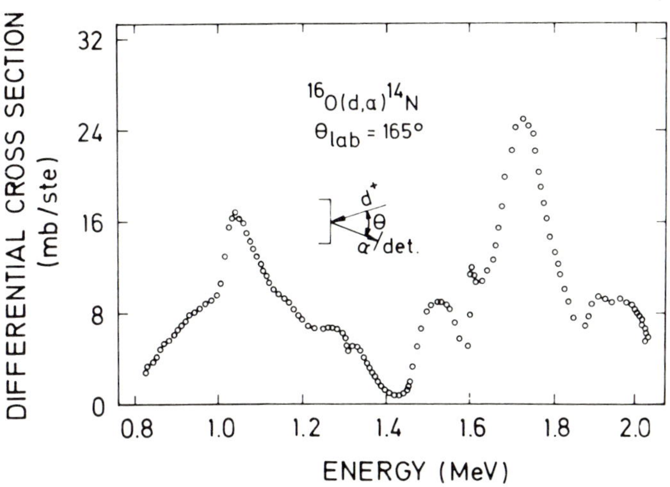


Fig. 11. Differential cross section for the $^{16}O(d,\alpha)^{14}N$ reaction [18]

carry an energy of 2.7 MeV if emitted into an angle of 165° with respect to the incoming beam (i.e., in backward direction). The differential cross section of the reaction for this angle is shown in Fig. 11 [18]. At 900 keV, it is approx. $6 \cdot 10^{-3}$ barn/sterad. The incoming deuterium beam loses, by electronic stopping, approx. 130 keV/µm (e.g. in nickel), so that only negligible variation over the depth of an implantation profile in the cross section is present. In this sense, the surface layer of interest is homogeneously excited by the deuterium beam, and the intensity distribution of the outgoing alpha particles around 2.7 MeV reflects the depth distribution of oxygen atoms in the sample, the electronic stopping for alpha particles in nickel being approx. 600 keV per µm (stopping power data taken from [19]). An application of this technique is shown in connection with a study of the effects of implanted oxygen on the formation of nickel silicide [20]: Fig. 12 shows the oxygen depth distribution for an unannealed implant into nickel on silicon, together with the annealed case where silicide formation has occurred. The oxygen depth distribution is derived from the energy distribution of the emitted alpha particles detected from a glancing angle with respect to the surface. These results are hardly obtainable by techniques using layer removal such as SIMS, because of artefacts due to the sputtering and secondary ion formation processes in the different materials.

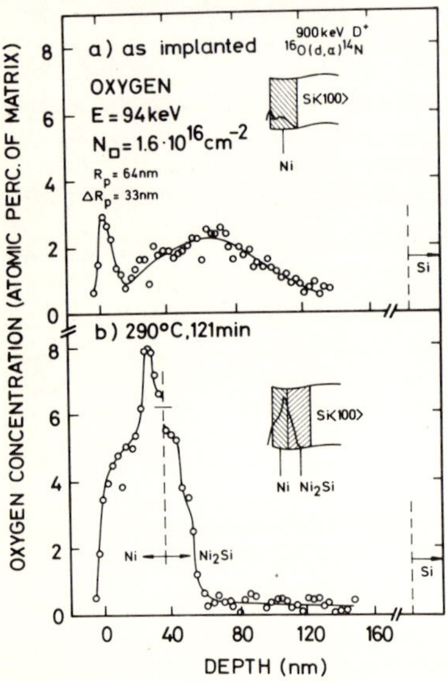


Fig. 12. Oxygen depth profiles as measured with NRA in a study of silicide formation [20]

There are quite a number of spontaneous nuclear reactions induced by energetic, light particles (protons, deuterons) which are useful for impurity analysis. A selection is given, for incident energies $E_0$ below 2 MeV (as accessible with one-stage Van de Graaff accelerators or small tandem accelerators), in the following table [18]. $E_1$ is the energy of the emitted particle.

The cross sections are of the order of millibarn/sterad, and refer to an emission angle of 150°. All reactions are exothermic; the outgoing particles generally carry a much higher energy than the exciting beam, so that the signals are essentially free of background, and are yield-limited provided a suitable system for energy discrimination is available.

Table 2. Ion-induced nuclear reactions

| Nucleus | Reaction | Q(MeV) | $E_o$(MeV) | $E_1$(MeV) | $d\sigma/d\Omega$(mb/sr) |
|---|---|---|---|---|---|
| $^2H$ | $^2H(d,p)^3H$ | 4.032 | 1.0 | 2.3 | 5.2 |
| $^2H$ | $^2H(^3He,p)^4H$ | 18.352 | 0.7 | 13.0 | 61 |
| $^3He$ | $^3He(d,p)^4He$ | 18.352 | 0.45 | 13.6 | 64 |
| $^6Li$ | $^6Li(d,\alpha)^4He$ | 22.374 | 0.7 | 9.7 | 6 |
| $^7Li$ | $^7Li(p,\alpha)^4He$ | 17.347 | 1.5 | 7.7 | 1.5 |
| $^9Be$ | $^9Be(d,\alpha)^7Li$ | 7.153 | 0.6 | 4.1 | ~ 1 |
| $^{11}B$ | $^{11}Be(p,\alpha)^8Be$ | 8.586 | 0.65 | 7.57($\alpha_o$) | 0.12($\alpha_o$) |
| | | 5.65 | 0.65 | 3.70($\alpha_1$) | 90($\alpha_1$) |
| $^{12}C$ | $^{12}C(d,p)^{13}C$ | 2.722 | 1.20 | 3.1 | 35 |
| $^{13}C$ | $^{13}C(d,p)^{14}C$ | 5.951 | 0.64 | 5.8 | 0.4 |
| $^{14}N$ | $^{14}N(d,\alpha)^{12}C$ | 13.574 | 1.5 | 9.9($\alpha_o$) | 0.6($\alpha_o$) |
| | | 9.146 | 1.2 | 6.7($\alpha_1$) | 1.3($\alpha_1$) |
| $^{15}N$ | $^{15}N(p,\alpha)^{12}C$ | 4.964 | 0.8 | 3.9 | ~ 15 |
| $^{16}O$ | $^{16}O(d,p)^{17}O$ | 1.917 | 0.90 | 2.4($p_o$) | 0.74($p_o$) |
| | | 1.05 | 0.90 | 1.6($p_1$) | 4.5($p_1$) |
| $^{18}O$ | $^{18}O(p,\alpha)^{15}N$ | 3.980 | 0.730 | 3.4 | 15 |
| $^{19}F$ | $^{19}F(p,\alpha)^{16}O$ | 8.114 | 1.25 | 6.9 | 0.5 |
| $^{23}Na$ | $^{23}Na(p,\alpha)^{20}Ne$ | 2.379 | 0.592 | 2.238 | 4 |
| $^{31}P$ | $^{31}P(p,\alpha)^{28}Si$ | 1.917 | 1.514 | 2.734 | 16 |

Energy spectroscopy of MeV particles, however, can hardly be performed with electrostatic equipment because of the high potentials necessary, and furthermore, because of the dependency of the charge state of the particles on their depth of origin in the solid. Therefore, a different concept of energy spectroscopy is used with MeV particles, based on the ionization chamber principle, i.e., energy is converted - regardless of the ionization state - into a free electrical charge in a semiconductor detector. The silicon surface barrier detector is similar to a photodiode and delivers charge pulses proportional to the energy of an absorbed particle. The spectrum of charge pulses, i.e. the energy spectrum, is recorded with a multichannel analyzer. As a consequence of this spectroscopic technique, nuclear reaction products are recorded simultaneously with lower-energy primary particles scattered back elastically from the solid. With high particle fluxes ($\geq 10^4\ s^{-1}$), the individual charge pulses may overlap in time and simulate particles of higher energy (pile-up effect). Since the cross section for elastic backscattering of the incident particles by matrix atoms can be larger by 2-3 orders of magnitude than the nuclear reaction cross section, the sensitivity of NRA for depth profiling can be severely restricted. Thin foils in front of the detector entrance window can selectively absorb the low-energy fraction, but depth information carried by the penetrating particles may be obscured due to energy straggling in the foil.

Elastic backscattering of the primary high-energy ions may be considered as a special nuclear reaction of the type A(x,x)A. Thereby the energy is conserved, i.e. Q = 0, and the electrostatic repulsion force between the positively charged nuclei of the incident particles or of the target atoms defines the interaction, rather than nuclear forces. The differential cross section for this interaction is described by the well-known Rutherford formula. The principle of the Rutherford backscattering technique (RBS) is shown in Fig. 13.

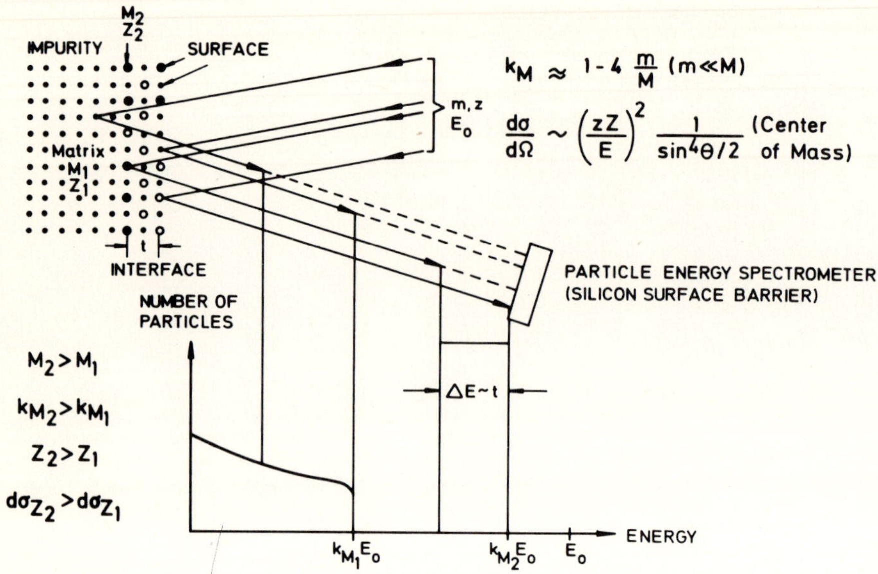


Fig. 13. Principle of Rutherford backscattering (RBS)

A parallel beam of light ions with a mass m and an atomic number z hits a solid surface with an energy $E_o$. The energy distribution of the reflected particles is measured with a semiconductor detector. An elastic collision with a (heavier) target nucleus results in an energy transfer, i.e., the reflected primary particle has a lower energy $k(M)E_o$ ($k(M)\leq 1$) as a result of a collision with a nucleus of the mass M. The larger the mass difference between the collision partners, the less energy can be exchanged, i.e., the energy loss of the impinging particle decreases with increasing M: $k(M_2)\geq k(M_1)$ for $M_2 \geq M_1$. If we consider a sample with matrix constituents $M_1$ which is homogeneously doped with heavier elements $M_2$ within a surface layer of depth t, an energy spectrum results with a structure as shown in Fig. 21. The highest possible energy is due to collisions with impurity atoms $M_2$ at the surface ($E=k(M_2)E_o$). If the same type of collision occurs at the interface between doped and undoped layers, a particle with lower energy will emerge from the surface, because of electronic stopping in transversing the doped layer twice. The energy difference $\Delta E$ between particles colliding with impurity atoms at the surface, or at the interface, is proportional to the depth t. If the impurity atoms are evenly distributed over this depth, a uniform distribution will result in the corresponding energy interval. If the depth distribution of impurities follows a characteristic profile, the same profile will be reflected in the energy distribution. At energies lower than $k(M_2)E_o-\Delta E$, a gap is seen in the energy spectrum extending to the value $k(M_1)E_o$, which corresponds to collisions with matrix atoms at the surface. This gap is present as long as the impurity-doped layer is thin enough and the mass difference between matrix and impurity atoms is large enough. For this case, an unambiguous identification of the count rate in a given energy interval with the corresponding chemical element is obtained. A continuum of counts extends from $k(M_1)E_o$ to zero energy, corresponding to collisions at deeper sites with matrix atoms until the total incoming energy is consumed by electronic stopping together with the collision energy transfer.

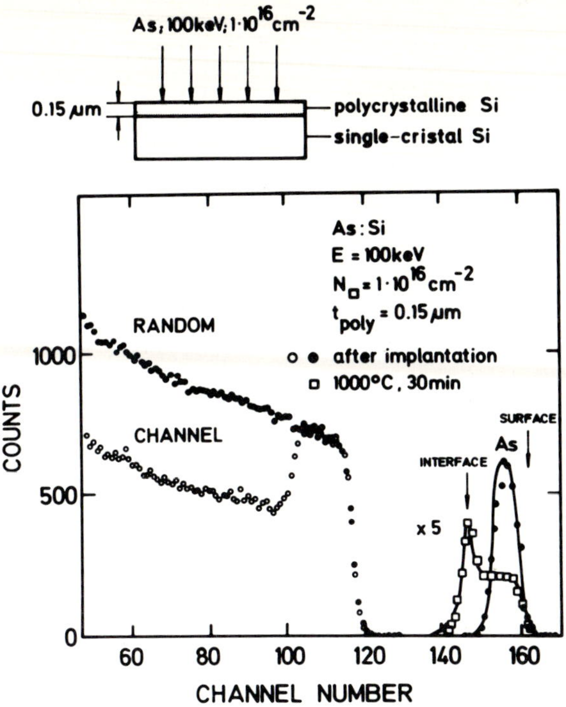


Fig. 14. As distribution in a poly/monocrystalline Si structure, as measured by RBS

While details of RBS will be given in Chapter 5, an example of its unique analytical capability is shown in Fig. 14 [21]. As has been implanted in a polycrystalline Si layer on a monocrystalline Si substrate, with the polysilicon layer serving as a diffusion source. We see the energy spectrum of backscattered helium particles (incident energy 1.6 MeV) as recorded with a multichannel analyzer calibrated so that the channel number on the abscissa is proportional to the energy. The right-hand part of the spectrum originates from collisions with the heavier As atoms, and the As depth distribution can be seen for the as-implanted and the annealed cases. After implantation, a Gaussian distribution is obtained, centered at the mean projected range. With annealing, the As redistributes, resulting in a constant concentration within the polysilicon layer, together with an accumulation at the interface and a sharply dropping tail into the monocrystalline substrate. The concentration value in the polysilicon layer can be quantitatively established without any standard by refering the count rate level to the matrix (left-hand portion of the spectrum). Since only nuclear interactions are involved, the respective yields have to be corrected for the well-known cross sections only, and no chemical effects have to be taken into account, i.e., a backscattering measurement of this type is intrinsically quantitative. A particularly interesting feature in this example is the As enrichment at the interface layer, which has been shown to depend on the pretreatment of the substrate before polysilicon deposition, and is probably due to the presence of a thin oxide layer. While the depth resolution of the method is not sufficient to indicate the top concentration, the area of the peak allows the determination of the areal arsenic density at the interface. Such a result can hardly be obtained by erosion techniques like SIMS, since both the sputtering rate and the secondary ion yield vary unpredictably when crossing the very interface.

Another interesting feature of RBS can be seen from the Si part of the spectrum: If the specimen surface is oriented at a random angle with respect to the incident beam, a continuously varying yield distribution is obtained, increasing slightly towards lower energies, the latter being due to the energy dependence of the cross section (see insert in Fig. 13). But orienting the substrate in a channel direction results in a sharp drop in the backscattering yield for the substrate portion of the Si signal. Thus, the thickness of the polycrystalline layer, which consists of randomly oriented crystallites, is apparent in the spectrum. Generally, by making use of this channeling effect, the crystalline quality of surface layers can be assessed, as will be shown later. From the above example, it is clear that RBS is only a sensitive technique as long as the impurities to be traced are considerably heavier than the matrix constituents. Otherwise, overlapping of the two parts of the spectrum would occur, leading to reduced sensitivity and ambiguities in interpretation. In this sense, NRA, which is sensitive to light elements only (as has been shown before), is complementary to RBS. In the high-concentration domain, however, RBS may be able to determine quantitatively the composition of surface layers with light element components, especially if the substrate contribution is partly suppressed by channeling, and if glancing angle observation of the scattered particles is used to reduce the probing depth. Fig. 15 [22] shows the formation of a thin (6 nm) nitride layer by low-energy implantation (2 keV) of nitrogen, with the aid of an ion milling machine. Furthermore, oxygen, fluorine and carbon signals are present, due to native surface oxide, HF etching and hydrocarbon contamination in the backscattering setup. (A similar study has been performed in [23], using AES for the composition of the surface layer.)

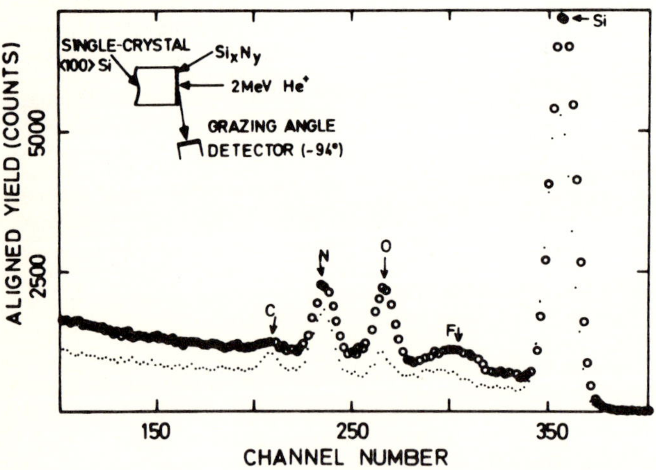


Fig. 15. Grazing angle/channeling backscattering spectra of nitrogen-implanted silicon (implantation energy 2 keV) [22]

To summarize, RBS is a non-destructive tool which is quantitative in concentration and depth scale, with a depth resolution limit of approx. 10 nm (grazing angle technique) and a minimum detectable impurity concentration of the order of $10^{18}$ $cm^{-3}$ for heavy elements. Through the use of channeling techniques, the crystalline quality of surface layers can be evaluated and background signals from a monocrystalline matrix can be effectively suppressed. Thus RBS may ideally complement SIMS measurements in many problems, especially if changing matrix compositions and high concentrations are involved.

The condition of a sufficiently large mass difference between the impurity atoms and the host lattice, with the impurity being the heavier species, has been shown to be essential for depth profiling using RBS. This condition does not hold for many very important systems in implantation work, e.g., P or B in Si, or for most of the problems with compound semiconductors and metal surface technology. To end this section on non-destructive analysis techniques based on homogeneous excitation, a method for boron depth profiling in silicon using a nuclear reaction induced by thermal neutrons is demonstrated. The boron isotope $^{10}B$, present with a natural abundance of 20%, absorbs thermal neutrons with an extremely high cross section of approx. $4 \cdot 10^3$ barns and decays spontaneously according to the reaction $^{10}B(n,\alpha)^7Li$. The reaction products are isotropic and monoenergetic, with 96% of the emitted alpha particles carrying an energy of 1.471 MeV. Energy spectroscopy of these particles is used for boron depth profiling [24]. The thermal neutrons pass a silicon wafer with negligible attenuation, so the reaction efficiency is constant with depth. The analysis has to be carried out in situ, i.e., during irradiation with a thermal neutron beam, conveniently obtained from the beam part of a research reactor. Fig. 16 shows a set of $^{10}B$ depth profiles in Si under different annealing conditions [25], obtained with a neutron flux density of $10^9$ thermal neutrons per cm² and s from the Grenoble research reactor. With this technique, a dynamic range of more than four decades is possible, with the measurements being perfectly quantitative also in the high-concentration region. A total implanted dose of $10^{11}$ $cm^{-2}$ is within the sensitivity of the method. For improved depth resolution, a glancing angle geometry for the alpha particle detection can be employed: With a take-off

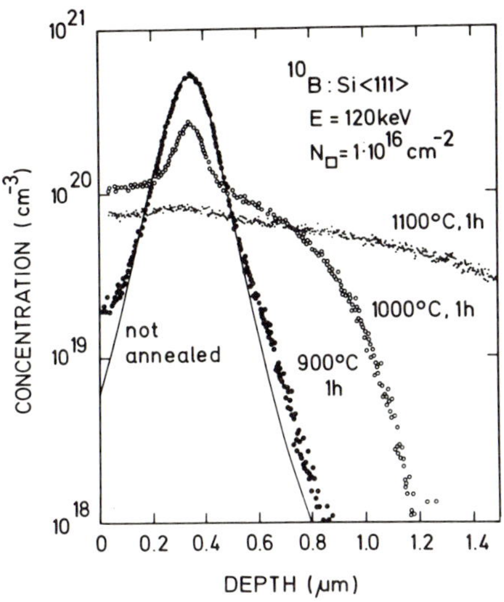


Fig. 16. $^{10}B$ depth profiles in Si using the $^{10}B(n,\alpha)^7Li$ reaction [25]

angle of 10 degrees with respect to the surface, the depth resolution is as good as 10 nm.

A different method based on the same principle is the detection of the emitted particles with the aid of the nuclear track technique: The samples are covered with organic resin foils, which record every single particle absorbed in the foil upon a subsequent chemical exposure treatment. The particle tracks after neutron irradiation and exposure are counted with the aid of an optical microscope. This technique has been used for bulk impurity determination of boron and lithium in various matrices [26] at a ppm level. Applications for depth profiling using bevelled samples should be possible.

### 3.2.2 Resonance Techniques

Again, ion beam induced nuclear reactions are considered. In Fig. 11, the cross section of the $^{16}O(d,\alpha)^{14}N$ reaction was shown as a function of the energy of the incident deuteron, and it was demonstrated that this excitation function varies only negligibly over an energy interval which corresponds to a typical implantation depth via electronic stopping. This led to the concept of homogeneous excitation. On the other hand, a variety of nuclear reactions exhibits sharp resonances with incident energy, with a resonance width small enough to probe an impurity profile by varying the incident energy over a small interval. A good example is the $^{27}Al(p,\gamma)^{28}Si$ reaction occurring at a proton energy of 992 keV, and showing a full width at half maximum of only 0.1 keV. This resonance width would correspond to a depth interval of only approx. 1 nm in nickel by taking into account electronic stopping for 1 MeV protons in this material. In practical applications, however, the achievable depth resolution is limited by the energy spread of the accelerator in the near-surface region and for deeper layers by the additional energy straggling imposed on the proton beam when transversing matter, through the statistical nature of the interaction processes. The reaction yields gamma quanta with energies up to 10 MeV, which

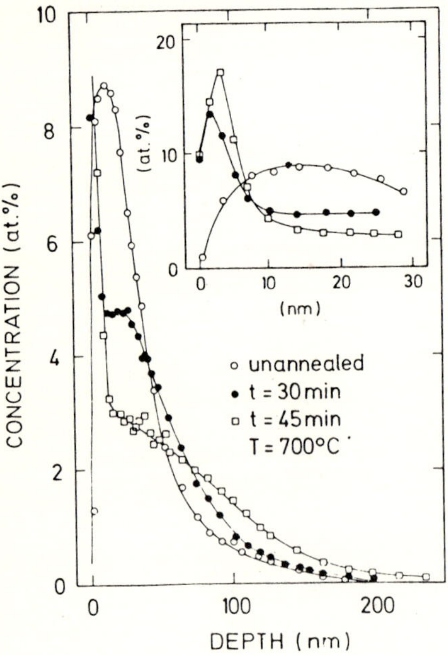


Fig. 17. Concentration curves of aluminum in nickel, as measured with a (p,γ) resonance technique. The insert shows surface enrichment after annealing [27]

can be counted using a scintillation counter and setting a suitable energy window for background suppression. Fig. 17 shows, as an application of this technique, Al depth profiles from implantation of $4.4 \cdot 10^{16}$ ions/cm² into polycrystalline nickel with different annealing conditions [27]. The proton beam energy resolution was about 0.4 keV, corresponding to about 3.5 nm in the depth scale.

There exists a great number of resonance reactions of the $(p,\gamma)$ type, which are compiled e.g. in [18]. Among the resonant reactions leading to particle emission, the $^{18}O(p,\alpha)^{15}N$ reaction occurring at a proton energy of 1167 keV has proven useful in oxidation studies [28]. The main advantage in using resonant techniques is the achievable depth resolution, which is determined by accelerator energy resolution (usually better than 2 keV) rather than by the energy resolution of a semiconductor particle detection system (typically 15 keV), as in the previous section for homogeneous excitation.

## 4. Secondary Ion Mass Spectroscopy (SIMS)

It has already been shown in section 3 that the SIMS method can give excellent results for impurity depth profiling (see esp. Figs. 3.2 and 3.3) with regard to sensitivity, dynamic range and depth resolution, provided that a low concentration of impurities has to be traced in a matrix with uniform sputtering characteristics. On the other hand, reference has been made on several occasions to the implications of the SIMS technique which are a consequence of both the sputtering process and the secondary ion formation. These implications may be dominant when analyzing multilayer structures and interfaces, regions of high impurity concentrations, or compound rather than elemental matrices. The crystallinity of the specimen can also lead to severe distortions of a SIMS measurement. As a matter of fact, effects inherent in this analytical technique may completely obscure even the simplest structures, as is demonstrated in Fig. 18 [29]. This is

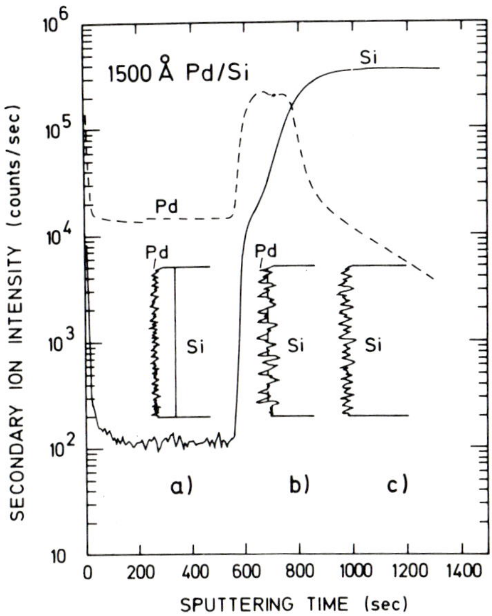


Fig. 18. SIMS multi-element depth profile 1500 Å Pd/Si. Region (a): surface totally within the Pd; region (b): surface intersecting both Pd and Si; region (c): sputtered down to the Si substrate. To be noted are: surface roughening, changes in secondary yield, and ion-induced atomic mixing. 5 keV $Ar^+$ with $O_2$ jet on sample, positive ion detection [29]

a depth profile of a 150 nm Pd/Si sample, showing the detected signals for Pd and Si as a function of sputtering time. Analysis conditions were: 5 keV $Ar^+$ primary ion bombardment, positive secondary ion detection, with an oxygen jet directed onto the sample to increase positive secondary ion sensitivity (see below). As schematically shown in part (a) of the figure, the polycrystalline Pd layer develops a rough surface as it is sputtered away. As the deepest areas between cones break through the Pd/Si interface, a sharp rise is detected in the Si signal. In region (b), both Pd and Si are exposed to the sputtering beam, resulting in only a slow rise in the Si signal until the cones have essentially been etched completely into the Si substrate, corresponding to part (c) of Fig.18. One can see how the artefact of sputter-induced surface roughening has thus completely obliterated the true shape of the depth profiles. A second artefact in SIMS is apparent from Fig. 18: the rise in the Pd signal in region (b) of the depth profile. This can be attributed to the presence of Si, which increases the amount of O sticking to the sample, which in turn increases the Pd ion yield (fraction of sputtered particles emitted as ions). This effect is less pronounced when using inert-gas bombardment without an O jet, but is still present to some degree due to a chemical yield enhancement effect attributable to Si alone. A third disadvantage of SIMS can be seen in region (c) of the Pd depth profile. The slow decrease in the Pd signal beyond 900 s sputtering time can be attributed to an atomic mixing of Pd into Si as a consequence of the ion bombardment.

Many phenomena connected with sputtering and ionization probability are still a subject of research, so that SIMS - in its general application - may still be considered as a semi-quantitative technique. These phenomena are beyond the scope of this presentation, which concentrates on the principles of the method together with the experimental techniques for application to ion implantation work. In order to select the most appropriate analytical conditions and to avoid misinterpretations of SIMS data, however, the intrinsic difficulties of this method have to be clearly recognized.

## 4.1 Sputtering and Secondary Ionization

In a SIMS depth-profiling experiment, the specimen surface is continuously eroded by sputtering. Most of the material leaves as neutral particles, but a small fraction is ejected as positive or negative ions in the form of single atoms or molecular compounds. These secondary ions are then focused into a mass spectrometer to provide an elemental analysis of the sample volume eroded by the primary ion beam. The total ion current of a particular mass M at the output of the mass spectrometer $I_M$ is the signal to be detected and processed. Thereby M is not necessarily an atomic or molecular mass already present in the specimen. It may be formed as well by a reaction with the primary beam (reactive sputtering) or with the residual gas atmosphere, or by a reaction with the matrix atoms during sputtering which leaves the material in a highly excited state. Strictly speaking, we monitor in general a molecular ion mass M with the molecule containing the atom A to be traced. This differentiation is important in the context of detection limits because there is in many cases an ambiguity with respect to M: $^{31}P^-$ and $^{30}SiH^-$ are ions with the same mass number 31 and the presence of hydrogen in the residual gas may limit severely the detection capabilities of P in Si. (Experimental techniques to overcome this problem - mass interference - will be discussed later.)

Under ideal conditions, the measured current $I_M$ at a given time is proportional to the concentration $c_X$ of the impurity X to be traced within the surface layer exposed at that particular time:

Let us assume an impurity concentration $c_X$ low enough not to affect the sputtering process, so that sputtering is completely determined by the matrix consisting of n atoms per unit volume, i.e. $c \ll n$. We define the sputtering yield S as the number of (matrix) atoms sputtered off by the impact of a single primary particle. Then with an impinging electrical current density $j_p$ of primary ions, each carrying the charge q, $y = Sj_p/q$ matrix atoms will be removed per unit area and time. y is called the sputtering rate. (The sputtering front will move with a velocity $v = Sj_p/qn$ into the solid.) By multiplication with $c_X/n$ we obtain the corresponding number of impurity atoms. A fraction f(X,M) of the freed impurity atoms will convert into ionized compounds with mass M and of these, a further fraction $\eta$ will be collected and transmitted through the mass analyzer to hit the detector. The resulting current $I_M$ is thus composed of factors describing the sputtering process, the formation of secondary ions and the detection characteristics of the instrument:

$$I_M = S\, j_p A/n\, f(X,M)\eta c_X\ . \qquad (4.1)$$

Here it is assumed that the primary ions cover homogeneously an area A which is fully within the acceptance field of the mass analyzing system. $Sj_pA/qn$ is the volume sputtered off per unit time, f(X,M) the probability for the formation of a molecular ion M from an impurity X, and $\eta$ the transmission of the mass analyzer.

The minimum concentration which can be measured, if all background effects are disregarded, is thus limited by the minimum detectable current and the transmission of the instrument, together with the material consumption rate. The latter factor is irrelevant for bulk analysis, but important for dynamic measurements, i.e. depth profiling: If the profile varies strongly with depth, a distortion can only be avoided by using a sufficiently large sampling area.

## 4.1.1 Sputtering Yield

For sputtering, inert-gas ions, mainly $Ar^+$ ions, are used because they give the highest sputtering yield and do not react with the specimen chemically. Other important sputtering ions are $O^+$ and $Cs^+$, because of the enhancement of the secondary ion yield (see next section) which is obtained for many elements in the presence of these atoms at the surface. The energy of the bombarding ions is generally between 1 keV and 20 keV. Below 1 keV, the sputtering yield of many materials may become very low, whereas on the contrary, there is no advantage to bombarding beyond approx. 20 keV energy, since the sputtering yield changes very little in this energy range and even tends to decrease at higher energies. When bombarding with reactive ions, e.g. oxygen, the surface layer is chemically altered by implantation and oxides are formed.

There is a pronounced dependence of the sputtering yield on the angle of incidence of the primary beam with respect to the surface normal: S rises steadily from normal incidence to an angle of about 60-70 degrees, then falls off rapidly. The increasing part of the angular dependence is

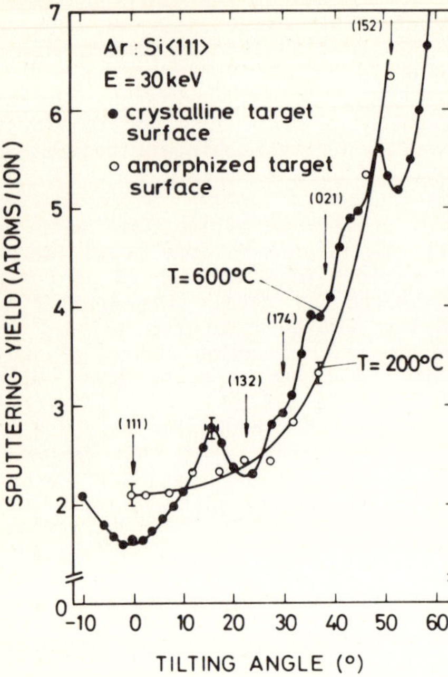


Fig. 19. The sputtering yield as a function of incidence [30]

shown in Fig. 19 [30] for 30 keV $Ar^+$ ions impinging on a (111) Si crystal rotated around the [112] axis. Furthermore, it can be seen from this figure that with sputtering, the surface layer is virtually amorphized by the damage produced by the bombardment. Only through sputtering at high temperatures (600°C), part of the damage anneals out instantaneously, so that channeling patterns appear in the secondary ion yield. For metals, the latter effect is more pronounced than for semiconductors, leading often to a microstructure with polycrystalline layers as in the example of Fig. 18.

When sputtering a compound matrix rather than a monoatomic one, different sputtering yields for the constituents have to be considered. This leads to a surface layer composition, approx. within the range of the sputtering ions, different from the bulk composition: The surface layer is enriched by the slower sputtering component under equilibrium sputtering conditions. In order to establish this equilibrium, a layer thickness corresponding to the range of the primary ions has to be sputtered off, e.g., 5-10 nm for a compound with a stopping power similar to silicon and 10 keV $Ar^+$ ions - data points for smaller depths are not quantitative. For very shallow profiles, substantially lower primary ion energies are necessary. The same compositional changes within the surface layer may occur with high-concentration impurity profiling, and especially with reactive sputtering at the surface and at interfaces.

### 4.1.2 Secondary Ion Emission

In Eq. 4.1, the factor f(X,M) has been defined as the probability that an impurity atom X in a given matrix will be found in a charged compound with the mass M after being sputtered off. f depends on the chemical nature of the atom X itself, e.g. its ionization potential, as well as on its surroundings in the surface layer which is structurally, and in the case of reactive sputtering and/or surface adsorption of residual gases chemically, altered. Furthermore, back-neutralization effects by electron emission from the surface have to be included. The questions involved are far too complicated to be discussed here, so that only a few features can be given, together with some experimental material. In ion implantation work in general, this complexity does not significantly impede quantitative depth profiling, because the empirical use of standards is routine.

For inert sputtering in a sufficiently clean vacuum ($\leqslant 10^{-6}$ Pa), compound formation can be neglected, so that the ionization process is basically due to the kinetic energy transfer of the primary ions. We find an ionization yield $f(X,X)=S^+$ (defined as the number of positively ionized atoms to total atoms) varying approximately inversely to the ionization engergies of the elements for the formation of positive ions. For 12 keV $Ar^+$ bombardment, the values range from $10^{-4}$(Te) to more than $10^{-1}$(Ga) as measured in the pure of elemental matrix. From the right to the left side of the periodic table of elements, i.e. with increasing electronegativity, the tendency of negative ion formation increases as well.

Figure 20 [31] shows the influence of oxygen adsorption at the surface on the count rate, and thus the ionization yield, for boron-doped silicon. An enhancement of two orders of magnitude is found for the $Si^+$ signal, and the $B^+$ signal is stronger by three orders of magnitude if the oxygen pressure allows for the adsorption of a monolayer under the given sputtering conditions. This so-called chemical emission is used in many SIMS experiments, either by directing an oxygen jet onto the analyzed surface with inert ion sputtering, or by sputtering with oxygen ions. Especially for electronegative ions, a similar yield enhancement is possible using a $Cs^+$ beam for sputtering. With reactive sputtering, as already mentioned, compound ions with constituents from the sputtering gas are formed besides the elemental ions, and may be advantageously used for analysis.

The energy distribution of secondary ions is characterized by a peak at low energies (~ 10 eV) and a tail towards higher energies, the slope of which depends upon the number of constituents of the ion: The larger the cluster, the more rapidly the energy distribution tails off, as shown in

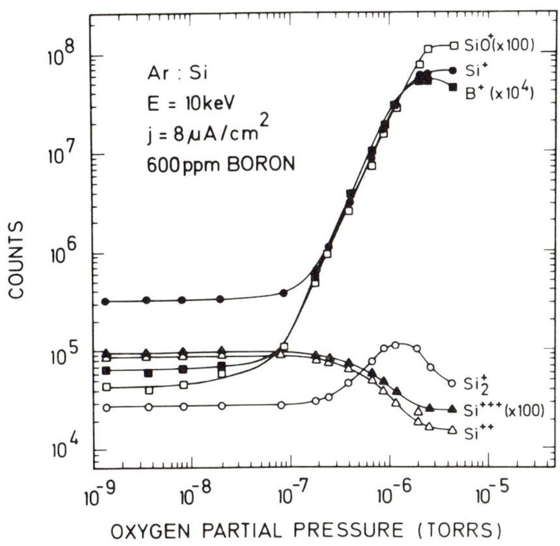


Fig. 20. Influence of oxygen adsorption on secondary ion yield [31]

Fig. 21 [32]. The energy is measured by variation of the target potential with a constant energy window at the mass spectrometer entrance. Accordingly, molecular ions of interfering mass can be suppressed by adequate setting of the energy window of the quadrupole mass spectrometer, as shown for $^{75}As^+$ and $^{30}Si^{29}O^+ (Si_2O^+)$ in Fig. 22 with the target potential set at -15 V, thus gaining a factor of 100 in detection sensitivity.

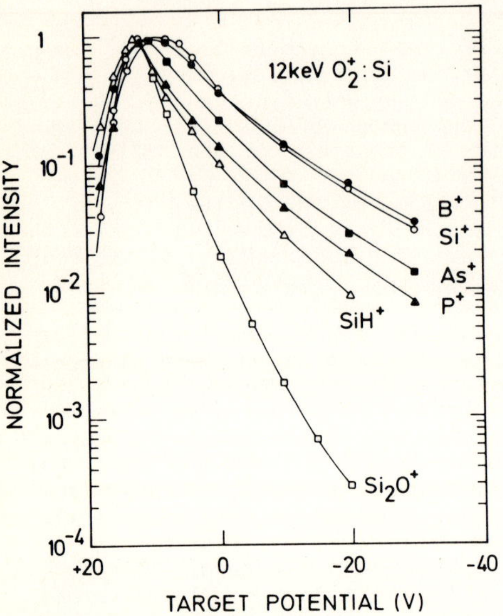


Fig. 21. Normalized intensity distributions of secondary ions emitted from silicon specimens homogeneously doped with either boron, phosphorus, or arsenic [32]

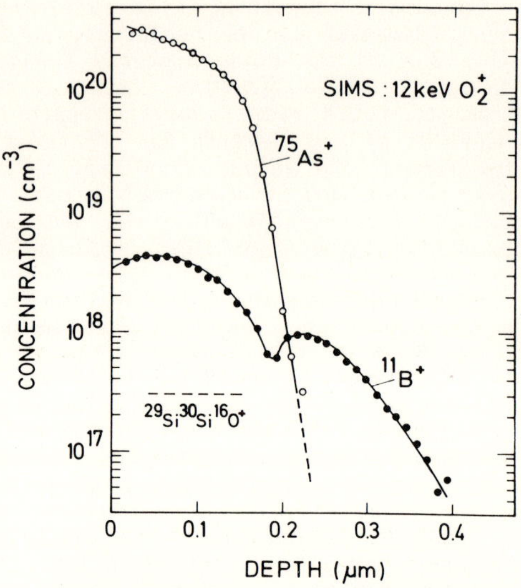


Fig. 22. Concentration profiles of boron and arsenic implanted-diffused in silicon. Target potential $V_T \approx -15$ V. The corresponding intensity level of $^{29}Si^{30}Si^{16}O^+$ is indicated for comparison [32]

## 4.2 Experimental Techniques

### 4.2.1 Instrumentation

Surface analysis by SIMS falls into two categories, as determined by the intensity of the primary beam: low-current density sputtering and high-current density sputtering. A low-current sputtering analysis results in a very small fraction of the surface layer being consumed for analysis, and approaches the requirements of a true surface analysis method; it is generally known as Static SIMS (SSIMS). For in-depth profiling, high-current density sputtering is mandatory (Dynamic SIMS). A modern SIMS configuration suitable for depth profiling of implanted impurities is typically based on the following principles:

(1) In the primary column, an ion gun produces positive or negative argon, oxygen or caesium ions with energies between 0.5 and 20 keV at current densities up to a few milliamps per $cm^2$. The ion beam is focused at the target to a spot size with a diameter of 10-50 μm, and rastered electrostatically over an area of a few hundred μm to ensure a flat-

bottomed sputtering crater. The ion gun consists of a duoplasmatron source (or a liquid metal source for Cs), electrostatic extraction, acceleration and focusing components. It is differentially pumped, and sometimes mass analyzed by a magnet. Electrostatic beam deflection may prevent neutrals from hitting the target.

(2) Secondary ions are extracted by a suitable potential (100 V - 5000 V) and energy filtered by an electrostatic analyzer (ESA). The ions transmitted through the ESA are subsequently mass analyzed either by a quadrupole mass filter or a magnetic prism. Ions of the selected mass, transmitted through the mass filter, are counted using a channeltron (a device similar to a photomultiplier). With a magnetic mass spectrometer, mass resolution $\Delta M/M$ can be as high as $10^4$; with a quadrupole mass filter, it is between $10^2$ and $10^3$.

(3) The raster gate technique is employed to avoid signal contributions from the wall of the sputtered crater where various depths are exposed to the sputtering beam. With this technique, the data are recorded only if the beam is striking the central flat portion of the crater as shown in Fig. 23. This is achieved by electronic gating.

(4) When investigating insulating surface layers, an electron gun floods the surface to prevent electrostatic charging of the specimen surface.

(5) Ultra-high vacuum conditions prevail in the analysis chamber; the specimens are conveniently introduced via a vacuum interlock system.

SIMS instrumentation ranges from custom-designed laboratory equipment to general-purpose commercially available instruments, sometimes including electron spectroscopic techniques such as AES. Although lateral mapping of impurity distributions is usually not the task in ion-implanted impurity depth profiling, two different basic instrumental concepts may be conveniently distinguished, according to their different approaches for the generation of lateral images:

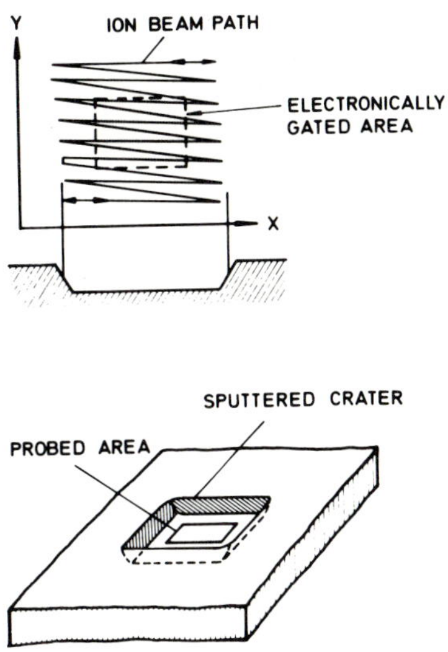


Fig. 23. Principle of the raster gate technique

(A) The Scanning Secondary Ion Microprobe where the area of interest is rastered by the primary beam, and image generation is analogous to the principle used in a SEM. Lateral resolution is determined by the minimum spot size of the primary ion beam (between 1 and 10 µm spot diameter with the most commonly used duoplasmatron ion sources, depending on current and energy). These instruments use either a quadrupole mass spectrometer [33] or a magnetic prism [34] for mass selection.

(B) The Secondary Ion Microscope or Secondary Ion Imaging Mass Spectrometer [35] with direct imaging. It generates, for one particular mass, a magnified image of the surface spot excited by the primary beam via ion-optical projection onto an image plate. The lateral resolution is determined by the ion-optical system, and can be as good as a few tenths of a µm. This instrument employs a double-focusing mass spectrometer using a magnetic prism. It can also be operated in the scanning mode for depth profiling.

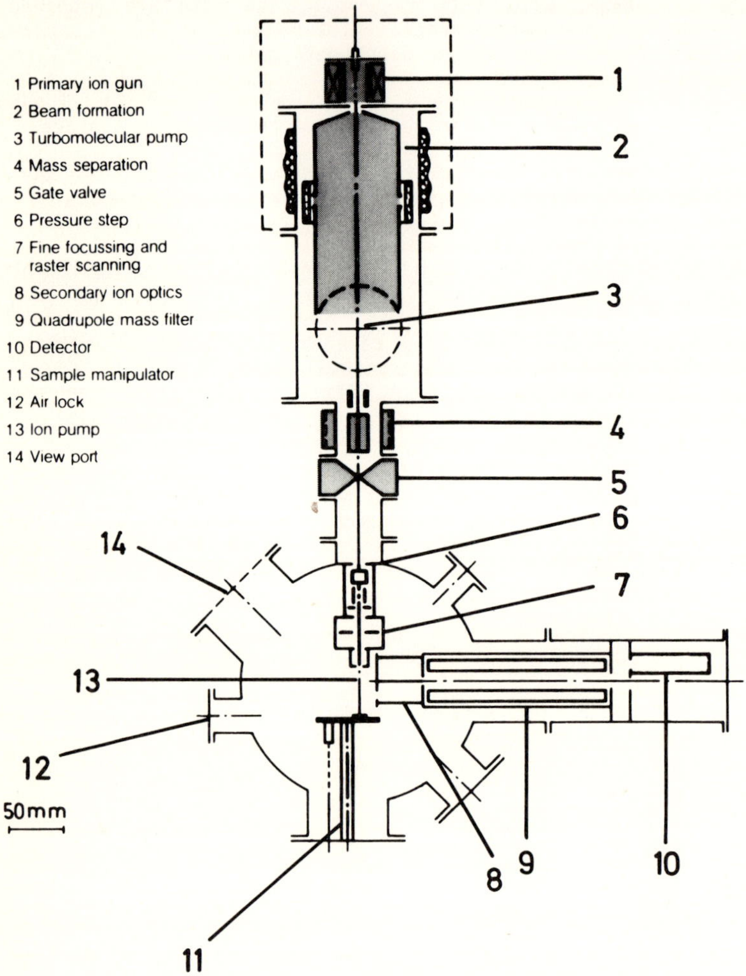


Fig. 24. A-DIDA scanning secondary ion microprobe, schematic [36]

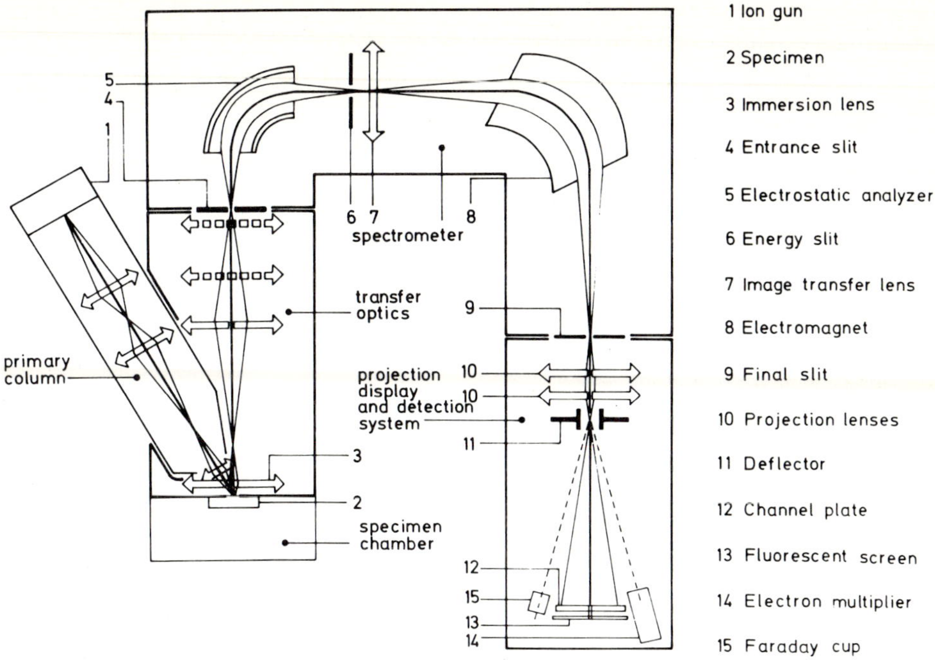


Fig. 25. I.M.S. 3f direct-imaging secondary ion microscope, schematic [37]

A schematic view of a commercial instrument of type (A), employing a quadrupole mass spectrometer, is given in Fig. 24 [36]. Fig. 25 shows the operating principle of an ion microscope of type (B) [37].

### 4.2.2 Depth Resolution, Dynamic Range, Sensitivity

It is very difficult, if not impossible, to present a general discussion of SIMS instrumentation performance. This is due to the great variety of instrumental hardware, operational settings and analytical requirements. Optimum SIMS performance is often a semiempirical compromise among a considerable number of factors. With in-depth profiling of impurities, important figures of merit are depth resolution, dynamic range and sensitivity of measurement. In the following, some important aspects will be mentioned in terms of these quantities.

Depth Resolution. While the thickness of the emission zone for secondary particles is only 0.5 - 5 nm, depending on the primary ion energy, the depth resolution obtainable with in-depth profiling is usually much worse. Apart from instrumental irregularities, such as inhomogeneities of the primary current over the sputtered area, the limiting factor is the sputtering process itself, leading to atomic mixing (recoil implantation, cascade mixing) even in the case of monocrystalline or amorphous perfectly flat surfaces. Surface irregularities and/or preferential sputtering with polycrystalline specimens introduce progressive distortions in the depth profile with increasing sputtering depth. Some of the effects mentioned can be partly suppressed by using low primary energy and off-normal beam incidence.

Dynamic Range. The dynamic range is defined as the range in concentration over which the ion of interest can be monitored accurately. The factors which affect the dynamic range of an analysis are [29]:

- crater edge effects
- neutral beam effects
- absolute count rate and background.

The influence of some of these factors on depth-profiling performance is schematically shown in Fig. 26.

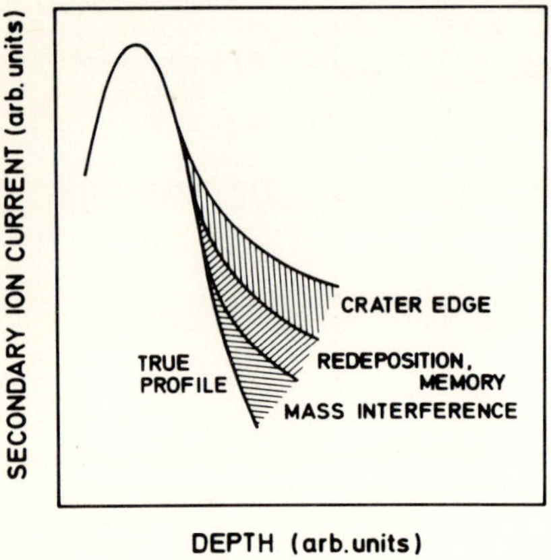


Fig. 26. Limitation of dynamic range by various factors

Crater edge effects can be effectively suppressed by choosing a sufficiently small area for signal pick-up with the raster gate technique, at the expense of overall yield. Another solution to overcome this problem would be structuring of the surface layer by photolithography, i.e., mesa etching of an area smaller than the area scanned by the primary beam, or selective implantation through an appropriate mask. With the same preparation technique, one can eliminate the influence of the neutral non-focused component of the primary beam which may strike the sample, thereby producing secondary ions from sections outside the sputtered area. An instrumental solution to the neutral primary ion problem is to use an electrostatic primary beam deflector [2].

Background can be due to mass interference, either from surface reactions with the residual gas atmosphere, or from matrix constituents and their molecular compounds. In the former case, improved vacuum conditions and/or faster sputtering will reduce the background; in the latter case, energy discrimination is sometimes possible, as has been shown in Figs. 21 and 22, but generally, a mass-resolving power is needed which is able to separate atoms and compounds with the same mass number through the small mass defect due to the nuclear binding energies. This is possible with a magnetic mass spectrometer, as employed in the direct-imaging SIMS microscope. Fig. 27 [38] shows a mass spectrum as recorded from hydrogenated, phosphorus-doped amorphous silicon with a mass resolution better than $6 \cdot 10^3$. The mass defects are clearly resolved, making depth profiling of P in a-Si:H possible, as shown in Fig. 28 [38].

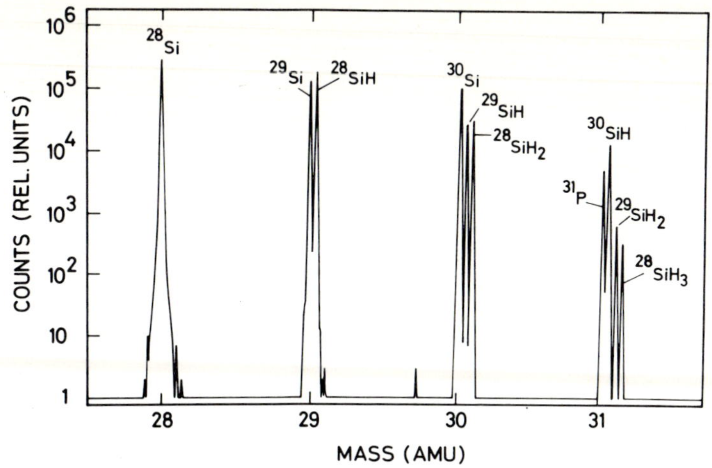


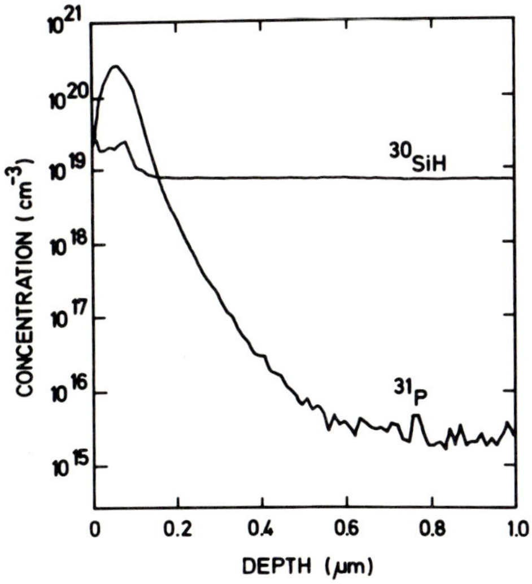


Fig. 27. High mass resolution ($m/\Delta m \geq 6 \cdot 10^3$) secondary ion spectrum for P in a-Si:H [38]

Fig.28. Depth profile of P implanted in a-Si:H, recorded with high mass resolution [38]

Another source of background signals in depth profiling may be due to redeposition of material sputtered from the crater wall onto the crater bottom. Furthermore, material sputtered onto the extraction lens may be redeposited onto the analyzed area (memory effect).

Sensitivity. The sensitivity in depth profiling is defined as the minimum detectable impurity concentration, and is largely set by the background, which in turn may be determined by the preceding profile measurement (e.g. memory effect). With increasing sputtering depth and increasing sputtering speed, a background reduction is possible. A further possibility for sensitivity gain may be obtained by subsequently reducing the analyzed area.

## 5. Rutherford Backscattering (RBS)

The principle of elastic backscattering of MeV light ions from surface layers has already been discussed in section 3, in the context of homogeneous excitations, and especially nuclear interactions (see Fig. 13). RBS is an elastic interaction, i.e., energy and momentum are conserved (backscattering kinematics); the interaction force is the electrostatic repulsion between the positively charged nuclei of the projectile and the target (Rutherford cross section); and finally, depth information is obtained by the energy loss of the incoming or the backscattered projectile on its essentially straight path in the solid to and from the scattering site (electronic stopping). These simple concepts, together with a set of reliable data, make RBS analysis transparent and quantitative, inasmuch as the analytical zone remains essentially unchanged by light-ion bombardment.

RBS analysis has been thoroughly treated by Chu et al. [39]. In this section, we follow in general their presentation and nomenclature, with emphasis on impurity depth profiling applications.

## 5.1 Concepts of RBS

### 5.1.1 Energy of Backscattered Particles

The backscattering geometry considered in the following is shown in Fig. 29: A particle (mass m) with energy $E_o$ incident at an angle $\theta_1$ with respect to the surface normal penetrates in a straight direction into the solid, until a nuclear collision occurs at a depth t. Thereby it is slowed down to an energy $E \leq E_o$ by the stopping power of the traversed layer. After the scattering event, the energy is further reduced to $k_M E$, due to kinetic energy transfer to the target atom (mass M) originally at rest. $k_M$ is called "kinematic factor", and is obtained by application of the laws of energy and momentum conservation:

$$k_M = \left(\frac{m\cos\theta + \sqrt{M^2 - m^2\sin^2\theta}}{M+m}\right)^2 \tag{5.1}$$

with the scattering angle $\theta = 180° - \theta_1 - \theta_2$. For $\theta = 180°$ and $m \ll M$, (5.1) reduces to:

$$k_M \approx 1 - 4\frac{m}{M} \tag{5.2}$$

which is a convenient formula for estimations. Using this formula, one obtains for the mass resolution $\delta M$:

$$\delta M = \frac{M^2}{4m} \cdot \frac{\Delta E}{E_o} \tag{5.3}$$

for ions scattered from the surface ($E=E_o$), with $\delta E$ being the energy resolution of the particle spectrometer. For the most frequently used case, i.e. backscattering of 1.5 MeV He ions (m=4) and an energy resolution of 15 keV, nearly independent of energy as typical for silicon surface barrier detectors, one obtains $\delta M \approx 0.5$ amu for Si and $\delta M \approx 25$ amu for Au; i.e. for heavy elements, mass resolution is essentially lost due to the small energy transfer. Using higher energies would increase mass resolution, but this is mostly not feasible, because of the resulting loss in depth resolution (see below).

The evaluation of energy losses due to electronic stopping is schematically shown in Fig. 30. A typical curve for the energy dependence of the stopping power dE/dx for a given material is drawn, and the energy path of

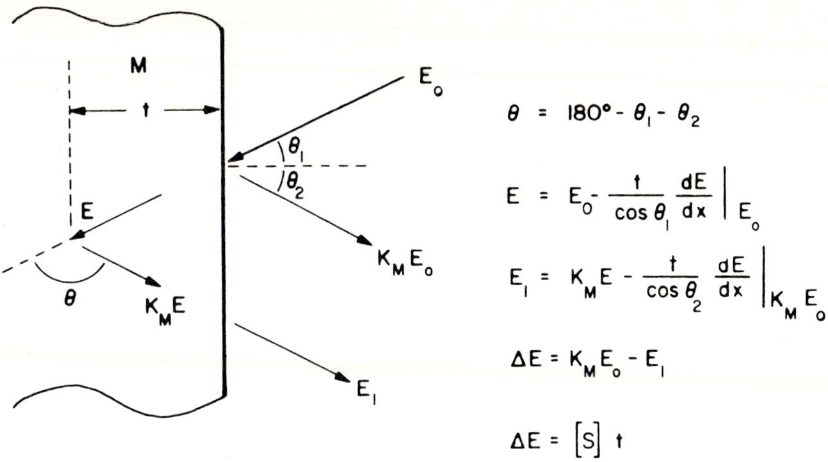


Fig. 29. Near-surface depth analysis by RBS

the particle is marked on its way from the surface to the scattering site ($E_0 \rightarrow E$) and from the scattering site again to the surface ($kE \rightarrow E_1$). In the near-surface approximation, the variation of dE/dx with energy is neglected within the incident or outgoing path, and dE/dx values are taken at the energy $E_0$ or $kE_0$. This approximation is valid for most ion implantation related problems, i.e., for profiling within less than one micron from the surface. By combining energy losses from electronic stopping and nuclear scattering, the backscattering energy loss factor $\lfloor S \rfloor$ is obtained, which allows the conversion of the total energy loss $\Delta E = E_0 - E_1$ of the backscattered particle to the depth of the scattering event:

$$\Delta E = [S]t \quad \text{with} \quad [S] = \frac{k_M}{\cos\theta_1} \frac{dE}{dx}\Big|_{E_0} + \frac{1}{\cos\theta_2}\Big|_{k_M E_0} \quad . \tag{5.4}$$

The depth resolution $\delta t$ is connected with the energy resolution $\delta E$ by:

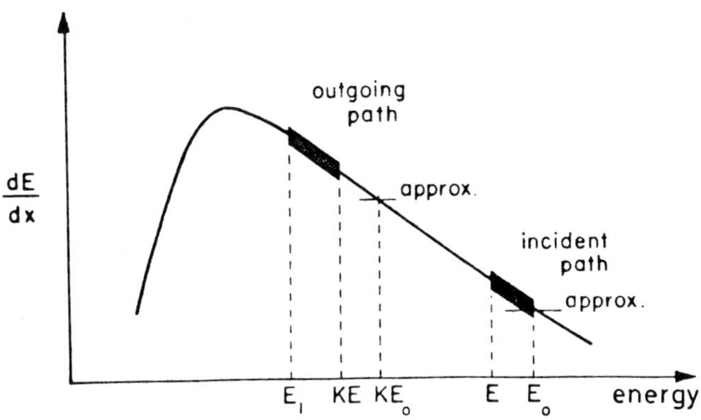


Fig. 30. Stopping power curve and energy path

$$\delta t = \frac{\delta E}{[S]} \quad . \tag{5.5}$$

Since both dE/dx values needed are in the descending portion of the stopping power curve in Fig. 30, [S] decreases with increasing energy. Assuming again an energy resolution independent of the energy, it can be seen that best depth resolution is obtained with maximum [S], i.e., the incident energy should be chosen so that the energy path of the particle in the stopping power curve is located near the top of this curve, which is around 0.5 MeV for particles in silicon and around 0.9 MeV in gold. In ion implantation, the impurity to be profiled is usually well known, so that the energy is adjusted to optimum depth resolution rather than mass resolution.

The stopping power of a given material is made up additively by the different constituent atoms (Bragg's rule), i.e. if the material is composed of $n_A$ atoms A per cm³ and $n_B$ atoms B, the following relation holds:

$$\frac{dE}{dx} = n_A \varepsilon^A + n_B \varepsilon^B \quad ; \tag{5.6}$$

$\varepsilon^A$ and $\varepsilon^B$ are termed the stopping cross sections of the atoms A and B, with the dimension eVcm². The quantitative strength of RBS is mainly due to this additivity, i.e. the absence of chemical effects, and the fact that ε varies slowly and smoothly with the element's atomic number, e.g.,

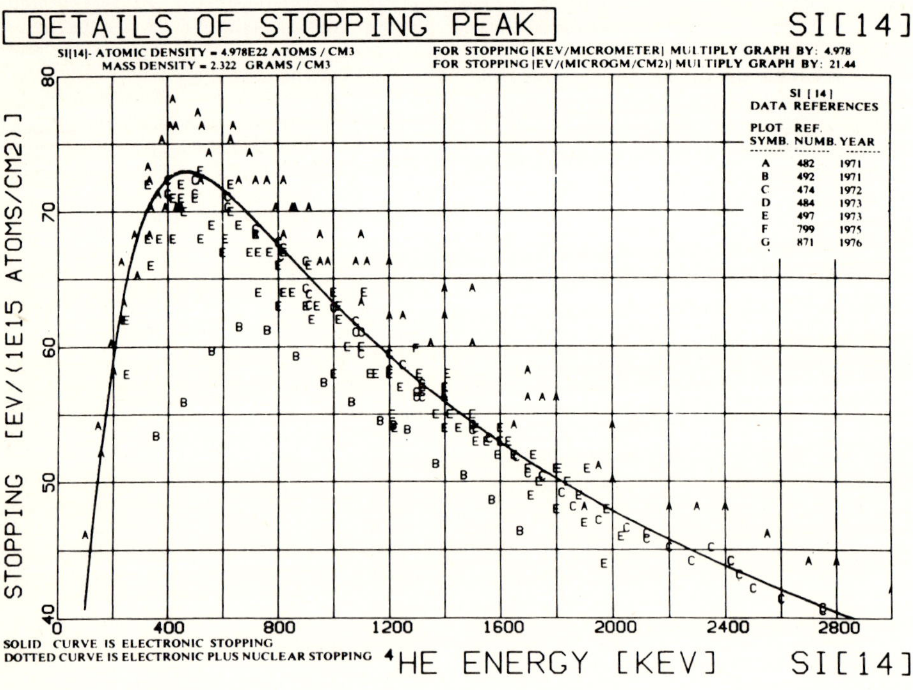


Fig. 31. Stopping cross section for ions in Si [19]

for 1 MeV helium particles, ε is $63 \cdot 10^{-15}$ eVcm² in Si, and $125 \cdot 10^{-15}$ eVcm² in Au. As a consequence, with an impurity concentration of the order of a few percent or below, the electronic stopping is completely determined by the matrix, and the depth scale is concentration independent. Values for ε have been extensively determined for most elements. A compilation and evaluation of the huge amount of data material has been published by Ziegler [19]. Fig. 31 is a reproduction of one page (silicon) of this work.

### 5.1.2 Differential Scattering Cross Section

The differential cross section has already been defined in Section 3.2.1. For Coulomb scattering as in RBS, it is given by:

$$\frac{d\sigma}{d\Omega} = \left(\frac{zZe^2}{4E}\right)^2 \frac{4\{\cos\theta + [1-(\frac{m}{M}\sin\theta)^2]^{1/2}\}^2}{\sin^4\theta[1-(\frac{m}{M}\sin\theta)^2]^{1/2}} \qquad (5.7)$$

Z and z are the atomic numbers for the target and the projectile nucleus, respectively, E the projectile energy immediately before scattering, and e the elementary charge. The first factor at the right side of Eq. 5.7 is not dependent on the scattering angle. It is 1 barn for 1.0 MeV helium scattering by silicon. For a typical backscattering geometry with $\theta = 165°$ and a solid angle of $10^{-3}$ sterad (i.e., a detector with a sensitive area of 50 mm² at a distance of 150 mm from the target), with a helium beam current of 0.1 µA at 1 MeV, we obtain a count rate of 8.5 counts per second for a monolayer of silicon ($1.4 \cdot 10^{15}$ atoms per cm²). If one percent of the silicon atoms is substituted by arsenic the count rate in the As peak would be 0.5 counts per second.

The simple Z² dependence of the cross section, and thus the backscattering yield, is a further strength of RBS, especially in multielement analysis. In the case of impurity profiling in a known matrix, this dependence allows for an absolute concentration calibration by comparing the respective yields.

## 5.2 Experimental Technique

The basic equipment for RBS and other experiments with energetic light ions (e.g. NRA) is a particle accelerator, usually a one-stage Van de Graaff generator or a tandem accelerator, a deflecting magnet and vacuum beam lines connected to one or more scattering chambers. A possible configuration is shown in Fig. 32. A MeV helium beam with an energy stability of approx. ± 2 keV and a current of typically 10-100 nA is used for most applications in implantation studies. The beam is confined by apertures in the beam line to a diameter of about 2 mm, and is guided by active slits. It may be focused to much smaller diameters with a magnetic quadrupole, or further subdivided by apertures. The scattering chamber is evacuated to approx. $10^{-4}$ Pa. It is equipped with the specimen holder, usually mounted on a two-axes goniometer and/or an xy stage. Furthermore, the scattering chamber contains the particle detector and a movable Faraday cup. The particle detector yields, for each absorbed particle, a current pulse, which is integrated in the preamplifier (attached to the chamber) to give a voltage pulse proportional to the energy of the particle. After signal processing in an amplifier/filter chain, the pulses are fed into a multi-

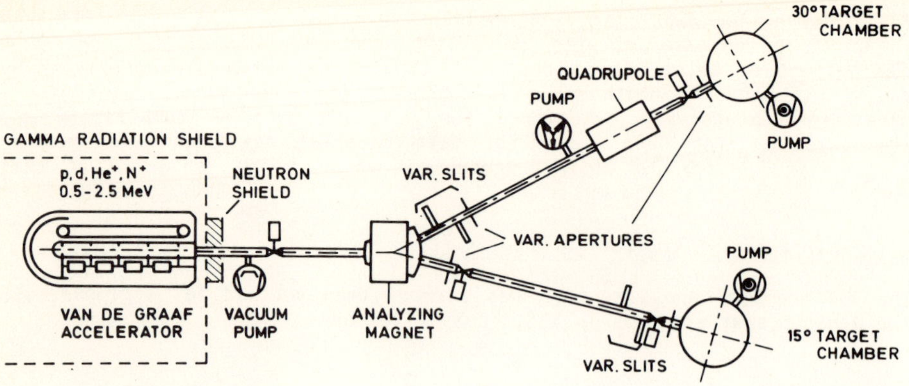


Fig. 32. Typical configuration for RBS, NRA and channeling measurements using energetic light ions

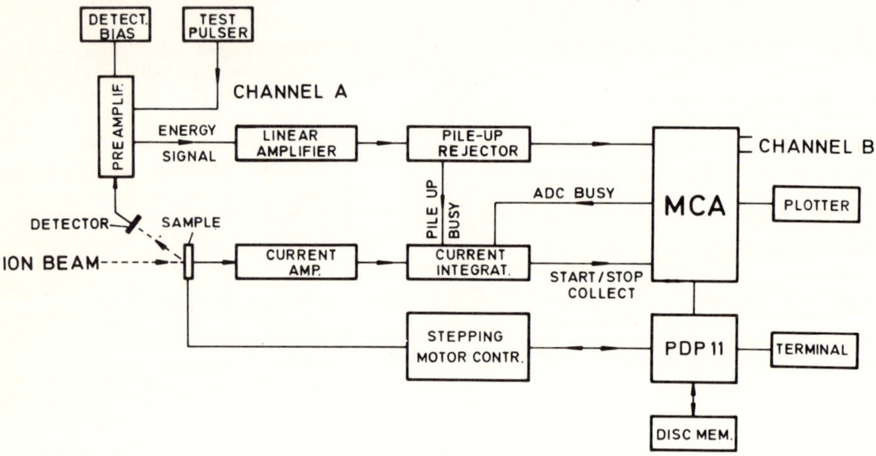


Fig. 33. Schematic arrangement for particle energy spectroscopy with RBS

channel analyzer (MCA), which transforms the voltage level by analog-digital conversion into a number (channel number) and displays the number of collected events for each channel, i.e., the energy spectrum is obtained at the CRT and is stored in a magnetic core or semiconductor memory for further processing. A more detailed schematic arrangement for RBS spectroscopy is shown in Fig. 33. The analysis time for RBS measurements is typically 20 min; pump-down time is of the order of 10 min.

## 6. Damage Evaluation by Channeling Techniques

Various analytical techniques allow investigations of the damage produced by ion implantation and its annealing behavior. MeV ion channeling, X-ray topography, transmission electron microscopy, optical methods, or electron-paramagnetic-resonance measurements are applied to extract information regarding classification and quantification of lattice disorder. Ion implantation generates at first a layer of randomly displaced lattice atoms in a depth somewhat shorter than the projected range of the ions, where maximum energy transfer occurs due to nuclear stopping. At high concentrations, the displaced atoms may form amorphous clusters, and amorphization of the crystal may extend up to the surface. Furthermore, the implanted ions occupy random positions with respect to the crystal lattice. These primary defects rearrange with annealing, which may already partly occur during implantation. This leads to a reconstruction of the lattice, with the impurities at substitutional lattice sites or at well-defined interstitial positions, or to the formation of extended damage configurations such as dislocations, stacking faults or other point defects. Extended damage structures are difficult to remove by annealing and are therefore referred to as residual damage. For high concentrations above thermal solubility, impurities precipitate to form clusters. Of the various methods mentioned above for the investigation of damage and impurity incorporation, only MeV ion channeling is discussed in this section. It is widely used for a semi-quantitative evaluation of primary-damage depth profiles and the determination of the substitutional fraction of impurities. For an analysis of extended defects, the application of X-ray topography or transmission electron microscopy is mandatory, although some limited information may be obtained from channeling studies as well. A discussion of optical methods (refractrometry, ellipsometry, absorption, etc.), or of paramagnetic resonance techniques, is beyond the scope of this contribution.

## 6.1 Channeling of Energetic Light Ions

In a RBS experiment with single-crystalline specimens, a sharp drop is observed in the backscattering yield (i.e. the total number of backscattered particles) if the axis of the analyzing ion beam is parallel to one of the various crystallographic axes of the crystal. This result can be qualitatively understood with the aid of a macroscopic lattice model (Fig. 34): the diamond-type crystal appears transparent if viewed from an axial (c) or a planar (b) direction compared with a random angle of view (a).

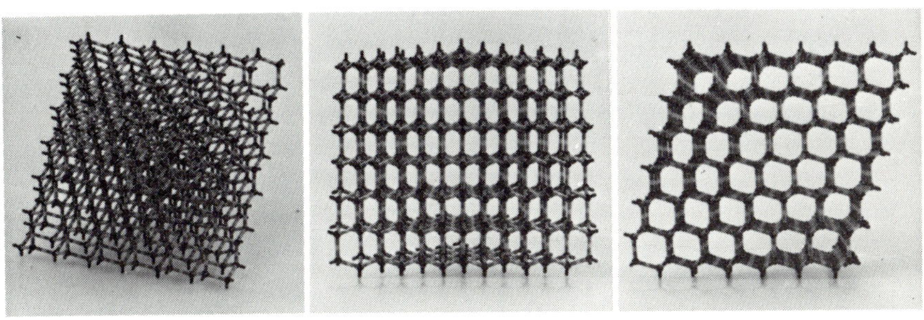

Fig. 34. Diamond-type lattice model viewed along random (a), planar (b), and axial (c) directions

The crystal symmetry is reflected in the angular pattern of the backscattering yield, as shown in Fig. 35. This pattern has been obtained by tilting a silicon crystal in subsequent steps around two perpendicular axes normal to the (110) direction, and recording the backscattering yield for 2 MeV helium particles for each angular position. Planar channeling directions show up as straight lines with lower yield. Their intersection

Fig. 35. Backscattering yield as a function of tilting around two perpendicular axes normal to the (110) direction

defines the (110) axial direction; areas with high backscattering yield represent random incidence.

The angular resolution in Fig. 35 is 0.1 degrees per point, from which a full width at half maximum of the (110) direction crater of approximately 0.8° can be obtained, i.e., if the direction of the incoming beam coincides within $\psi_c = \pm 0.4°$ with the (110) direction, the backscattering yield is reduced by at least a factor of two compared to random incidence. $\psi_c$ is often referred to as the critical angle for channeling. This somewhat surprisingly slow dependence of the backscattering yield is explained by a steering effect of the lattice atom rows ,as schematically depicted in Fig. 36. Particles entering the channel with an angle $\psi \leq \psi_c$ are totally reflected from the potential walls made up of the superimposed potentials of the linear row defining the channel boundaries. For $\psi \geq \psi_c$, the particles are able to leave the channel, resulting in a random trajectory. A rough estimation for $\psi_c$ can be obtained using Lindhard's formula:

$$\psi_c = \left(\frac{2Z_1Z_2e^2}{Ed}\right)^{1/2} \approx 0.3 \left(\frac{Z_1Z_2}{Ed}\right)^{1/2} , \quad \psi_c \text{ in degrees} \qquad (6.1)$$

with the incident energy E in MeV, the atomic numbers of target $Z_2$ and projectile $Z_1$, and the atomic spacing d along the channel in Å. The minimum backscattering yield for axial channeling is between 5 and 10 percent of the random yield with perfect crystals, and depends highly on the surface conditions.

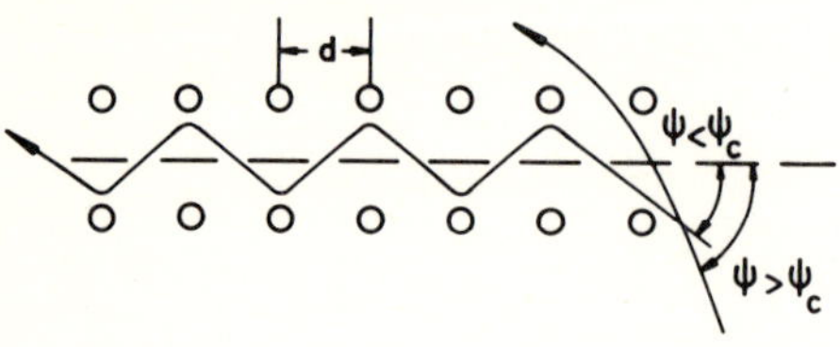


Fig. 36. Trajectories of a channeled and a random particle

## 6.2 Channeling Spectra and Lattice Disorder

In order to interpret channeling spectra, i.e. RBS spectra from a single crystal oriented in a channeling direction, different contributions must be distinguished giving rise to backscattering events, as schematically depicted in Fig. 37:

- Direct backscattering from surface atoms.
- Direct backscattering from displaced atoms.
- Forward scattering from displaced atoms, with a scattering angle larger than the critical angle (dechanneling), and subsequent backscattering from regular lattice sites.

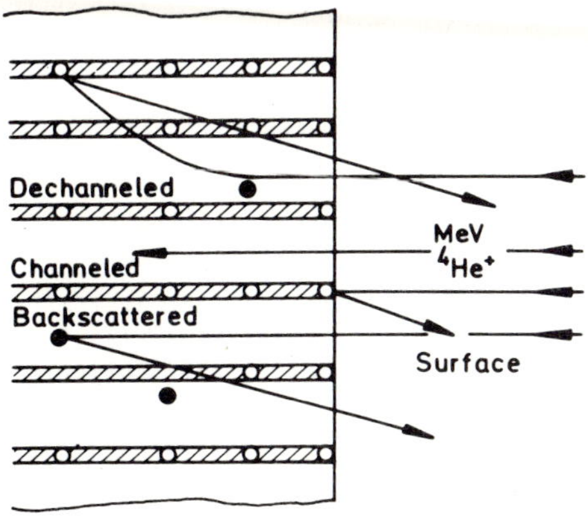


Fig. 37. Different types of events contributing to backscattering

Therefore, atoms on irregular lattice sites or on the surface have a twofold effect on the channeling spectrum. Apart from direct backscattering, their dechanneling ability accounts for additional backscattering events from lattice atoms. These double collision events generally appear at a lower energy than the direct backscattering contribution (dechanneled fraction). This situation is schematically shown in Fig. 38 [39]. In the spectrum for the crystal with disorder, a superimposition of direct backscattering events and the dechanneled fraction is observed, leading to an increased yield at energies which would correspond to depth values greater than the extension of the damage for the random-incidence case. Since the dechanneled fraction accumulates with the depth of the traversed damaged layer, a direct determination of the damage concentration from the backscattering yield is only possible near the surface.

For deeper layers, a quantitative evaluation is possible; however, this is complicated and rarely performed in ion implantation work, especially with semiconductors where any kind of damage within the sensitivity of channeling techniques (in the order of a few percent of displaced atoms) is detrimental for device performance. The emphasis is clearly on the evaluation of effective annealing procedures from the viewpoint of the lowest possible thermal impurity redistribution, together with that of maximum activation, i.e., lattice incorporation of the impurities.

As an example, Fig. 39 [40] shows a typical investigation of thermal annealing behavior for a high-dose implantation of As into Si. The following conclusions can be drawn from the sequence of the channeling spectra:

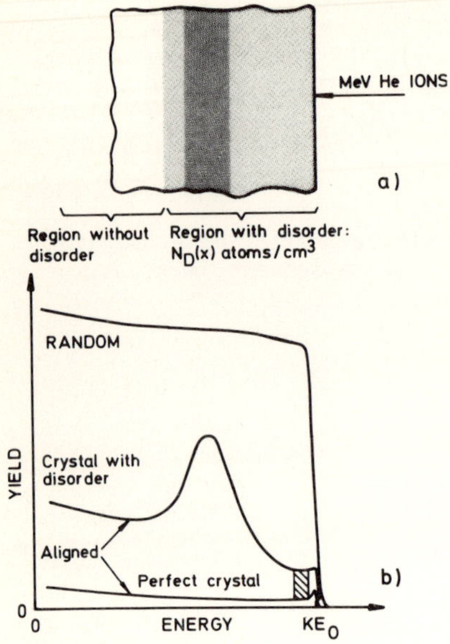


Fig. 38. (a) Schematic and (b) random and aligned spectra for mega-electron-volt $^4$He ions incident on a crystal containing disorder. The aligned spectrum for a perfect crystal without disorder is shown for comparison. The difference (shaded portion) in the aligned spectra between disordered and perfect crystals can be used to determine the concentration of displaced atoms at the surface [39].

After implantation, a disorder peak with random yield extends from the surface to a depth of 0.2 μm, indicating a virtually amorphous layer. Recrystallization takes place after annealing at or above 600°C, and starts from the bulk side of the implanted layer, since the single-crystalline bulk acts as a seed for nucleation of solid phase epitaxy. After annealing at 700°C, the amorphous peak induced during implantation disappears, but a second and more stable peak appears at the former interface, which can be attributed to the formation of extended defects such as stacking faults or dislocation loops. This damage has a marked influence on the dechanneled fraction, and disappears only with annealing at 900°C, where the channeling spectrum is essentially equal to the unimplanted case, with about 80 percent of the As atoms incorporated on lattice sites (or on interstitial sites shadowed by the (111) rows, see section 6.3).

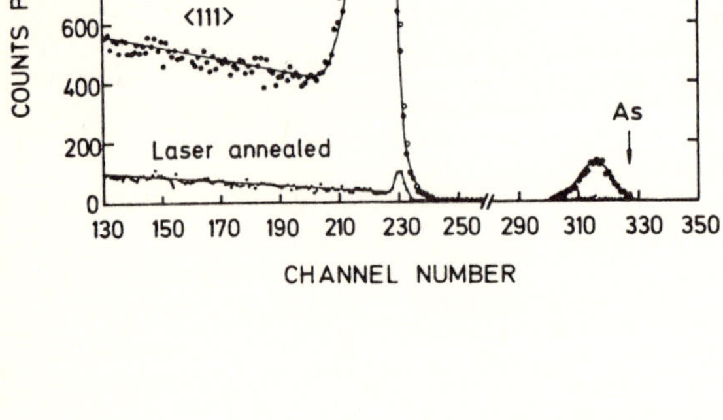


Fig. 39. Aligned and random backscattering spectra for (111) Si samples annealed thermally at temperatures between 600 and 900°C. For clarity, the arsenic peaks are shown only in random and channeling directions after 900°C annealing [40].

As a general rule, primary damage, such as single displaced atoms or amorphous clusters, can be quantitatively evaluated from the direct backscattering contribution in the energy spectrum (damage peak). Extended damage configurations introduce a complex relationship between wide and small-angle scattering which is characteristic for the particular type of extended defect involved [41]. The effect of extended damage is generally more pronounced in the dechanneled fraction, as can be seen from Fig. 40 [42]. A series of (100) channeling spectra is shown for a silicon sample implanted with phosphorus. After implantation, bright bands of color appeared at the surface, indicating an inhomogeneous temperature dependence. The channeling analysis of Si at different points clearly demonstrates the presence of a separate disordered layer below the heavily-disordered or amorphous layer near the surface, which is attributed to the formation of extended damage. Contrary to the primary damage, it is nearly independent of the temperature distribution, and completely dominates the dechanneled fraction.

Evaluation of extended defects from channeling spectra is a complex task and still a subject of research. It has been carried out recently [43] for depth profiling of stacking faults and dislocations, by analyzing axial and planar channeling data from the dechanneled part of the spectrum at different energies. For identification of the damage type, TEM micrographs on thinned samples were recorded.

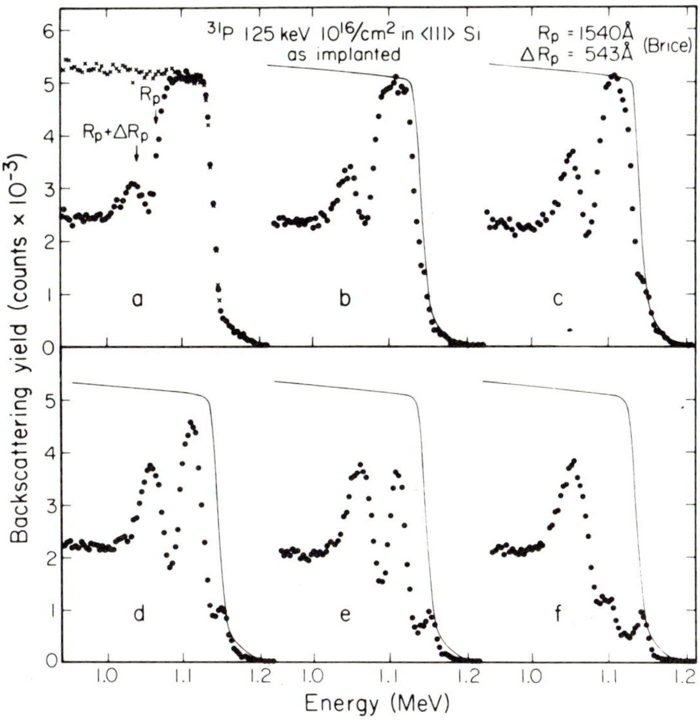


Fig. 40. 2 MeV $^4$He channeling spectra from phosphorus-implanted silicon. The spectra were taken at different positions, exhibiting different temperatures during implantation.

## 6.3 Lattice Location

By irradiation of a crystal along different main crystal axes, it is possible to obtain information on the lattice location of impurity atoms which are off regular lattice sites. This is shown in Fig. 41 [44] for silicon implanted with thallium. By comparing the areas under the Tl peaks, it can be concluded that approximately 40% of the Tl atoms are located on substitutional sites, and another 40% on the tetrahedral interstitial sites in Si. The area of the peak for the (111) direction (20%) corresponds to the number of atoms located at random sites.

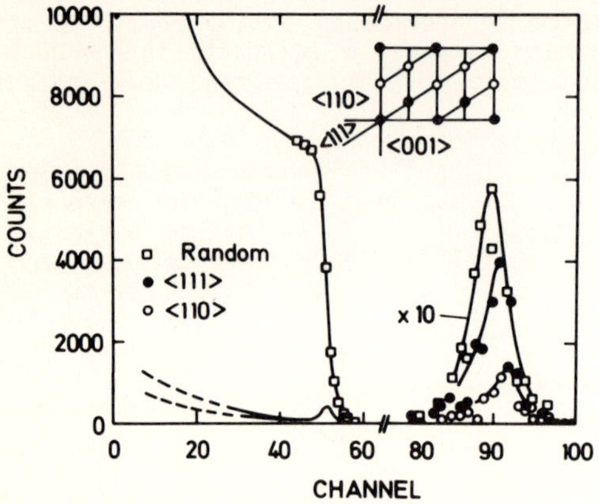


Fig. 41. Energy spectra of helium ions backscattered from a thallium-implanted crystal [44]

## References

1. Ion Beam Modification of Materials, R.E. Benneson, E.N. Kaufmann, G.L. Miller, and W.W. Scholz eds., North Holland Publ. Co., Amsterdam 1981.
2. K. Wittmaack and J.B. Clegg, Appl.Phys. Lett. 37, 286 (1980).
3. P. Williams and C.A. Evans, Jr., Appl.Phys. Lett. 30, 560 (1977).
4. J.B. Clegg, Surface and Interface Analysis, Vol. 2, No. 3, Heyden and Sons Ltd., 1980.
5. W.K. Stuckey in Systematic Materials Analysis IV, J.H. Richardson and R.V. Peterson eds., Academic Press, New York 1976.
6. J.W. Colburn, E. Taglauer, and E. Kay, J. Appl.Phys. 45, 1779 (1974).
7. V.G. Isotopes Ltd, Publ. 02.484 (1982).
8. M. Maier, D. Bimberg, H. Baumgart, and F. Phillip in Secondary Mass Spectroscopy SIMS III, A. Benninghoven, J. Giber, J. László, M. Riedel, and H.W. Werner eds., Springer-Verlag, Berlin 1982.
9. H. Oechsner and E. Stumpe, Appl.Phys. 14, 43 (1977).
10. G. Blaise, Surf. Sci. 60, 65 (1976).
11. K.R. Williamson, W.M. Theis, S.S. Yun, and Y.S. Park, J. Appl.Phys. 50, 8019 (1979).
12. H. Ryssel et al., to be published.
13. T. Bereznai, F. DeCorte, and J. Hosk, Radiochem. and Radioanal. Lett. 17, 279 (1974).
14. A.P. Janssen and J.A. Venables, Ninth Int. Congr. on Electr. Microscopy, Toronto 1978, Vol. I, p.520.
15. J.A. Borders in Site Characterization and Aggregation of Implanted Atoms in Materials, A. Perez and R. Coussement eds., Plenum Publ. Co., New York 1980.

16. M.T. Thomas, D.R. Baer, H.R. Jones, and S.M. Bruemmer, J.Vac.Sci. Technol. 17, 25 (1980).
17. M. Baron, A.L. Chang, J. Schreurs, and R. Kossowksy, Nucl. Instr.& Meth. 182/183, 531 (1980).
18. Ion Beam Handbook for Material Analysis, J. W. Mayer and E.Rimini eds. Academic Press, New York 1977.
19. The Stopping and Ranges of Ions in Matter, Vol. 3 and 4, H.H. Anderson, J.F. Ziegler eds. Pergamon Press, New York 1977.
20. D.M. Scott and M.-A. Nicolet, Nucl. Instr. & Meth. 182/183, 665 (1981).
21. H. Ryssel, F. Iberl, M. Bleier, G. Prinke, K. Haberger, and H. Kranz, Appl.Phys. 24, 197 (1981)
22. T. Chiu, H. Bernt, and I. Ruge, J. Electrochem.Soc. 129, 408 (1982).
23. R. Hezel and N. Lieske, J. Electrochem. Soc. 129, 379 (1982).
24. J.F. Ziegler, G. W. Cole, and J.E.E. Baglin, J. Appl.Phys. 43, 3809 (1972).
25. H. Ryssel, K. Müller, K. Haberger, R. Henkelmann, and F. Jahnel, Appl.Phys. 22, 35 (1980).
26. L.J. Pilione and B.S. Carpenter, Nucl. Instr. & Meth. 188, 639 (1981).
27. J. Hirvonen, Appl. Phys. 23, 349 (1980)
28. G.Amsel and D. Samuel, J.Phys. Chem. Solids 23, 1707 (1962).
29. C.W. Magee, R.E. Honig, and C.A. Avans, Jr., in [8], p. 172.
30. H.E. Rosendaal in Sputtering by Particle Bombardment I, R. Betrisch ed., Springer-Verlag, Berlin 1981.
31. J. Maul and K. Wittmaak, Surf. Sci. 47, 358 (1975).
32. K. Wittmaack, Appl.Phys. Lett. 29, 552 (1976).
33. K. Wittmaack, Proc. 7th Intern. Vac.Congr. and 3rd Conf. Solid Surf., R. Dobrozemsky et al., eds. Vienna 1977, p. 2573.
34. B.L. Bentz and H. Liebl in [8], p. 30.
35. J.M. Gourgout in [33], p. A-2755.
36. ATOMIKA Technische Physik GmbH, Ionprobe A-DIDA 3000, Technical brockure, Munich 1980.
37. CAMECA, Ion Microanalyzer ims 3f, technical brochure, Courbevoie 1980.
38. J.M. Gourgout, Chem. Phys. 9, 286 (1979).
39. W.-K. Chu, J.W. Mayer, and M.-A. Nicolet, Backscattering Spectrometry, Academic Press, New York 1978.
40. M. Takai, P.H. Tsien, S.C. Tsou, D. Röschenthaler, M. Ramin, H. Ryssel, and I. Ruge, Appl.Phys. 22, 129 (1980).
41. J. Hory and Y. Quéré, Rad. Effects 13, 57 (1972).
42. L. Csepregi, E.F. Kennedy, S.S. Lau, and J.W. Mayer, Appl.Phys. Lett. 29, 645 (1976)
43. B. Gruska and G. Götz, Phys.Stat.Sol.(a) 67, 129 (1981).
44. S. Namba and K. Masuda, Advances in Electronics and Electron Physics 37, 263 (1975).

# Annealing and Residual Damage

Siegfried Mader

IBM General Technology Division
Hopewell Junction, New York 12533, USA

## Abstract

This chapter describes aspects of ion implantation damage which are important for Si process technology. Primary damage consists of atomic displacements and amorphization of Si (except for B implantation). Annealing restores crystallinity and induces electrical activation of implanted dopant ions. It can also cause the formation of residual defects with well-defined crystallographic nature, for example stacking faults and dislocation loops. During prolonged annealing these defects change their sizes and configurations in response to climb forces. The climb forces are related to indiffusion of the implanted profile and to surface oxidation. Residual dislocations also respond to mechanical stresses which arise from the geometries of masking patterns used for defining active device areas.

## 1. Introduction

The most widespread application of ion implantation is localized doping of semiconductor wafers during planar device fabrication. The more conventional doping methods, which are alternatives to implantation, employ thermal indiffusion of dopant atoms from source layers or from the vapor phase. Compared to in-diffusion, implantation allows a more precise control over the amount of impurities which are introduced into the substrates. This is because we can - in principle - measure the electrical charge deposited by an ion beam directly, whereas for in-diffusion we have to rely on highly temperature-dependent thermodynamic driving forces and kinetics. However, the price is high: implantation cannot be done without radiation damage. Ions which are propelled into the substrate crystal collide with substrate atoms and displace them from their lattice sites in large numbers. The success of semiconductor device fabrication has in part been due to our ability to grow perfect crystals of the substrate material. Thus, it is not surprising that the success of ion implantation doping depends on the restoration of the damaged crystals, at least in critical junction regions.

In this chapter we will describe some aspects of implantation damage and its annealing in Si. In modern technology one uses junction depths of 0.5 µm or less, and the trend is to shallower junctions. For doping with high concentrations (of the order of $10^{20}$ $cm^{-3}$) it has become customary to select implant depths which are smaller than the desired junction depth and to thermally diffuse the implanted profile to the final position. This predeposition and drive-in scheme avoids junction positions in the regions of primary implant damage. But it does not guarantee a restoration of crystal perfection. Frequently secondary or residual defects evolve from the primary damage. Lower doping concentrations ($\leq 10^{18}cm^{-3}$) can be implanted directly into the desired depth position; they are less likely to produce residual defects.

In the following sections we will first list methods for the characterization of damage and defects and briefly describe the primary damage structure. Then we deal with the events during thermal annealing and with the evolution of residual defects.

In 1972 Gibbons reviewed damage production and annealing [1], and in 1973 Dearnaley et al. published a monograph with a large section on semiconductor applications [2]. These references contain most of the relevant concepts. In the intervening years the literature proliferated with many detailed observations and measurements. This can be seen from bibliographies on implantation, e.g. [3], which are books in themselves. Convenient summaries of the damage and defect aspects of our subject can be found in recent conference proceedings of the Materials Research Society [4] and of the Royal Microscopical Society [5].

## 2. Characterization of Damage and Defects

A very suitable tool for characterization of implantation damage is transmission electron microscopy (TEM). There are two reasons for the good match between object and methodology: 1. the scale of distances between damage features is of the order of nm to µm and most of them are in the range of TEM resolution; 2. the defects are present in shallow layers just below the surface of the implanted crystal. These layers are accessible to TEM by simply etching away the substrate wafer.

TEM images arise from variations of Bragg diffraction at distorted crystal lattice planes. The methods for interpretation of diffraction contrast are well established [6]. They are particularly useful for characterization of crystallographic features of secondary defects which can form during heat treatment of the primary damage structure.

X-ray topography also images distortions of lattice planes. Its limited lateral resolution renders it less sensitive for implantation damage. Optical microscopy of defects decorated by chemical etches (Sirtl etch, Jenkins-Wright etch) also suffers from limited lateral resolution. Closely spaced defects do not give rise to individually distinguishable etch figures and a dense defect structure may remain unnoticed. The situation is somewhat better for surface examination with a scanning electron microscope (SEM) after a very light chemical etch.

Another well-established and frequently used method for characterizing implant damage is Rutherford backscattering (RBS) [7]. Its popularity among practitioners of implantation stems - in part - from the similarity of tools. For both implantation and RBS one needs accelerators and ion beams; in the latter case one uses $H^+$ or $He^+$ and energies in the MeV range.

RBS is particularly useful for exploring primary implantation damage which is less accessible to TEM. Backscattering can measure the number of atoms which are displaced from regular lattice sites. It is unsurpassed for detecting the presence of an amorphous layer and for measuring its thickness. It can also determine profiles of implanted ions which are heavier than Si and it can ascertain whether they occupy substitutional lattice sites. In a wafer with residual defects after heat treatment RBS analysis shows dechanneling yields which are caused by crystallographic defects. Correlations between these dechanneling effects and microscopic defect structures are emerging [8], [9].

Point defects are created in abundance during implantation, and a powerful method for their analysis is electron paramagnetic resonance (EPR), compare [10]. Amorphous Si gives rise to a fingerprint in the EPR spectra; this was used to study implantation-induced amorphization, e.g. in [11].

## 3. Primary Implantation Damage

Detailed theories and tabulations exist for the ranges of implanted ions in solid targets [12], [13], [14]. But the understanding of what happens to the Si target itself is less detailed. Brice [13] tabulated depth distributions of the energy which is deposited in the crystal.

An ion slowing down and coming to rest in the target loses energy by electronic stopping and by nuclear collisions. Both loss mechanisms contribute to the primary implantation damage. Collisions with energy transfers above a threshold of about 15 eV displace Si atoms from their lattice sites and create a Frenkel pair. The recoiled Si atoms can act as projectiles for secondary collisions and generate displacement cascades. Electronic excitations enhance the diffusion of point defects which can exist in several charge states. The travel of recoiled atoms, thermal diffusion in the beam-heated target, and electronic-enhancement effects contribute to migration and agglomeration of point defects during the implantation process.

Accumulation of displacement damage can lead to amorphization of the Si structure. There is a critical dose for the formation of an amorphous layer which increases with decreasing mass of the implanted ions and with increasing target temperature. In a target at room temperature $5 \cdot 10^{14}$ $P^+/cm^2$ or $10^{14}$ $As^+/cm^2$ produce amorphous layers, whereas $B^+$ ions do not amorphize a Si target.

The nuclear stopping power increases with increasing projectile mass. For light ions like $B^+$, this translates into a trail of well-separated primary recoils in the wake of the implanted ion. But heavy ions produce closely spaced recoils, and their secondary cascades touch each other or overlap. The results are amorphous zones with dimensions of the order of 10 nm. When these zones completely fill the volume of the implanted layer, this layer becomes amorphous. It can be shown that this happens when the energy deposited into collisions exceeds $6 \cdot 10^{23}$ $eV/cm^3$ or 12 eV for each target atom [11].

Amorphization begins in a layer at the depth of the maximum collision energy deposition (slightly less than the projected range $R_p$) and spreads towards the surface and towards deeper positions in the target. The interface with the single crystal target is not a well-defined plane due to the statistical nature of ion penetration. Beyond the interface we expect a considerable concentration of Si interstitials which have diffused out of the damage clusters during implantation.

## 4. Thermal Annealing

In order for a dopant atom to be electrically active as donor or acceptor it has to be incorporated into a lattice site of the Si structure. The resistivity of as-implanted material is very high because only few implanted atoms happen to come to rest in a substitutional site. Furthermore, after implantation there are more structural defects than dopant atoms and most of these defects have deep trapping and compensating states. Therefore, thermal annealing is necessary to remove the defect states and to incorporate all implanted ions in lattice sites.

A large arsenal of annealing methods is available. One way of categorizing these methods is the time of exposure to high temperature. This time ranges from microseconds for pulsed laser annealing, milliseconds for CW laser annealing, seconds for heating with incoherent light and thermal radiation, to minutes and hours for conventional furnace annealing. The short-time methods have been very actively investigated in the past few years, and most of the results are conveniently available in conference proceedings [15], [16]. However, in actual device fabrication, where reproducibility and throughput are important, furnace annealing dominates.

In pulsed laser annealing the surface of the wafer melts locally while the remainder of the wafer remains at ambient temperature. The very rapid resolidification usually restores a perfect crystal. There is not enough time for any process which requires diffusion in the solid, such as precipitation of a supersaturated dopant concentration.

The methods of millisecond to second duration heat the implanted layer without melting and allow activation by thermal diffusion. The diffusion distances are short, of the order of a few nm, and implanted profiles are not significantly broadened. Amorphous layers grow back into crystalline structures. But the duration of the diffusion tends to be insufficient for complete restoration of crystalline perfection. The interstitials which had accumulated in the crystalline region adjoining the amorphous layer cannot diffuse out of the crystal, but they form agglomerates. Figure 1 shows an example in an As-implanted wafer. It had been exposed for 10 sec to a graphite heater at 1250°C in an apparatus similar to that described by [17]. There are many small interstitial prismatic dislocation loops. (They are platelets of extra Si material which is in perfect register with the Si crystal. The thickness of a platelet is that of a 1/2 {110} plane. The boundary is a complete edge dislocation with Burgers vector a/2 <110>).

The remainder of this section deals with furnace annealing. As is to be expected, the effects of heat treatments are different for implantations which had amorphized the substrate and for B implantations.

Consider first amorphized layers. They regrow epitaxially onto the single-crystal substrate with well-defined growth kinetics. The interface velocity V was elegantly determined by Csepregi et al. [18], [19], using RBS. It can be described by

$$V = V_0 \exp(-2.35 \text{ eV}/kT) \tag{1}$$

where $V_0$ depends on the crystallographic orientation of the interface and on the type of doping. For an isothermal annealing at 550°C the interface velocity is 8.5 nm/min for (100) and about 1 nm/min for (111) interfaces in the case of undoped (Si-implanted) wafers and 60 times higher for P-and

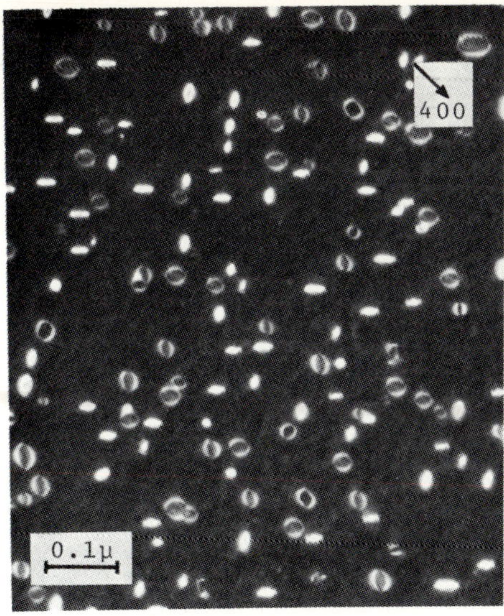


Fig. 1. Interstitial prismatic dislocation loops after regrowth of amorphous Si. Implantation: $5 \cdot 10^{14}$ $As^{+}/cm^{2}$ at 140 keV, annealing: 10 sec exposed to graphite heater at 1250°C

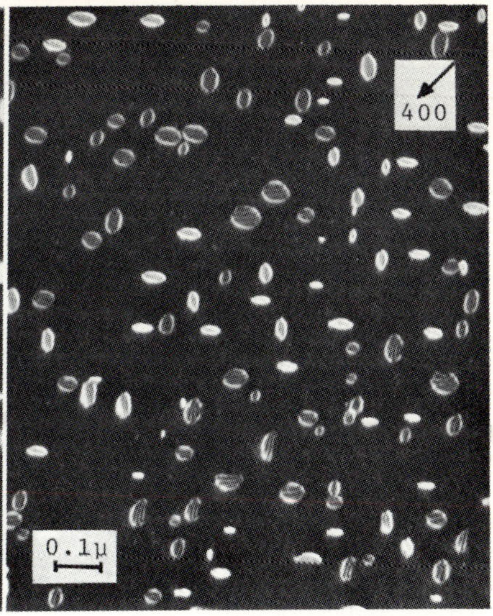


Fig. 2. Frank loops with extrinsic stacking faults after regrowth of amorphous Si. Implantation: $2 \cdot 10^{14}$ $Sb^{+}/cm^{2}$ at 150 keV, annealing: 15 min at 950°C

As-implanted wafers. These kinetics are obeyed over a very large range of temperatures and growth rates including the regime of very rapid heating by CW laser annealing [20].

In a typical furnace annealing at 900°C to 1000°C the amorphous layer regrows during heat-up. For (001) wafers a single-crystal layer is restored except at the position of the original amorphous-crystalline interface. In this region we have found - in all cases examined in our laboratory - extrinsic defects, i.e. defects which can be ascribed to agglomeration of interstitials. They usually have the configurations of dislocation loops similar to Fig. 1; but other configurations occur also. Figure 2 shows Frank loops where the excess Si atoms are incorporated on {111} planes and form stacking faults. In TEM pictures they show characteristic contrast of parallel fringes. They were observed after annealing of an Sb implantation.

In interfaces with {111} orientations additional imperfections occur: microtwins on inclined {111} planes. They propagate through the whole thickness of the regrowing layer. Their nucleation was discussed by Washburn [21].

The driving force for the regrowth of an amorphized layer is very large, and dopant atoms are swept into lattice sites even for peak concentrations which are larger than the solubility at the regrowth temperature. After a regrowth annealing of, e.g., 30 min at 600°C the implanted dose is electrically active.

Electrical activation can be checked by measuring the sheet resistance $\rho_s$. In a first approximation $\rho_s$ is inversely proportional to the electronic charge q, average mobility $\mu$ and implanted dose N:

$$\rho_s = 1/q \cdot \mu \cdot N \ . \qquad (2)$$

The mobility $\mu$ depends strongly on the concentration of doping atoms, and values of $\rho_s$ have been computed using known mobility data [2]. After regrowth of an amorphous layer, $\rho_s$ measurements usually yield values in the range expected by such computations. This indicates that the defect structures of Figs. 1 and 2 do not significantly degrade the mobility.

The situation is different for B implants (and P implants below the critical dose for amorphization). For low doses ($\leq 10^{13}$ cm$^{-2}$) one observes a monotonous increase in the fraction of activated dopant atoms with increasing annealing temperature. However, with higher doses an initial increase of the activated fraction reverses in the temperature range of 500°C to 800°C and complete activation can only be achieved above 900°C.

During annealing of B implants a large and complex variety of crystallographic defects are formed. In the temperature range of reverse annealing, rod-shaped defects dominate. They are elongated in <110> directions. An example is shown in Fig. 3. With increasing temperature the rod structure coarsens and, in addition, elongated interstitial dislocation loops and Frank loops appear, Fig. 4. The structures of the rod defects are not completely known; there appear to be several types. Wu and Washburn [22] showed that their shrinking is controlled by the diffusivity of B in Si. This implies that they contain B or are boron precipitates. The existence of precipitates and their dissolution during heat treatment above 800°C further implies that, in this case, the concentration in solid solution - and therefore the electrical activity - is limited to the equilibrium solubility of B at the annealing temperature. Similar results were recently reported for P [23].

Fig. 3. Microstructure after B implantation: Rods elongated in <110> directions. Implantation: $5 \cdot 10^{14}$ $B^+/cm^2$ at 150 keV, annealing: 15 min at 750°C

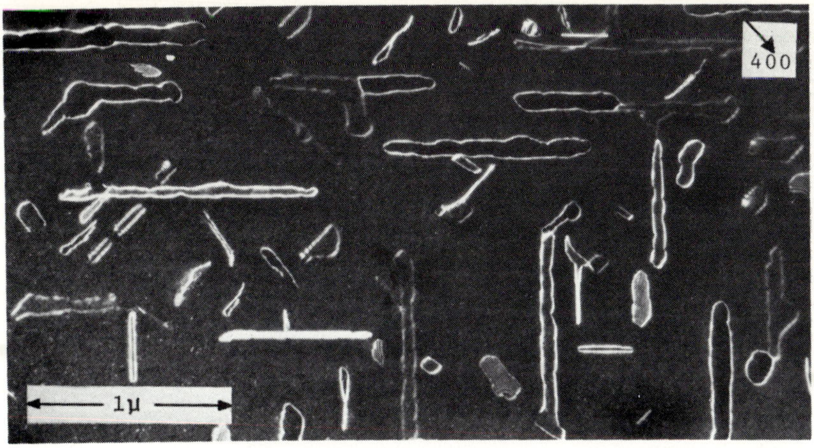


Fig. 4. Microstructure after B implantation. Elongated interstitial dislocation loops and Frank loops. Implantation: $5 \cdot 10^{14}$ $B^+/cm^2$ at 150 keV, annealing: 20 min at 950°C

So far we have considered heat treatments mainly from the viewpoint of electrical activation. An equally important concern is the broadening and in-diffusion of the profiles. For doping by predeposition and drive-in the primary function of the annealing cycle is to diffuse the junction to a desired depth position. The necessary times and temperatures can be preselected with good accuracy from diffusion models or process models [24].

During drive-in the defects described so far react with one another and form a variety of different arrangements. This will be discussed in the next section. Eventually, with prolonged drive-in, the defects tend to shrink and to disappear. Shrinking occurs by dislocation climb which is primarily controlled by self-diffusion of Si. Since the activation energy of self-diffusion is always larger than that of impurity diffusion, the ratio of the rate of defect elimination to the rate of dopant in-diffusion increases with increasing temperature. Therefore, it is desirable to choose as high a temperature as possible - within the limits of achieving the junction depth of the device design - in order to remove as many defects as possible.

## 5. Residual Defects

After an implanted profile is fully activated and diffused to the desired depth position, it is quite possible that defects are still present in the doped layer. They are referred to as residual defects.

A layer with dislocation loops or stacking faults is not an equilibrium structure because the defects carry with them strain energy, core energy, and stacking fault energy. After sufficiently long annealing times we would expect that they had disappeared or reorganized themselves into a cross grating of misfit dislocations in those cases where doping changes the Si lattice constant. This is indeed the case. Residual defects after a specific drive-in cycle are just intermediate states of a dislocation structure which is evolving towards its equilibrium. We will try to describe this evolution for the case of As implantation in (001) wafers.

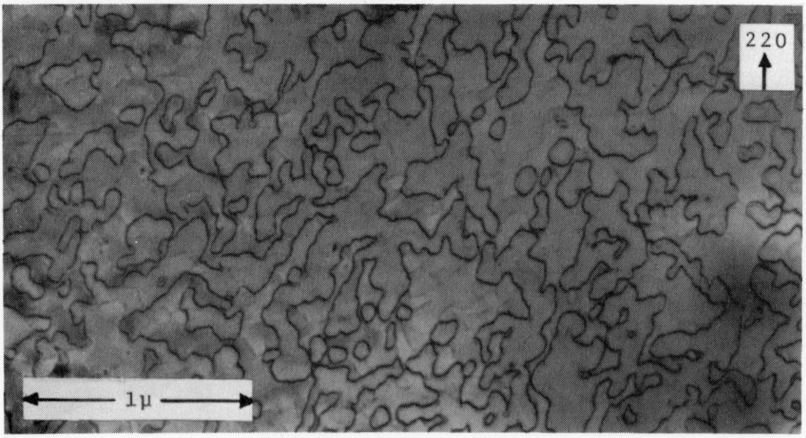


Fig. 5. Winding dislocation loops with Burgers vector parallel to the wafer surface. Implantation: $1 \cdot 10^{16}$ $As^+/cm^2$ at 40 keV, annealing: 1 hr at 850°C

During heat-up in a furnace small interstitial prismatic dislocation loops form, similar to the ones in Fig. 1. Most of them have Burgers vectors which are inclined to the wafer plane. During further heat treatment the loops grow until they touch one another. Two touching dislocation segments then undergo a Lomer reaction which reduces their line energy. It can be shown that this reaction transforms all of the prismatic loops into

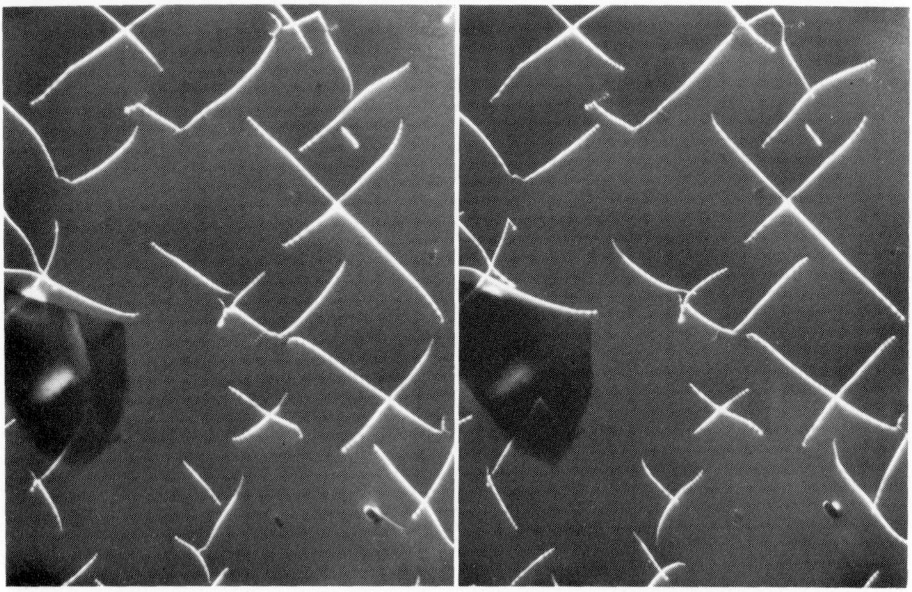

Fig. 6. Half loops of edge dislocations, stereo pair. Implantation: $5 \cdot 10^{15}$ $As^+/cm^2$ at 50 keV, annealing: 10 min at 1000°C. Average depth penetration of half loops: 150 nm

two sets of larger loops [25]. They now have long meandering outlines and Burgers vectors parallel to the wafer plane. This structure is shown in Fig. 5.

With continued heat treatment a second stage occurs where the edge dislocation segments climb normal to the wafer surface in the direction defined by absorption of interstitials (or emission of vacancies). This leads to the formation of edge dislocation half loops with straight segments parallel to the surface and end segments turning up to the surface, Fig. 6. The extra half planes are contained between the surface and the dislocation lines. The edge segments initially climb deeper into the crystal to penetration depths of 100 to 200 nm. Then they reverse their advance, climb back to the surface and disappear [26].

The second stage is greatly retarded for implantations through a screen oxide. It appears that recoiled oxygen atoms precipitate along the dislocation lines and stabilize the structure with winding loops, Fig. 7. On the other hand the second stage is enhanced and the edge segments penetrate deeper for drive-in in oxidizing atmosphere. Frequently a drive-in cycle begins with a short oxidation which is intended to cap the crystal surface and to prevent out-diffusion of the implanted impurities.

Expansion and shrinking of interstitial loops (and half loops) is controlled by non-equilibrium concentrations of point defects. Several factors contribute to the imbalance of point defects during thermal annealing. Not all of them are due to the implantation process. In fact, only the small interstitial loops in the wake of the amorphous regrowth can be ascribed to the primary implant damage. They had nucleated at a position where we had expected to find a super-saturation of Si interstitials as a result of the implantation process.

The growth of these loops into the structures of Figs. 5 and 6 takes place while simultaneously the implanted profile broadens and diffuses into the crystal. The diffusivity of dopant atoms is always larger than the self-diffusion of Si. This leads to a pile-up of extra atoms (or an

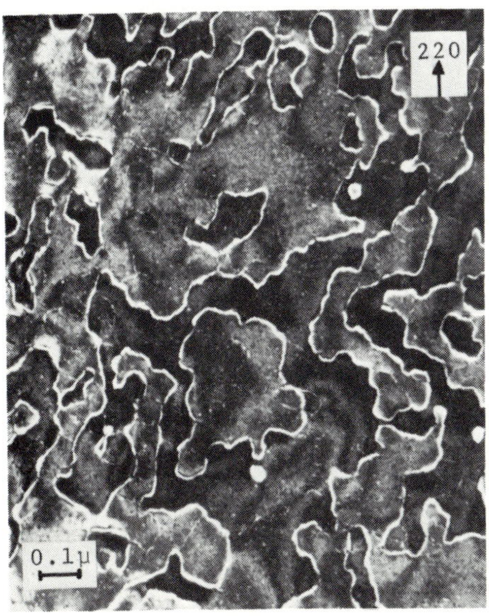


Fig. 7. Winding dislocation loops decorated with small precipitate particles. Implantation: $5 \cdot 10^{15}$ $As^+/cm^2$ at 50 keV through 25 nm $SiO_2$; drive-in: 60 min at 1000°C

undersaturation of vacancies) in front of the steepest gradient of the profile, which tends to enlarge interstitial dislocation loops at that location. (This concept is similar to the one discussed by Gösele and Strunk [27]). When diffusion proceeds into a perfect crystal no dislocation loops form because, presumably, the imbalance of point defects is not large enough to overcome the nucleation barrier. But in the present case nucleation had already occurred as a consequence of implantation damage, and in-diffusion merely enlarges existing loops. Once the diffusion front has passed beyond the depth position of the loops or half loops, the climb force reverses and aids the shrinking and elimination of the defects.

Oxidizing heat treatment creates an additional super-saturation of Si interstitials, as is known from the formation of oxidation stacking faults and oxidation enhanced diffusion [28]. This enhances the growth phase of our half loops and tends to retain dislocations in the crystal even after all of the volume of primary damage has been consumed by oxidation.

With phosphorus implantation the half loops of the second stage tend to join together and form a cross grating or a network of edge dislocation. This structure remains in the crystal; it is stabilized by the lattice misfit between the phosphorus doped layer and the substrate. Figure 8 shows the development of this structure. Its formation is strongly enhanced by annealing under oxidizing conditions, as was shown by Tamura [29].

Double implantation e.g. of P and Sb greatly reduces the density of residual dislocations for equivalent doping concentrations and junction depths. P and Sb introduce lattice misfits of opposite signs which compensate each other. In addition, the double implants modify primary damage and in-diffusion characteristics [30].

Residual defects after drive-in do not always have the configuration of complete dislocations. Figure 9 shows defects in a layer doped by B implantation. They are Frank loops with extrinsic stacking faults, and they occur between the position of the primary damage and the driven-in junction. Since stacking faults contain extra atoms on {111} planes, they also accommodate lattice misfit between the B doped layer and the substrate.

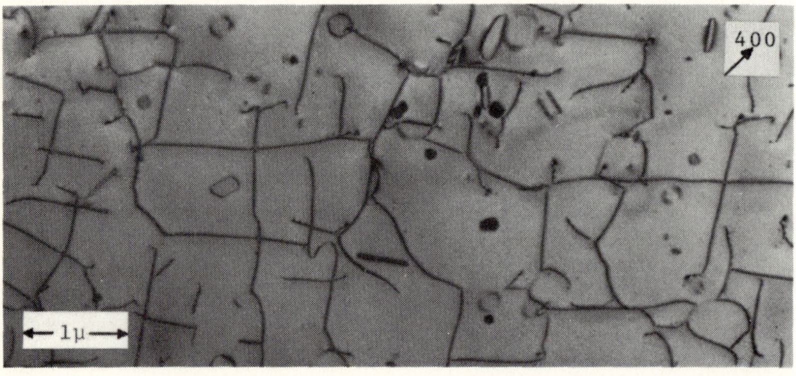


Fig. 8. Dislocation half loops developing into a cross grating of misfit dislocations. Implantation: $3\cdot10^{15}$ $P^+/cm^2$ at 50 keV and $10^{15}$ $P^+/cm^2$ at 200 keV, annealing: 2 hrs at 900°C and 30 min at 950°C

Fig. 9. Extrinsic stacking faults in layer between 200 nm and 400 nm below surface (within doped layer, below primary damage). Implantation: $5\cdot10^{15}$ $B^+/cm^2$ at 40 keV, annealing: 30 min at 970°C

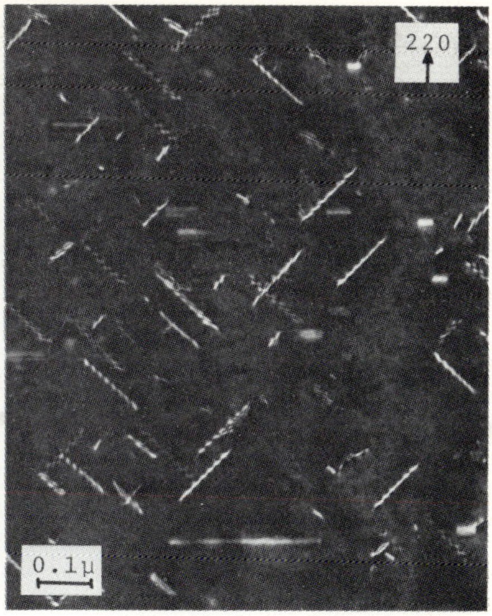


Fig. 10. Rod defects elongated along <110> directions, formed during $H^+$ bombardment at 50 keV in wafer at 575°C, about 400 nm below surface. Dose rate $10^{13}$ $H^+/cm^2$ sec

We conclude this section with a few observations made after implantation into hot targets. In this case the implanted species was $H^+$ and the purpose was not doping of the substrate but production of Frenkel pairs to induce radiation-enhanced diffusion [31]. In targets with temperatures above 500°C rod-shaped defects were found; they are shown in Fig. 10. They are elongated in <110> directions and dissociated into narrow ribbons on {113} habit planes. On these planes they form extrinsic stacking faults. These defects are yet another configuration of agglomerated Si interstitials. They have been observed in a variety of cases where displacement damage occurs [32].

For radiation-enhanced diffusion to be effective the point defects must have an opportunity to diffuse distances of the order of a micron before recombining or being trapped at immobile agglomerations. This happens at target temperatures above 600°C. The diffusion distances can be qualitatively appreciated in Fig. 11, where the $H^+$ beam had produced a super-saturation of point defects in the volume from which the TEM foil was prepared. Interstitial dislocation loops had nucleated and grown to sizes of about 1 μm diameter, most of them larger than foil thickness in Fig. 11. The scalloped outlines of the edge dislocations reflect the high super-saturation of interstitials in this particular case, compare [33].

We have seen that Si interstitials can agglomerate in a variety of configurations: prismatic dislocation loops, Frank loops with stacking faults on {111} planes, and {113} stacking fault ribbons. One might ask what happens to the other partner of a displacement collision, the vacancy. Vacancy agglomerates have been rarely observed in semiconductors. They

Fig.11.

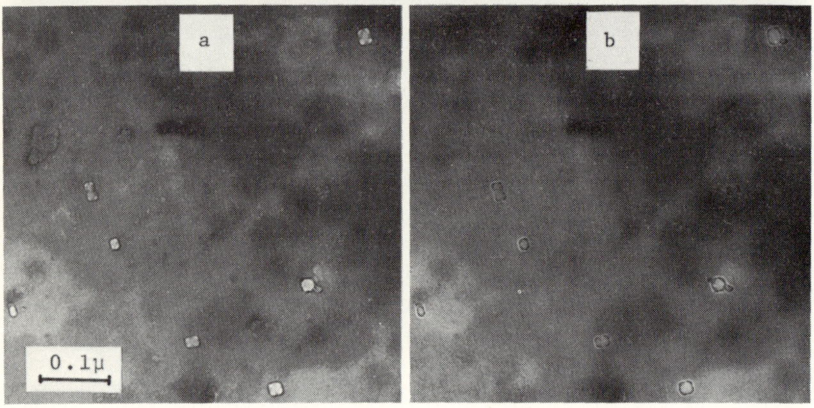


Fig.12.

Fig. 11. Edge dislocations absorbing interstitials during $H^+$ bombardment at 150 keV in wafer at 720°C, about 1.5 μ below surface. Dose rate $3.5 \cdot 10^{13}$ $H^+/cm^2 sec$

Fig. 12. Cavities (vacancy agglomerates) in same wafer and same depth as Fig. 11. a: 220 nm under focus, b: 220 nm over focus

occur when displacements are spread out in space and not concentrated in amorphous zones. They form three-dimensional cavities without distorting the Si lattice. Therefore, they are far less conspicuous in TEM images, but they can be made visible by out-of-focus contrast [34]. An example is shown in Fig. 12 which is taken from the same $H^+$-bombarded wafer as Fig. 11. The cavities can grow to sizes of 10 to 20 nm and they tend to develop facets on {111} planes. In our example the distances between cavities (and thus the diffusion distances of vacancies) in Fig.12 are much smaller than the distances between edge dislocations which absorb interstitials in Fig. 11.

## 6. Effects of Residual Defects

Implantation damage and residual dislocations act as nucleation sites for the precipitation of impurities in a wafer. They can be put to good use as gettering agents [35]. With backside implantation one can remove fast diffusing impurities from the device side of a wafer [36].

This property becomes, of course, detrimental when a dislocation threads through a junction and attracts unwanted impurities into the depletion region. Even undecorated dislocations degrade a device, although one can never be completely sure that a dislocation has not attracted some impurities.

The electrical behavior of dislocations on a microscopic scale is not completely understood. They act as recombination centers, which is evident from their SEM contrast in electron-beam-induced current (EBIC) mode [37]. When they cross a junction they cause a soft reverse characteristic and leakage currents with superlinear dependence on the applied reverse voltage. This was very clearly shown for the case of bipolar transistors with implanted bases by Ashburn et al. [38] and for implanted emitters by Bull et al. [39].

But why could dislocations be near the depletion region after a properly executed drive-in? In the previous section we had shown that conditions can be chosen such that dislocations climb away from the junction or disappear altogether. This is true for experiments with blanket wafers. However, when devices are fabricated implantations have to be restricted to local areas. They are defined by windows in masking films or by oxide patterns recessed into the Si surface. The pattern edges usually are sources of mechanical stresses in the substrates. Dislocations react to mechanical stress by glide motion. Even transient dislocations, such as interstitial half loops during As drive-in, can respond to mechanical stress and glide away from the implanted areas into regions where they are cut

Fig. 13. Penetration of dislocation from As- implanted area (left) into masked area (right). Implantation: $5 \cdot 10^{15}$ $As^+/cm^2$ at 50 keV, annealing: 60 min at 1000°C

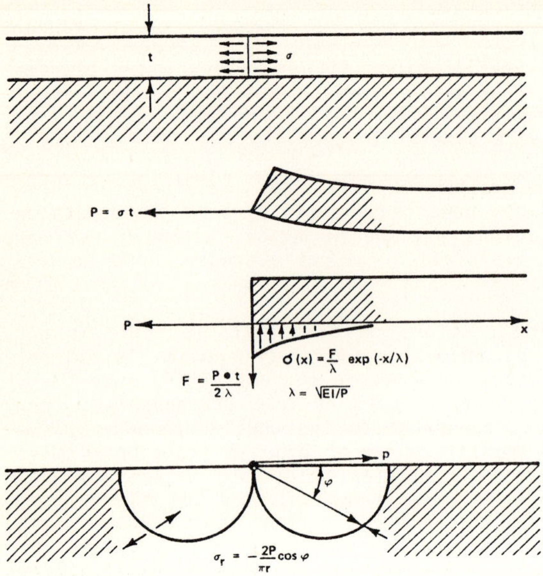


Fig. 14. Analysis of stress induced in the substrate by masking film with intrinsic tension

off from the driving forces for climb which would have eliminated them in the implanted area.

An example is the so-called side wall penetration in implanted emitters [26], [39 to 41], shown in Fig. 13. The area on the left had been implanted with As while the area on the right had been masked by 80 nm $SiO_2$ and 160 nm $Si_3N_4$. After drive-in, the implanted area is free of defects while half loops are present at the edge of the window and under the masked area. Slip traces show that the dislocations had indeed moved out of the implanted area.

In this particular case a mechanical stress analysis was possible [26]; it is outlined in Fig. 14. The masking film had a known amount of intrinsic tensile stress σ. After etching the window, the edge exerts a surface traction on the substrate whose main component is a tangential force $P = 2.25 \cdot 10^5$ dyns/cm$^2$. It transmits a stress distribution into the substrate which is indicated in Fig. 14. The shear stress components of this distribution on planes parallel to the substrate surface exert glide forces on dislocations which are present during drive-in annealing. Inside the implanted areas the dislocations are attracted to the mask edge, and under the masking film they are pushed away from the edge. Contours of constant shear stress underneath the mask are shown in Fig. 15, together with coordinates of individual dislocations which were observed in the masked area. They are located around the contour $\tau = 1.7 \cdot 10^8$ dyns/cm$^2$. This value can be identified with the critical shear stress for glide on {001} planes at 1000°C. Dislocations will not glide into regions where the stress is lower than this value.

Figure 16 shows an example of an As-implanted area which was bounded on both sides by a thick oxide recessed into the substrate surface by 0.5 μm. Drive-in under slightly oxidizing conditions did not completely eliminate the dislocation half loops. But they are all confined well within the doped region; no glide or penetration took place. In this case the defi-

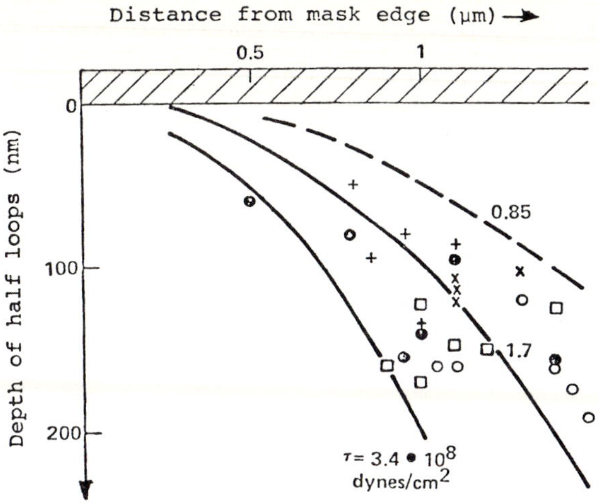


Fig. 15. Contours of constant shear stress and positions of dislocations outside implanted area for conditions of Fig. 13

ning oxide exerted a small compression on the implanted volume. An examination of the associated shear stress components shows that they do not induce glide of dislocations which have the nature of edge dislocations with Burgers vectors parallel to the substrate plane.

The two examples show the importance of mechanical stresses for understanding the effects of residual dislocations which are free to respond to shear stresses. Stacking faults (such as the ones in Fig. 9) are enclosed by sessile dislocations. They cannot move when they are exposed to mechanical stress.

So far, in this section, we have emphasized that defects should not be in the depletion region of a junction. Residual defects which are away

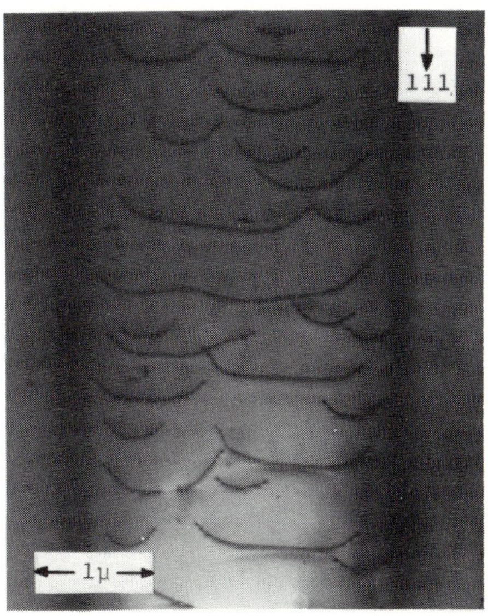


Fig. 16. Dislocation half loops in region bounded by recessed oxide on left and right. TEM foil tilted 35° to show half loop outlines. Implantation: $2 \cdot 10^{15}$ $As^+/cm^2$ at 70 keV, annealing: 45 min at 1000°C

from the junction, closer to the implanted surface, are not always electrically harmful. They can even improve the junction quality by local gettering. However, there is one application where defects must be completely eliminated. This is when the implantation doped wafer is to be a substrate for deposition of an epitaxial layer. Dislocations which terminate at the surface, as in Figs. 6 and 16, replicate themselves through a growing epitaxial layer. Moreover, out-diffusion of dopant material is greatly enhanced along the dislocation lines. If the buried doped layer is the subcollector of a planar transistor, the out-diffusion along the dislocations can reach into the emitter region and give rise to electrical pipes. Buried layers, therefore, have to be more thoroughly annealed than the minimum drive-in of a very shallow junction. Fortunately this is compatible with the use of buried layers, where the desirability of a small sheet resistance translates into a deep junction before epitaxial deposition.

Acknowledgement

I am grateful to my colleagues J.R. Gardiner, C.T. Horng, I.E. Madgo, B.J. Masters, A.E. Michel and R.O. Schwenker for many discussions and for interesting specimens from a variety of projects.

References

1. J.F. Gibbons, Proc. IEEE, 60, 1062 (1972).

2. G. Dearnaley, J.H. Freeman, R.S. Nelson and J. Stephen: Ion Implantation, North-Holland, Amsterdam 1973.

3. A.H. Agajanian: Ion Implantation in Microelectronics, IFI Plenum, New York 1981.

4. Defects in Semiconductors, ed. by J. Narayan and T.Y. Tan, North-Holland, New York 1981.

5. Microscopy of Semiconducting Materials, 1981, ed. by A.G. Cullis and D.C. Joy, The Inst. of Phys., Bristol 1981.

6. Modern Diffraction and Imaging Techniques in Material Science, ed. by S. Amelinckx, R. Gevers, G. Remaut, and J. Van Landuyt, North-Holland, Amsterdam 1970.

7. W.K. Chu, J.W. Mayer, and M.A. Nicolet: Backscattering Spectrometry, Academic Press, New York 1978.

8. G. Foti et al., ref. 5, p. 79.

9. D.K. Sadana and J. Washburn, ref. 5, p. 301.

10. J.W. Corbett et al., ref. 4, p. 1.

11. J.R. Dennis and E.B. Hale, J. Appl. Phys. 49, 1119 (1978)

12. J.F. Gibbons, W.S. Johnson, and S.W. Mylroie: Projected Range Statistics, Halstead Press, Stroudsburg 1975.

13. D.K. Brice: Ion Implantation Range and Energy Deposition Distribution, IFI Plenum, New York 1975.

14. U. Littmark and J.F. Ziegler: Handbook of Range Distributions for Energetic Ions in All Elements, Pergamon Press, New York 1975.

15. Laser and Electron Beam Processing of Materials, ed. by C.W. White and P.S. Peercy, Academic Press, New York 1980.

16. Laser and Electron-Beam Solid Interactions and Materials Processing, ed. by J.F. Gibbons, L.D. Hess, and T.W. Sigmon, North-Holland, New York 1981.

17. R.T. Fulks, C.J. Russo, P.R. Hanley, and T.I. Kamins, Appl. Phys. Lett. 39, 604 (1981).

18. L. Csepregi, E.F. Kennedy, T.J. Gallagher, J.W. Mayer, and T.W. Sigmon, J. Appl. Phys. 48, 4234 (1977).

19. L. Csepregi, E.F. Kennedy, J.W. Mayer, and T.W. Sigmon, J. Appl. Phys. 43, 3906 (1978)

20. S.A. Kokorowski, G.L. Olson, and L.D. Hess, J. Appl. Phys. 53, 921 (1982).

21. J. Washburn, ref. 4, p. 209.

22. W.K. Wu and J. Washburn, J. Appl. Phys. 48, 3742 (1977).

23. D. Nobili, A. Armigliatu, M. Finnetti, and S. Solmi, J. Appl. Phys. 53, 1484 (1982).

24. D.A. Antoniadis and R.W. Dutton, in Process and Device Modelling for Integrated Circuit Design, ed. by F. van de Wiele, W.L. Engl and P.G. Jespers, Noordhoff, Leyden 1977, p. 837.

25. S. Mader and A.E. Michel, Phys. Stat. Sol. (a) 33, 793 (1979).

26. S. Mader, J. Electron. Mater. 9, 963 (1980).

27. U. Gösele and H. Strunk, Appl. Phys. 20, 265 (1979).

28. S.M. Hu, ref. 4, p. 333.

29. M. Tamura, Phil. Mag. 35, 663 (1977).

30. A. Schmitt and G. Schorer, Appl. Phys. 22, 137 (1980).

31. B.J. Masters and E.F. Gorey, J. Appl. Phys. 49, 2717 (1978).

32. T.Y. Tan et al., ref. 4, p. 179.

33. M. Kiritani, Y. Machara, and H. Takata, J. Phys. Soc. Japan 41, 1575 (1976).

34. M.R. Ruehle in Radiation-Induced Voids in Metals, ed. by J.W. Corbett and L.C. Ianniello, AEC Information Services 1972.

35. T.E. Seidel, R.L. Meek, and A.G. Cullis, J. Appl. Phys. 46, 600 (1975).

36. H.J. Geipel and W.K. Tice, Appl. Phys. Lett. 30, 325 (1977).

37. A. Ourmaza et al., ref. 5, p. 63.

38. P. Ashburn, C. Bull, K.H. Nicholas, and G.R. Booker, Solid-St. Electron. 20, 731 (1977).

39. C. Bull, P. Ashburn, G.R. Booker, and K.H. Nicholas, Solid-St. Electron. 22, 95 (1979).

40. M. Tamura, N. Yoshihiro, and T. Tokuyama, Appl. Phys. 17, 31 (1978).

41. T. Koji, W.F. Tseng, J.W. Mayer, and T. Suganuma, Solid-St. Electron. 22, 335 (1979).

# Part IV

# Appendix: Modern Ion Implantation Equipment

# Evolution and Performance of the Nova NV-10 Predep™ Implanter

G. Ryding

Eaton Corporation, Ion Implantation Division
Beverly, MA 01915, USA

## 1. Introduction

Nova Associates was founded in September 1978 with the goal of developing a new generation of high-current (10mA), low-energy implanters. The basic performance objectives and machine configuration were established in November 1978 and have not changed significantly since (see Section 2). However, the system and subsystem designs have evolved along the expected learning curve. Sixty systems have now been shipped and a pattern of actual performance characteristics is emerging.

This report summarizes the most important of these characteristics and reviews the major design features of the NV-10.

## 2. Performance Specifications

The present performance characteristics of the NV-10 are summarized in Table 1 together with the original design goals which are included for comparison.

In almost all cases the system performance has gradually been upgraded, and in particular, the fundamental characteristics of "dose control" and "throughput" now exceed the original objectives by a significant margin. These characteristics and other features of the system will be discussed in detail in the following sections.

## 3. Dose Control

The ability to control dose in terms of both absolute value and uniformity is the single most important characteristic of any implanter. A new concept of dose control was developed for the NV-10 and has been described in detail [1].The advantages of this technique are summarized in Table 2.

Unfortunately not all of these advantages were fully realized in the early systems. In particular it was found that doping level was a function of system pressure. In most cases system pressures were sufficiently reproducible from batch to batch but in other cases large variations in outgassing rates led to unacceptable doping reproducibility.

The magnitude of this effect was totally inconsistent with the effect of beam neutralization or any other known phenomenon (see Section 11. ) Consequently, a comprehensive study of Faraday designs was initiated. Beam signals were recorded as a function of target chamber pressure for several

Table 1. NV-10 Specifications

| System Characteristic | | | 1978 Design Spec. | 1982 Performance Spec. |
|---|---|---|---|---|
| ENERGY[a] | | | 10-60 KeV | 5-80 KeV |
| BEAM CURRENT | P,As | 60-80 KeV | 10 mA | 12.5 mA |
| | | 40 KeV | 10 mA | 10.0 mA |
| | Sb | 40-80 KeV | 5 mA | 6.0 mA |
| | B | 40-80 KeV | 5 mA | 4.0 mA |
| THROUGHPUT[b] (wafers $hr^{-1}$) | P,As up to 3E15 | 3 inch | 300 | 235 |
| | | 100 mm | 200 | 350 |
| | | 125 mm | - | 175 |
| | 1E16 | 3 inch | 150 | 190 |
| | | 100 mm | 100 | 125 |
| | | 125 mm | 70 | 88 |
| | B up to 1E15 | 3 inch | 300 | 330 |
| | | 100 mm | 200 | 230 |
| | | 125 mm | 150 | 165 |
| | 1E16 | 3 inch | 80 | 80 |
| | | 100 mm | 50 | 50 |
| | | 125 mm | 30 | 33 |
| DOPING UNIFORMITY | 1σ across 125mm wafer | | 0.75 | 0.5 |
| DOPING REPRODUCIBILITY | 1σ wafer to wafer | | 0.5 | 0.5 |
| ION MASS RANGE | | | 1-150 | 1-130 |
| ION MASS RESOLUTION | M/ΔM fwhh | | 65[c] | 75 |
| WAFER COOLING | | | 140°C @ 600W | 100°C @ 1200W |
| MACHINE SIZE | | | 94"(W) x 185"(l) x 76"(h) | |
| MACHINE WEIGHT | | | 17000 lbs. | |
| FACILITIES | (max requirements) | Power | 30 KVA | |
| | | Air | 100 psi, 4 cfm | |
| | | $N_2$ | 5 psi, 0.14 cfm | |
| | | $H_2O$ | 40 psi, 5 gpm | |
| | | Exhaust | 1000 cfm | |

a. The NV-10-160 has an energy range 10-160 KeV.
b. All throughputs are achieved with either manual or automatic wafer handling.
c. Optional resolving apertures can be used to provide resolutions up to 100.

different Faraday designs using a calorimeter technique for absolute calibration. Silicon wafers were also implanted under the varied conditions so that relative dose could be determined from four-point probe sheet resistivity measurements.

Preliminary details of this work have been published [5] and will be further discussed at this conference [6]. Typical results are shown in Figs.1-5 where it is seen that the Faraday suppression scheme has a marked influence on performance. It was concluded that signal variations with pressure are primarily a result of residual gas ionization in the entrance region of the Faraday. The magnetic suppression design gives vastly improved performance as shown in Fig.4 and as confirmed by the sheet resistivity data of Fig.5.

Table 2. Advantages of the NV-10 Dose Control System

1. The beam signal is not measured on the wafer surface and dose error due to variation in surface composition and secondary effects [2,3,4] is eliminated.

2. Graphite has been selected for the ion impact surface within the Faraday to further minimize secondary effects and increase reproducibility.

3. Geometric effects of the spinning disk (velocity gradient $\propto R^{-1}$) are automatically compensated for.

4. The "slot technique" provides on-line control of doping uniformity.

5. Beam signal "commutation" from a rotating shaft is avoided.

6. Water or freon cooling is not required as part of the dose control Faraday and electrical leakage problems are avoided.

7. The stationary Faradays provide noise-free signals which permit precise control even at low doses ($<10^{12}$ ions $cm^{-2}$).

8. The use of two independent Faradays to determine "estimated" and "actual" implant times serves as a powerful check on the consistency of the system.

9. The entire beam path in front of the wafers is free of electrostatic fields. The migration of free electrons is uninhibited and surface charging is minimized.

10. If necessary the wafers can also be flooded with low energy electrons from "an electron shower" (optional). This system is not part of the dose control Faraday. Consequently it does not require electrical isolation and can not influence the Faraday dose control signal.

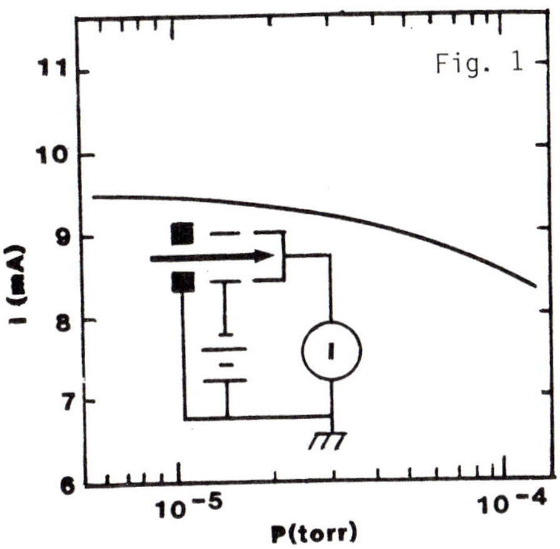


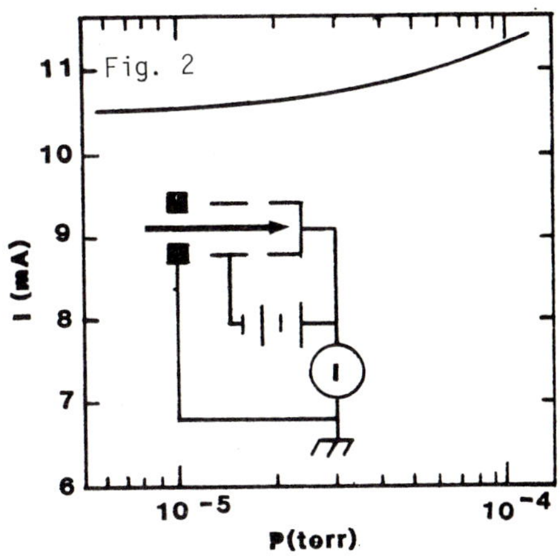


Figure 1. Faraday # 1. Beam signal vs pressure for 10 mA, 60 keV, $As^+$ beam

Figure 2. Faraday # 2. Beam signal vs pressure for 10 mA, 60 keV, $As^+$ beam

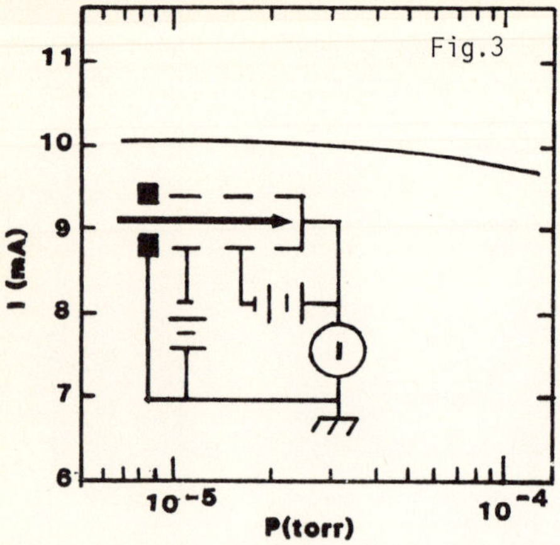


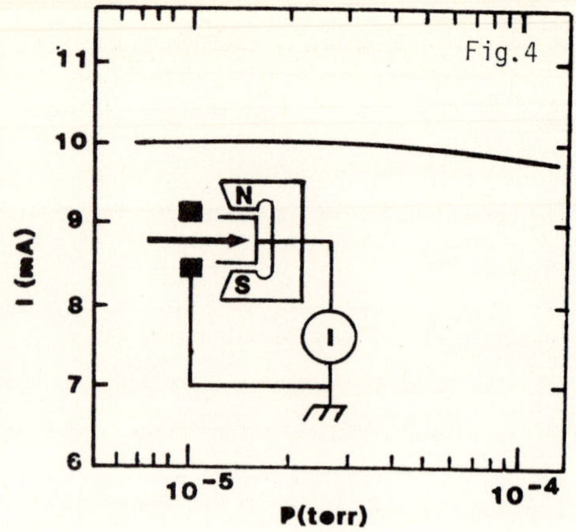


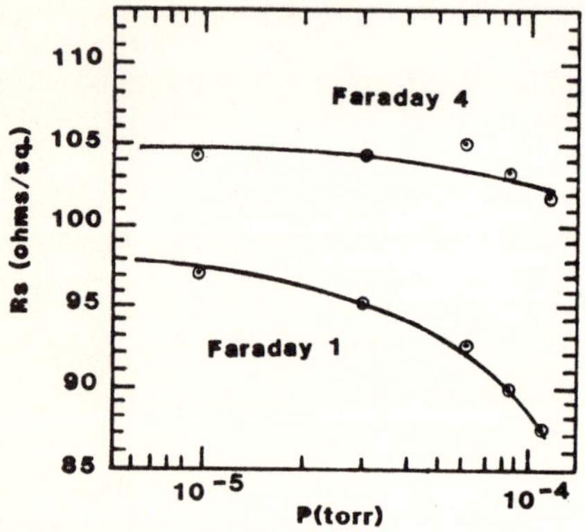


Figure 3. Faraday # 3. Beam signal vs pressure for 10 mA, 60 keV, $As^+$

Figure 4. Faraday # 4. Beam signal vs pressure for 10 mA, 60 keV, $As^+$

Figure 5. Sheet resistivity vs pressure for Faradays 1 and 4. $As^+$ 80 keV, $10^{15}$ ions $cm^{-2}$

The magnetic design is now standard on all NV-10 systems and is being retrofitted on earlier machines as required. Since the measured beam intensity and dose accuracy is independent of pressure over such a broad range, a major advantage is achieved in that implantation can be initiated earlier in the pump down cycle and throughput is increased (see Section 4.1). Furthermore the effects of photoresist outgassing are minimal over rather broad limits. This is particularly important since positive resist outgassing rates approach $10^{-1}$ torr $lsec^{-1}$ at beam powers of 1 kwatt, and target pressures reach $10^{-4}$ torr during the first scan of the beam over the disk. A further advantage is that the dangerous suppression voltages and breakdown complications of conventional Faradays are avoided.

In summary it is believed that the use of a "slot-technique" in conjunction with magnetic suppression of the Faradays will establish a new standard in terms of dose accuracy, reproducibility and uniformity. To quantify this we are undertaking a detailed study of machines undergoing final test in the factory so that each is characterized before shipping. In addition we are working with several customers on selected process requirements. Preliminary results of this work are as follows:

Figures 6, 7 and 8 show sheet resistivity measured by a four-point probe method as a function of implanted dose for B, P and As respectively

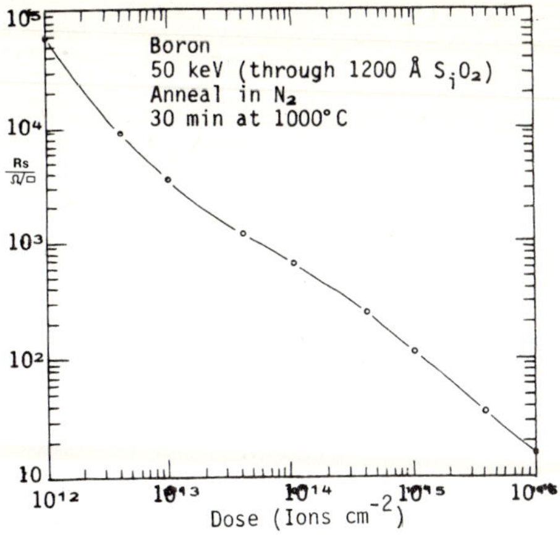


Figure 6. Sheet resistivity vs dose for 50 keV $B^+$

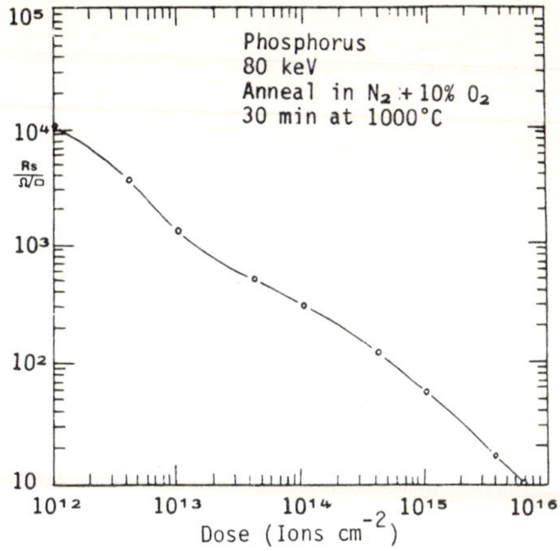


Figure 7. Sheet resistivity vs dose for 80 keV $P^+$

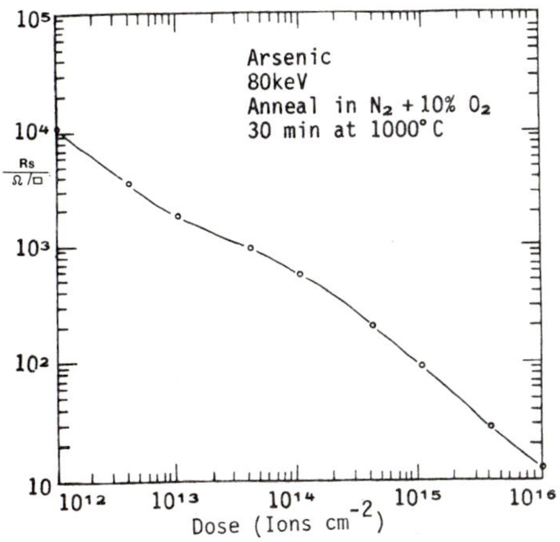


Figure 8. Sheet resistivity vs dose for 80 keV $As^+$

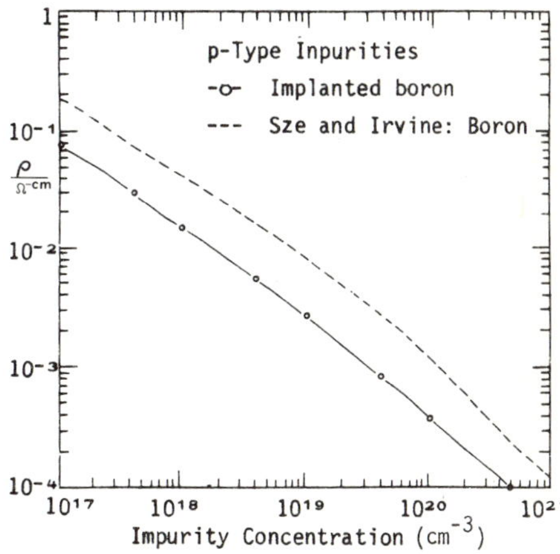


Figure 9. Bulk resistivity vs carrier concentration for p-type impurities

in 1 to 10 Ω-cm phosphorus doped, Czochralski <100> silicon. In order to compare these results with previously established data, sheet resistivity was converted to bulk resistivity using a simple model which assumes a uniform depth profile after annealing. The results are displayed in Figs.9 and 10. It can be seen that, in regard to slope and shape, the implanted data is in excellent agreement with the data of Sze[7]and Irvine[8]. It should be noted that in the case of boron (Fig.6) sheet resistivity increases more rapidly than expected as the dose is reduced below 5 x $10^{12}$ ions/$cm^2$. This is explained by the relatively high background concentration which yields about $10^{16}$ holes/$cm^3$ at low boron doses after anneal. Similar implants into silicon with a background concentration of $10^{14}$ impurities/$cm^3$ show no "tailing effect" at low boron doses.

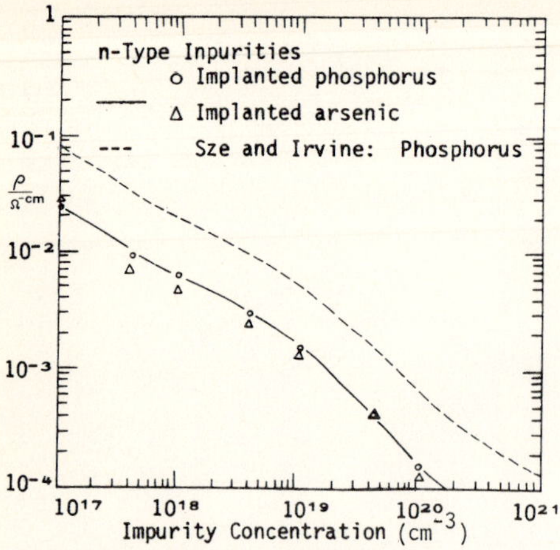


Figure 10. Bulk resistivity vs carrier concentration for n-type impurities

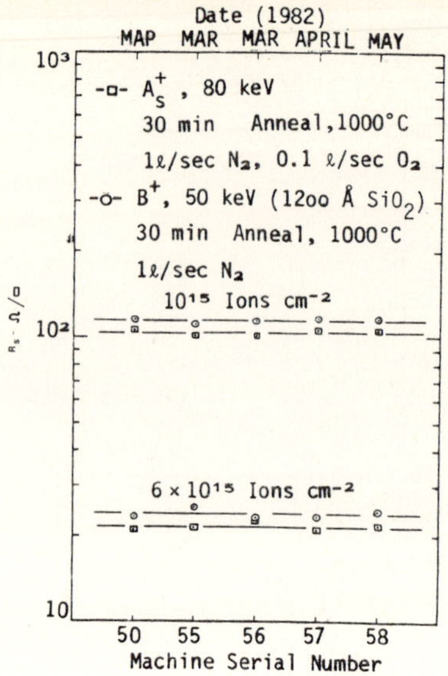


Figure 11. Sheet resistivity tracking chart

In all cases the measured uniformity across each wafer, including four-point probe measurement errors, gave "one sigma" values consistently below 1%. Finally the tracking chart shown in Fig.11 indicates that a consistent pattern is being established from day to day and machine to machine. Full details of this work will be published in the near future[9].

## 4. Throughput

The throughput of any implanter depends on the duration of each implant and the cycle time taken between implants when the implanted wafers are removed from the beamline and replaced by the subsequent batch to be implanted. The measured characteristics of the NV-10 are summarized as follows:

### 4.1 Implant Time

The implant time for each batch is given by the following equation:

$$T = 1.6022\ 10^{-13}\ D\ I^{-1}\ A \qquad (1)$$

where $T$ = the batch implant time in seconds
$D$ = the dose in ions $cm^{-2}$
$I$ = beam current on target in $\mu A$
and $A$ = the total implant area in $cm^2$

The total implant areas and batch sizes for the NV-10 are listed in Table 3.

| Table 3. NV-10 Implant Areas and Batch Sizes | | |
|---|---|---|
| Wafer Size | Implant Area $cm^2$ | Batch Size |
| 3 inch | $1.680\ 10^3$ | 18 |
| 100 mm | $1.969\ 10^3$ | 13 |
| 125 mm | $2.217\ 10^3$ | 10 |
| 150 mm | $2.425\ 10^3$ | 8 |

## 4.2 The Wafer Handling Sequence Between Implants

The sequence in which implanted wafers are replaced by the next batch of wafers is summarized in Table 4.

The pressure setpoint at which implant is initiated is normally set at $2 \times 10^{-5}$ Torr and pump down times and throughputs have been based on this number. However, we have shown that accurate dose control is obtained at much higher pressures (see for example the data discussed in Section 3.1) and the process engineer has the option of adjusting the setpoint to higher values if higher throughputs are required. In this case the disk "run-speed" servo time can be the "gating" item and the rotary drive servo system has recently been modified to give faster acceleration and settling time.

## 4.3 Throughput Values

The system throughput can be readily calculated for any dose, beam current and wafer size combination using Equation (1) in conjunction

| Table 4. NV-10 Timing Sequence Between Implants | |
|---|---|
| Action | Time (Seconds) |
| Implant Ends | 0 |
| "Flag Faraday" Closed then Gate Valve Closed | 3 |
| End Station Vent | 12 |
| Door Opens, Disk Stops Rotating | 10 |
| Disk Interchange | 30 |
| Door Closes, (Disk Acceleration Starts) | 10 |
| Endstation Roughing to 200 Microns | 25 |
| High Vacuum Pump to $2\ 10^{-5}$ Torr and Disk Run Speed O.K. | 30 |
| Implant Starts | 0 |
| Total Elapsed Time | 120 seconds |

Table 5. NV-10 Throughput

| Wafer Size | Beam (mA) | Dose (ions $cm^{-2}$) | Throughput (wafers $hr^{-1}$) |
|---|---|---|---|
| 3 inch | 12.5 | up to 3E15 | 350 |
| | | 5E15 | 280 |
| | | 1E16 | 190 |
| | | 2E16 | 115 |
| | 4 | up to 1E15 | 340 |
| | | 3E15 | 200 |
| | | 5E15 | 140 |
| | | 1E16 | 80 |
| 100 mm | 12.5 | up to 2.6E15 | 250 |
| | | 5E15 | 185 |
| | | 1E16 | 125 |
| | | 2E16 | 74 |
| | 4 | up to 8E14 | 250 |
| | | 3E15 | 130 |
| | | 5E15 | 90 |
| | | 1E16 | 50 |
| 125 mm | 12.5 | up to 2E15 | 200 |
| | | 5E15 | 135 |
| | | 1E16 | 88 |
| | | 2E16 | 50 |
| | 4 | up to 6.5E14 | 200 |
| | | 3E15 | 90 |
| | | 5E15 | 60 |
| | | 1E16 | 35 |

with the parameters listed in Tables 3 and 4. Typical throughput values are shown in Table 5.

It is emphasized that these values are conservative for three reasons:

a. In most cases beam current specifications are frequently exceeded (see Section 6.1).

b. The 120-second time between implants is reduced to ~105 seconds if a higher vacuum setpoint ($\sim 4 \cdot 10^{-5}$ Torr) is used.

c. The table values have typically been "rounded down" from the calculated values by ~3%.

## 5. Uptime

Implanter throughput specifications are meaningless unless they are complemented by good machine reliability. Unfortunately, machine reliability depends not only on basic design features but also on intangible factors such as operator training, preventative maintenance quality, spares availability, service response, documentation accuracy and service accessibility. Although these factors are difficult to quantify, the information on overall reliability is so important that in November, 1981, Eaton/Nova implemented a weekly "uptime" reporting scheme for all domestic implanters.

In summary, performance of each system is reported by the customer using the following definitions:

a. Scheduled Hours - The total hours for which production operations of the NV-10 are required or expected. It is normally forecasted on Monday morning at the time the "Uptime Report" is given but can subsequently be adjusted to allow for increases or for decreases in work load or facilities failures.

b. Uptime Hours - Total time of successful operation in the reporting period.

c. Uptime Percent - $100ba^{-1}$

d. Runtime Hours - Read directly from source elapsed time meter.

e. Number of Failures - Total number of distinct failures.

f. MTTR (mean time to repair) - $(a-b)\ e^{-1}$

g. MTBF (mean time between failures) - $be^{-1}$

h. Preventive Maintenance Hours - Maintenance performed according to a prescribed schedule (see Section 9). Service complications (overrun hours) are deducted from uptime hours and treated as a distinct failure.

i. Emergency Maintenance Hours - Maintenance required during scheduled hours. During unscheduled hours, emergency maintenance does not detract from Uptime Hours, but is tracked for purposes of calculating MTTR.

j. Deferred Maintenance - Maintenance required to correct malfunctions which do not prevent successful operation of the system. When this is performed by customer decision during unscheduled hours, it does not detract from uptime hours, but is tracked for purposes of calculating MTTR.

Report forms are provided to the customer and data is collected each week on Monday morning. On a monthly basis, the customer receives copies of the weekly reports and a graph charting the cumulative and weekly uptime percentage of his system as compared to the total system data base. Customer confidentiality is maintained throughout the reporting scheme.

So far the customer involvement has been very encouraging with an 80% participation. System cumulative uptimes have ranged from 70% to 95% with an overall average of 85%.

## 6. Beam Current and Source Performance

Because of the unusually short distance between source and wafer, the NV-10 operates with beam transmission efficiencies close to 100%. Consequently the

Table 6. Source Parameters for $As^+$ and $P^+$

| | | | |
|---|---|---|---|
| Energy | 40 keV | 80 keV | 80 keV |
| Beam current | 10 mA | 10 mA | 15 mA |
| Arc current | 1.7 A | 1.5 A | 2.3 A |
| Arc voltage | 60 V | 60 V | 60 |
| Extraction power supply current | 22 mA | 18 mA | 28 mA |
| Filament life | > 40 hr | > 40 hr | ~ 40 hr |

Table 7. Source Parameters for $B^+$

| | $BF_3$ | | Enriched $BF_3$ | |
|---|---|---|---|---|
| Energy | 40 keV | 80 keV | 80 keV | 80 keV |
| Beam current | 4 mA | 4 mA | 4 mA | 4 mA |
| Arc current | 6 A | 4.5 A | 4.0 A | 6 A |
| Arc voltage | 60 V | 60 V | 60 V | 60 V |
| Extraction power supply | 33 mA | 28 mA | 24 mA | 33 mA |
| Filament life | ~ 16 hr | ~ 25 hr | ~ 30 hr | ~ 16 hr |

Table 8. Vaporizer Performance Summary

| | |
|---|---|
| Capacity (As) | ~ 30 g |
| Charge lifetime at 10 mA | ~ 60 hr |
| Heat-up time to 400°C | ~ 5 min |
| Cool-down time to 200°C | ~ 5 min |
| Temperature stability | ± 0.5°C |
| Temperature Uniformity | ± 2°C |
| System "tune-up" time to 10 mA | <15 min |

specification for $As^+$ and $P^+$ is easily met or exceeded. Indeed, many implants have been performed in the 15-20mA range without over-stressing the system.

A few "critical" parameters generally serve as a barometer of an implanter's efficiency, and typical operating values for the NV-10 are listed in Table 6.

In the case of boron, beam specifications are less conservative and a clean source, good alignment and careful tuning are required. Typical operating values are shown in Table 7. In order to improve the boron situation we recommend the use of "enriched" $BF_3$. As can be seen, it is possible to obtain 6mA of beam at the higher energies but the reduced source lifetime makes these conditions impractical for long production runs.

Performance of the vaporizer is summarized in Table 8, and a typical heating and cooling cycle is shown in Fig.12.

This fast thermal response enables the machine to be brought up from a cold start to specified currents of $P^+$ and $As^+$ in less than 15 minutes.

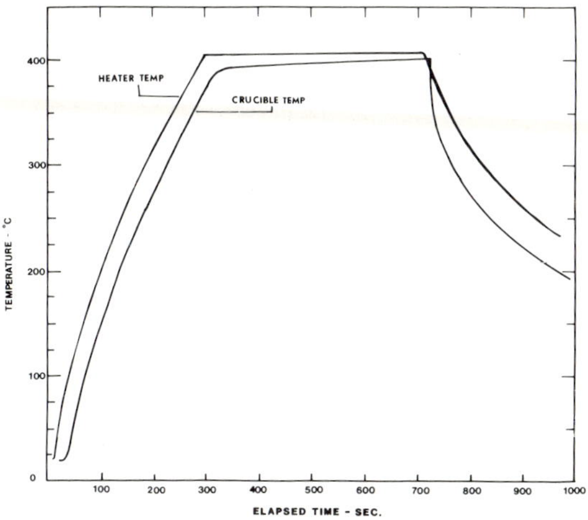


Figure 12. Vaporizer temperature vs time

## 7. Energy

Commercial implanters do not have an impressive history of reliable operation at maximum rated voltage. Problems traditionally fall into the following categories.

a. The accelerating electrode region deteriorates as a result of vapor and sputtered deposits, the vacuum degrades, and vacuum arc-overs become more violent and frequent. Operator errors can exaggerate these problems.

b. The vacuum arc-overs, which have very fast rise-times, stress the H.V. power supply and lead to frequent failure. Other electronic components in the implanter can also fail as a result of this noisy environment.

c. External breakdown can result from insufficient safety margin and poor machine maintenance.

Unfortunately, voltage ratings have often been based on factory tests under relatively favourable conditions as opposed to the conditions encountered in production when a system has operated for prolonged periods on a 3-shift basis.

The NV-10 source region is designed to operate at 100kV. In particular the source bushing can be operated at 100kV and the source is enclosed within a fixed corona shield to minimize stress between the source and extraction electrodes. A "double gap" extraction arrangement was evaluated but did not offer any advantages over the adjustable single gap presently used. Particular attention has been given to good pumping in this region, and all critical assemblies (source and electrodes in particular) are easily demountable for convenient maintenance and cleaning.

Regardless of design precautions, the source extraction region operates under the severe conditions of high pressure, high temperature, vapour and sputter deposits. Source "glitches" which trigger vacuum arc-overs are inevitable. Consequently the H.V. extraction supply must be capable of withstanding this stress.

Unfortunately the commercial supplies available were found to be unacceptable in the early NV-10 systems and in June, 1979, Nova embarked on a comprehensive power supply development program. It resulted first in power supply upgrade kits which have been retrofitted on all machines, bringing the reliability to acceptable levels. Following this a proprietary supply was developed specifically for use in the NV-10. An "overkill" philosophy was adopted and a wide safety margin was designed into all critical components. Particular attention was given to its ability to survive repetitive short circuits of the output, and indeed this remains an important part of system testing.

The resulting supply, which was introduced at machine serial #37, has a basic rating of 100kV and 50mA. The maximum requirement of the NV-10 is 80kV, and even with boron the extraction current is generally below 33mA (see Tables 6 and 7).

The ability of the source to run constantly at 80kV and the reliability of the H.V. supply itself are major features of the NV-10 system.

## 8. Wafer Cooling

Device designers are interested in the use of photoresist for ion beam masking in order to reduce the number of process steps in device fabrication. Unfortunately photoresist can only tolerate a limited temperature rise before flow, blistering or cracking results in unacceptable degradation of the patterns. In particular most positive resists deteriorate at temperatures greater than approximately 110°C. Consequently beam power, and therefore system throughput can be severely restricted by wafer temperature considerations.

| Table 9. Methods Used to Evaluate Wafer Cooling |
|---|
| 1. Vacuum Test Stand: A test wafer is illuminated by a high intensity lamp through a quartz window. thermocouples banded to wafers monitor the temperature rise. |
| 2. Thermal stickers: Small adhesive pads containing compounds which change color at various melting temperatures are attached to wafers during implant. |
| 3. I.R. Detector: An infrared detector mounted in the end station monitors wafer temperature during implant. |
| 4. Photoresist wafers: Patterned production wafers are inspected under a microscope after implantation in order to evaluate cooling efficiency. Ease of stripping after implant has also been investigated. |

As a result of this, wafer cooling has been a major concern at Nova and there has been an ongoing development project since the company was founded. A review of this work is to be published at this conference[10]. In summary the methods used to evaluate different cooling techniques are listed in Table 9.

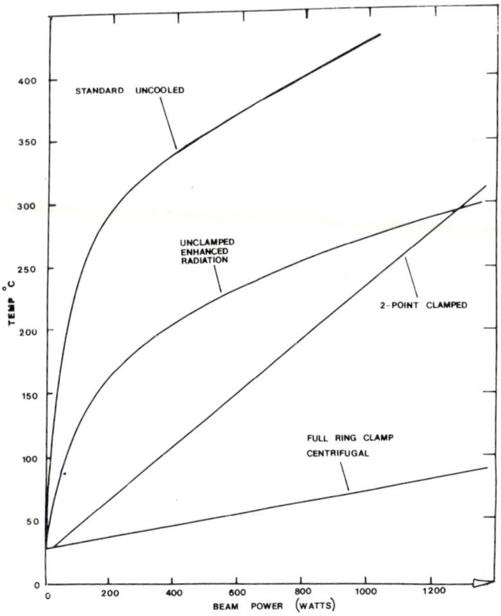


Figure 13. Wafer temperature vs beam power for 80 KeV, $10^{16}$ ions $cm^{-2}$ implants

The performance of different designs is displayed in Fig.13, which shows the temperature rise of 100mm wafers as a function of incident beam power on the disk for a dose of $10^{16}$ ions $cm^{-2}$. The different cooling methods are briefly described as follows:

## 8.1 Uncooled Disks

Wafers are placed on a simple aluminum pedestal inclined at $7^{0}$ to the plane of the disk so that centrifugal force holds them in place. In this case heat loss from the wafer is essentially by radiation only and depends on the emissivity of the wafer and its surroundings.

## 8.2 Radiation-Enhanced Cooling

In order to improve the cooling by radiation, surfaces surrounding the wafers are treated so that their absorption coefficients are increased. Specifically the aluminum pedestals are coated with alumina and the inside of the target chamber facing the wafers is coated with a black epoxy film. This configuration is available as an option.

## 8.3 2-Point Clamp Technique

In this system wafer cooling is enhanced by conductive cooling from the backside of the wafer. Each pedestal has a slight cylindrical curvature and is covered with a thin layer of high-conductivity silicone rubber. The wafer is pressed against this pliable surface with a spring-loaded arm which makes contact at just two diametrically opposite points on the edge of the wafer. In this way the amount of wafer surface shadowed by the clamp is kept to a minimum. Unfortunately the degree of backside surface contact achieved with this clamp design is limited by the stiffness of the wafer, and only partial success was achieved. As shown in Fig.13, beam power with the 2-point clamp is limited to approximately 300 watts.

## 8.4 Full-Ring Centrifugal Clamp

In order to further enhance conductive cooling, a full circumferential ring clamp system was designed and is now available as a standard option for all wafer sizes. In this case the wafer is pressed evenly around its edge over a slightly convex rubber surface. The clamping action is provided by the centrifugal force of the spinning disk as shown schematically in Fig.14. In this way stiff clamp springs are avoided and loading and unloading of the wafers is greatly simplified. A typical disk is shown in Fig.15 and, as can be seen from Fig.13, wafer temperatures do not exceed 90°C at beam powers up to 1200 watts. Using the full-ring clamps, successful use in terms of pattern integrity and post-implant stripping has already been demonstrated for the following resist types and beam conditions:

| | | |
|---|---|---|
| a. | Hunt 204 | 15mA, 80 keV, $10^{16}$ ions $cm^{-2}$ |
| b. | Shipley 1470 | 10mA, 80 keV, 2 $10^{16}$ ions $cm^{-2}$ |
| c. | KTI II | 18mA, 80 keV, $10^{16}$ ions $cm^{-2}$ |

## 9. Preventive Maintenance

It is important to know the preventive maintenance requirements of any implanter, since correct service procedures can significantly improve machine performance and reliability. A pattern of NV-10 service requirements has emerged and is summarized briefly in Table 10. The actual intervals between each service requirement naturally depend on the mode of operation of the

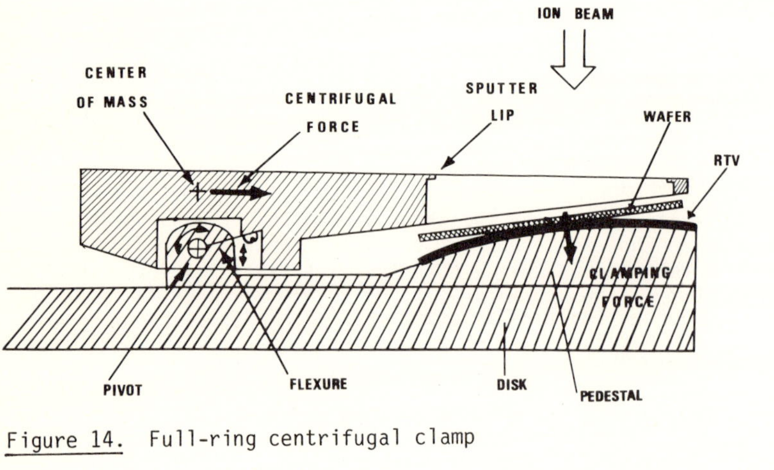


Figure 14. Full-ring centrifugal clamp

Figure 15. Wafer disk with full-ring centrifugal clamps

| Table 10. Preventive Maintenance Schedule | | |
|---|---|---|
| Interval Source Hours (total time) | Service Procedure | |
| 20 hr (daily) | 1. Update Log: | System Pressures, Timer Panel Readouts |
| | 2. General Inspection: | Doors, Panels, Gas Box, Source, etc. |
| 40 hr (2 days) | 1. Source Service as required: | Filament, Vaporizer Charge, Clean Arc Chamber, etc. |
| | 2. Oil Change: | Source Roughing Pump RP1 (see 3.7) |
| 120 hr (weekly) | 1. Beam-line Service: | Source, Bushing, Source Housing Electrodes, Beamguide, Strike-plate, Flag Faraday |
| | 2. Oil Change: | RP1, 2, 3 and 4 |
| | 3. Cryopumps: | Regenerate P2, P3 |
| | 4. Exchange Heads: | Check alignment, Lubrication |
| 3000 hr (Semi-annual) | 1. Subsystem Calibration: | Vac, Beam, Dose Controllers |
| | 2. Oil Change: | Source Diffusion Pump P1 |
| | 3. Cryopumps: | Clean Heads, Service Compressors |
| 6000 hr (annual) | 1. Cryopumps: | Replace adsorbers, Charcoal Arrays |
| | 2. Clean vacuum foreline, roughing system | |
| | 3. Inspect sliding seal assembly | |

system and the extent to which it is being stressed. However, we recommend conservative intervals for "worst-case" operation, since it is generally accepted that planned downtime is far preferable to unscheduled downtime during critical production runs.

The most critical service procedure is probably the beamline service which is recommended every 120 hours of beam operation. Typically this is performed at the weekend when the system is running 3-shift operation. It involves a thorough cleaning and inspection of all beamline components which either generate or intercept the ion beam. Although each component is designed as an easily removable subassembly, full service can take as much as 6-8 hours.

Consequently we have extended the well-accepted philosophy of a spare ion source and have generated a recommended list of spare beamline subassemblies called "Fast Maintenance Spares" as listed in Table 11. Use of these spares enables most of the beamline service to be done off-line in the maintenance area during normal operation of the implanter. The weekly service can then be performed in under 4 hours, a time largely dictated by the regeneration of the cryopumps.

| Table 11. Fast Maintenance Spares |
|---|
| 1. Ion Source Assembly |
| 2. Source Bushing |
| 3. Extraction Electrode Assembly |
| 4. Flag Faraday Assembly |

The frequency of oil change in the source roughing pump depends dramatically on whether the source is operating with a vaporizer or gas feed. $BF_3$ is the worst case, whereas the gas load is essentially zero with the vaporizer. A closed-loop oil filtration system has been added to the source roughing pump on all recent systems, and we expect a major reduction in oil change frequency as a result of this improvement.

## 10. Process Control (Datalock$^{TM}$)

Wafers are all too easily destroyed in an implanter if the operator selects an incorrect process parameter. Unfortunately, test results of the implantation process are not available until several steps later in the manufacturing cycle. Consequently incorrect implanter adjustment can persist without detection, and thousands of valuable wafers can be inadvertently destroyed.

The complete NV-10 control system is a distributed network of microprocessor-based subsystems and is shown schematically in Fig.16. Each unit operates on a "stand-along" basis with local intelligence, self-calibration and test diagnostics, so that the operator interaction is simple and the probability of error is minimized. For example the Mass/Energy (AMU) control subsystem[11]performs the following basic functions automatically on command from the operator:

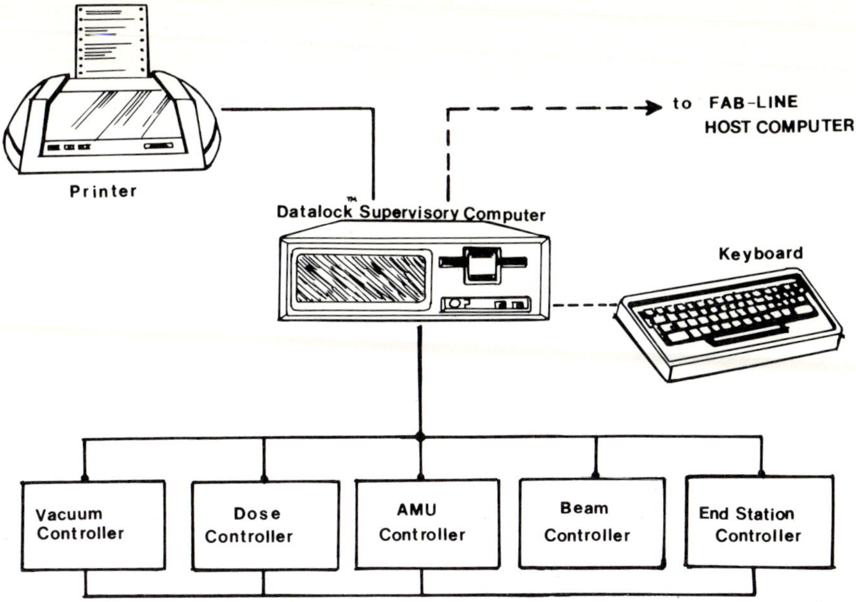


Figure 16. Datalock T.M. control system schematic

a. Energy Control

b. Analyzer Control

c. Ion Mass (AMU) Calculation

d. Beam Peak Search

e. Fine-Tune

f. Mass Resolution Measurement

g. Spectrum Recording

Each control subsystem is in constant communication with a supervisory system (Datalock TM) which consists of a microcomputer, touch-sensitive CRT display, floppy disk, printer and a detachable keyboard. In this way the "Datalock" provides continuous on-line monitoring of all machine parameters.

In the normal mode of operation the process engineer establishes by means of the keyboard a complete, password-protected recipe corresponding to a particular process code number (1-49). Each machine parameter is given an acceptable minimum/maximum value as illustrated in Fig.17. In production the operator recalls this recipe by means of the touch-sensitive screen which displays both the actual machine values together with the previously established limits. Parameters which are invalid are identified using an "inverse video" display, and prompted in this way, the operator must adjust the system for the correct values before an implant can be started. Each parameter is monitored throughout the implant, and the operator is alerted by error messages if any deviates from its limits. The floppy disk drive provides for mass storage of the actual implant data, which can be recalled at any time. Furthermore at the conclusion of each implant a printout is provided on lint-free paper showing all parameter values at the start, mid-point and end of each implant as shown in Fig.18.

PROCESS 31 12:46 30-Apr-82

| PARAMETER | UNITS | MIN | MAX |
|---|---|---|---|
| Dose | IONS/CM2 | 8.00E+15 | 8.00E+15 |
| Energy | V | 7.95E+04 | 8.05E+04 |
| AMU | AMU | 3.10E+01 | 3.10E+01 |
| # Scans | | 6.00E+00 | 1.00E+01 |
| Wafer Size | INCH | 4.00E+00 | 4.00E+00 |
| P1 Source | TORR | 1.00E-08 | 5.00E-05 |
| P2 Beam | TORR | 1.00E-08 | 1.00E-05 |
| P3 Disk | TORR | 1.00E-08 | 1.00E-04 |
| I Beam | A | 9.00E-03 | 1.10E-02 |
| I Arc | A | 1.00E+00 | 2.00E+00 |
| V Arc | V | 5.50E+01 | 6.50E+01 |
| I Extract | A | | 2.50E-02 |
| I Fil | A | 1.00E+02 | 2.20E+02 |
| V Fil | V | | 7.50E+00 |
| I Analyzer | I | 4.00E+01 | 5.00E+01 |
| I Source Mag | A | 8.50E+00 | 1.10E+01 |
| T Vaporizer | C | 3.50E+02 | 4.25E+02 |
| Axis 1 | | 3.00E+02 | 9.00E+02 |
| Axis 2 | | 3.00E+02 | 9.00E+02 |
| Axis 3 | | 4.50E+02 | 9.00E+02 |

Figure 17. Datalock:Process recipe example

BATCH NUMBER: WA 14-26 PROCESS NUMBER: 31
TIME & DATE: 30-Apr-82 13:16
IMPLANT DURATION, ESTIMATED: 4.1 MINUTES
ACTUAL: 4.3 MINUTES
NUMBER OF IMPLANT INTERRUPTIONS: 0

| IMPLANT PARAMETER | IMPLANT START CONDITIONS | MID IMPLANT CONDITIONS | IMPLANT FINISH CONDITIONS | |
|---|---|---|---|---|
| DOSE | 8.00E15 | 8.00E15 | 8.00E15 | IONS/CM2 |
| ENERGY | 79.9 | 79.9 | 79.9 | KEV |
| AMU | 31.0 | 31.0 | 31.0 | AMU |
| I BEAM | 9.50E+00 | 9.69E+00 | 1.00E+01 | MA |
| PRESET SCANS | 8 | 8 | 8 | |
| WAFER SIZE | 4 | 4 | 4 | INCHES |
| P1 SOURCE | 1.3E-5 | 1.3E-5 | 1.3E-5 | TORR |
| P2 BEAM | 6.0E-7 | 5.1E-7 | 1.4E-6 | TORR |
| P3 DISK | 1.7E-5 | 9.6E-6 | 2.3E-5 | TORR |
| I ARC | 1.46 | 1.47 | 1.48 | A |
| V ARC | 63 | 63 | 62 | V |
| I EXTRACT | 19.4 | 19.6 | 19.6 | MA |
| I FIL | 138 | 136 | 136 | A |
| V FIL | 3.78 | 3.79 | 3.94 | V |
| I ANALYZER | 42.0 | 42.0 | 42.0 | % |
| I SOURCE MAG | 9.5 | 9.5 | 9.5 | A |
| T VAPORIZER | 406 | 406 | 406 | C |
| AXIS 1 | 601 | 601 | 601 | |
| AXIS 2 | 415 | 415 | 415 | |
| AXIS 3 | 464 | 464 | 464 | |

Figure 18. Datalock: Implant data printout

The Datalock computer can also communicate with the host computer of the process line. The electrical connector is a standard EIA RS-232-C, and the message protocol, datalink protocol and character structure adhere to the proposed standard known as E2 SEMI Equipment Communications Standard (SECS)[12]. This link to the host computer provides for wafer inventory control and ensures that each process step is performed in a correct sequence. As the system evolves it will also be possible to directly download each implant recipe from the host computer.

The Datalock system greatly simplifies and improves the reliability of implanter set-up, and with it, the possibility of implanting wafers with an incorrectly adjusted process parameter has been virtually eliminated.

## 11. Neutral Beams

The subject of "neutral beams" and "neutral traps" in high-current implanters has become an emotional issue and is frequently a distraction during machine selection. Beam neutralization is an atomic collision phenomenon whereby the incident ion collides with a residual atom or molecule, captures an electron and thereby returns to a "neutral" atomic configuration as shown in Fig.19. This "charge exchange" or "electron-capture" process has been well recognized in most implanters and particle accelerator designs. The degree of beam neutralization is proportional to the residual gas pressure and the beam path length, and under most conditions the neutral fraction $F_0$, generated along a section of beamline, is given by:

$$F_0 = 3.3\ 10^{16}\ P\ L\ \sigma \qquad (2)$$

where $P$ = Pressure (torr)
$L$ = Path Length (cm)
and $\sigma$ = the collision cross-section ($cm^2$)

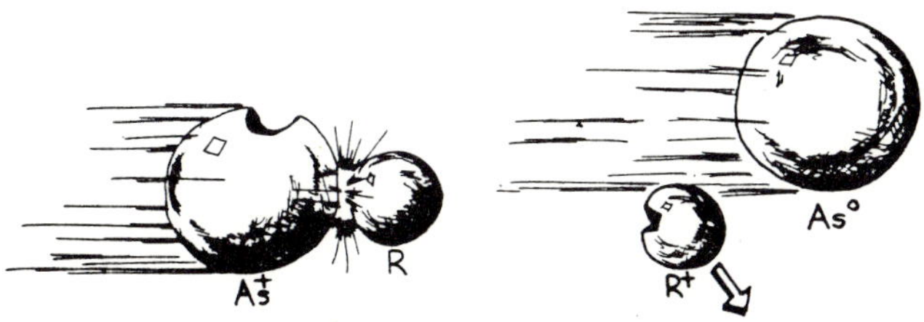


Figure 19. Charge exchange (electron-capture) schematic for $As^+$ ion collision with residual atom

Neutral formation can result in the following undesirable effects:

a. Dose error. The Faraday does not detect the neutral particles and the wafers are therefore subjected to excess doping.

b. Dose non-uniformity. With beam scan systems, the neutrals do not respond to deflection fields and "hot-spots" or "stripes" can result.

c. Depth profile errors. If acceleration of the ions involves two discrete stages, as in most conventional 200kV implanters, "energy contamination" can occur. Particles which undergo charge exchange before the final acceleration stage and yet still reach the end-station, arrive at the target with the wrong velocity and result in shallow profile tails (electron capture) or deep profile tails (electron loss).

d. Secondary particle effects. Neutral particles do not respond to deflection fields and can strike the beamline in unexpected places resulting in sputtering, heating and evaporation. Secondary particles so generated are undesirable companions to the primary ions if they are allowed to reach the wafer.

The design of the NV-10 will now be discussed in regard to these four effects and with reference to the beamline schematic shown in Fig.20.

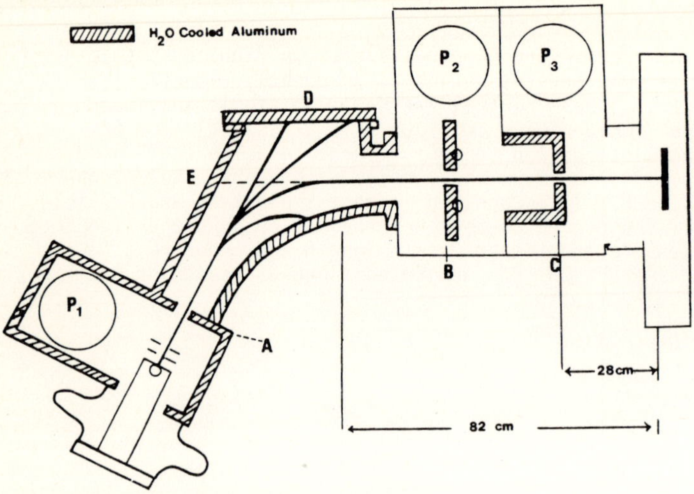


Figure 20. NV-10 beamline schematic

11.1 Dose Error

The beamline between the magnet and the wafers consists of two independently pumped regions. The beam-resolving aperture, which has a conduction of ~10 l/sec, is the only vacuum connection between the two and is located 28cm in front of the wafer. As a result, the pressure increase resulting from endstation vacuum cycling and wafer outgassing only occurs over the last third (28cm) of the total beamline from the magnet to the wafers (82cm). It should also be noted that this distance from the magnet to the wafer is comparable to the distance between the neutral offset and the wafer in electrostatic scan machines.

A typical endstation pressure plot for an 800-watt beam striking a full disk of positive photoresist-covered wafers is shown in Fig.21. For comparison, a similar plot using blank silicon wafers is shown in Fig.22. Analysis of the outgassing products by means of a residual gas analyzer indicates that nitrogen is the dominant component. Using a typical electron-capture cross-section[13] value of $4 \times 10^{-16}$ cm$^2$ and an average pressure during implant of $3 \times 10^{-5}$ torr, which is the worst case for short implants with photoresist, it can be seen from equation (2) that the neutral beam fraction formed in the target region is ~1%. In the region between magnet and resolving aperture the pressure is invariably below $10^{-6}$ torr and the neutral fraction generated here is ~0.1%. These estimates are entirely consistent with the Faraday data of Fig.4 and the sheet resistivity data of Fig.5 discussed in Section 3.1.

11.2 Dose Non-uniformity

The NV-10 uses the "mechanical scan" concept so that all incident particles are distributed uniformly over the wafer, independent of their charge. This immunity to charge exchange effects is a particularly attractive feature of mechanical scanning[14].

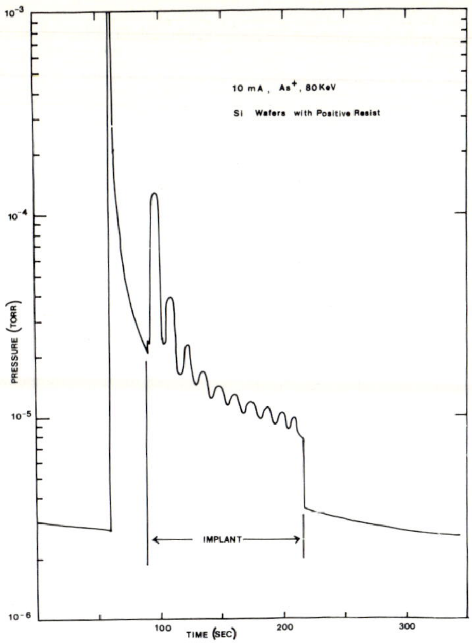


Figure 21. End station pressure vs time for 10 mA, 80 keV, $As^+$ in photo-resist-covered Si

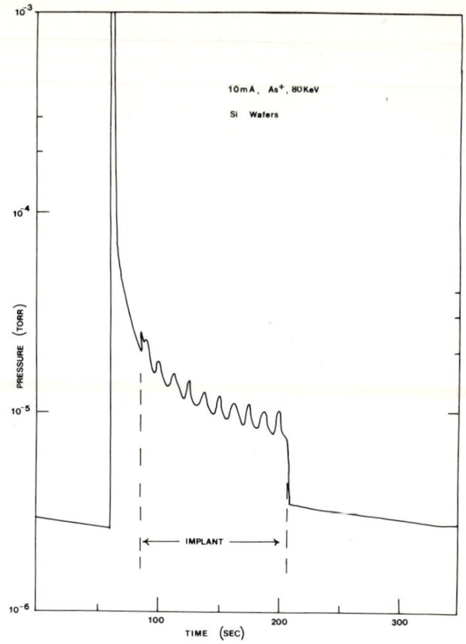


Figure 22. End station pressure vs time for 10 mA, 80 keV, $As^+$ in bare Si

## 11.3 Depth Profile Errors

Ions in the NV-10 are accelerated in a single stage at the ion source and then analyzed. Consequently, in normal operation, all particles arrive at the target with the same velocity even if charge exchange occurs after analysis. A typical implant profile obtained by SIMS analysis is shown in Fig.23 (provided by courtesy of G. Jung and

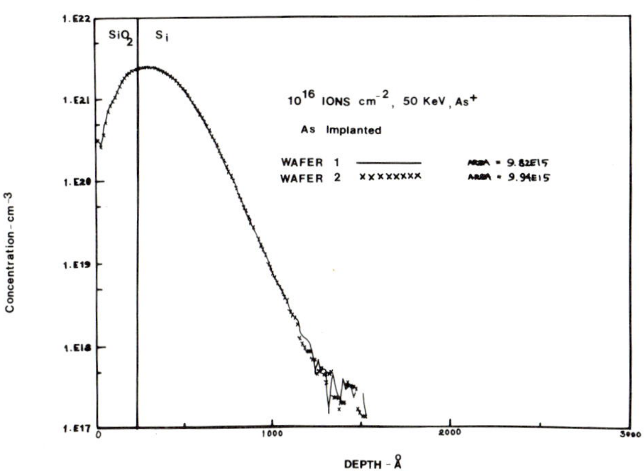


Figure 23. SIMS profile for 50 keV, $10^{16}$ ions $cm^{-2}$, $As^+$ implant

R.H. Kastl of I.B.M. East Fishkill). In the NV-10, energy contamination can only occur when doubly or triply charged ions are used to extend the energy range of the system[13].

### 11.4 Secondary Particle Effects

As shown in Fig.20, the NV-10 beamline is constructed so that all primary, satellite and neutral beam components are intercepted by water-cooled aluminum. The dominant satellite components strike a simple replaceable strike plate D, which is recessed as shown so as to minimize secondary effects. Furthermore, a water-cooled collimating aperture at B greatly reduces the solid angle in which secondary particles can directly reach the wafer. In fact, direct line-of-sight access to the wafer is only possible from the distant region E, which is never subjected to beam bombardment.

For the reasons discussed above, it is concluded that "neutral beam effects" in the NV-10 are negligibly small. The ability to keep the beamline short and use the analyzer as a neutral trap is one of the main reasons for selecting a mechanical scan design in a high-current implanter. Indeed, none of the eight designs of mechanical scan systems[15] (Western Electric PR30, IBM "Taconic", Hughes, Eaton/Nova NV 10-80 and 10-160, Ulvac IM-80H, Hitachi, Applied Implant Technology Model III-A and III-X, and Varian Model 200-1000) use a separate neutral trap, because none is needed.

## 12. Contamination

The primary ion beam itself is also a source of secondary particles which can reach the wafer. Contamination by sputtering occurs to some extent in all implanters and has been discussed in the literature[16,17]. In the NV-10 the entire beamline and wafer disks are constructed of aluminum, a relatively passive material, and the use of heavy metals (Fe, Cr, Cu etc.) has been avoided. The closest beam-defining aperture to the wafers is located 28cm in front of the disk and is graphite-faced. Consequently it is believed that the wafer fixturing and the disk itself are the only significant sources of secondary particles. In this regard every attempt has been made to minimize beam-exposed surfaces in front of the wafers. The most difficult situation is of course when a wafer clamp is required for wafer cooling (see Section 8.1.4). In this case we have added a small lip to the front surface of the clamp as shown in Fig.4 so that the portion of the clamp in contact with the wafer is shadowed from the beam.

Secondary particles are a particular problem when the beam sputters dopant materials which were implanted in the surfaces during previous use of the machine. This phenomenon is often referred to as the "memory effect". In regard to this effect the automatic disk interchange scheme [18] used on the NV-10 offers a particularly attractive feature. The double disk arrangement enables disks to be easily changed without reducing throughput and if required, the memory effect can be minimized by the use of "dedicated" disks for each dopant type.

## 13. Wafer Handling

The first sixty NV-10 units are operated with manual loading and unloading of the wafers. A fully automatic cassette-to-cassette wafer handling

Figure 24. The AT-4 automatic wafer handling system

system (Model AT-4) has recently been developed. This unit, which is now available as a standard option, is shown in Fig.24 and will be described in detail at this conference[19].

## 14. Summary

The most important performance characteristics of the Nova NV-10 Implanter have been discussed in the preceeding sections, and whenever possible, actual field operating data has been presented. By ordinary implanter standards the NV-10 is now a "mature" product. However, ongoing customer feedback inevitably fuels the design review process and leads to continued evolution of the system. In particular every effort is made to simplify designs and improve reliability.

New equipment designs are also being introduced. Development of a 160 keV version (Model NV-10-160) is well underway for delivery later this year. An infrared detector system which enables wafer temperatures to be monitored throughout the implant has recently been developed. At the same time the control systems and software are also evolving, and as the Datalock system assumes greater responsibility for machine set-up, the ultimate objective of complete implanter automation will be approached.

## Acknowledgement

This paper summarizes the research and development efforts of a large team of enthusiastic scientists and engineers. Although unable to acknowledge them all individually, the author would like to express appreciation to Peter Rose and David Hopkins, whose inspiration helped to found and fund Nova Associates.

## References

1. G. Ryding and M. Farley, Inst. Phys. Conf. Ser. 54 (1981) 100
2. P.L.F. Hemment, Inst. Phys. Conf. Ser. 54 (1981) 77
3. D.M. Jamba, Nucl. Inst. and Meth. 189 (1981) 253
4. S. Matteson, D.G. Tonn, and M.A. Nicolet, J. Vac. Sci. Technol., 16 (1979) 882.
5. G. Ryding, M. Farley, M. Mack, K. Steeples and V. Gillis, 10th International Conf. of Electron and Ion Beam Science and Technology. Montreal (1982).
6. G. Ryding, this conference.
7. S.M. Sze, Physics of Semiconductor Devices, 2nd Ed., Wiley Interscience (1982) 32.
8. J.C. Irvine, Bell System Tech. J. 41 (1962) 387.
9. K. Steeples and G. Ryding, to be published.
10. M. Mack, this conference.
11. G. Ryding, T. Maymay, V. Benveniste and M. Farley, 10th International Conf. of Electron and Ion Beam Science and Technology. Montreal (1982)
12. Semi, 625 Ellis Street, Suite 212, Mountain View, CA USA.
13. J.H. Freeman, Inst. Phys. Conf. Ser. 28 (1976) 340.
14. A.B. Wittkower, Solid State Technology (1982)
15. G. Ryding, Nucl. Inst. and Meth. 189 (1981) 239.
16. P.L.F. Hemment, Vacuum 29 (1979) 439.
17. E.W. Haas, H. Glawischnig, G. Lichti and A. Bleier, J. Electronic Mater 7. (1978) 525
18. G. Ryding and A. Armstrong, Nucl. Inst. and Meth. 189 (1981) 319.
19. A. Armstrong and G. Ryding, this conference.

# Ion Implantation Equipment from Veeco

W.A. Scaife and K. Westphal

Veeco Industrial Equipment Div.
Austin, Texas, USA

## 1. Introduction

Veeco/AI, a division of the Veeco industrial-equipment group, has recently introduced a number of improved and redesigned ion implantation systems. The major goals were to serve the requirements of both development and production, using the same basic system with modified end stations, with a high-quality standard to make the systems reliable tools with a minimum in downtime. In the following paragraphs, we present a summarized description of our developments.

## 2. A Versatile Ion Implantation System

The Veeco model 2100 MPR ion implantation system (Fig. 1) is designed specifically to maximize wafer-handling flexibility, process flexibility, process yield, and maintenance ease. It appeals especially to the facility that must utilize its implanters for R & D, process development, pilot-line and full-level production, yet does not wish to dedicate separate systems to each function.

A unique carousel end station (Fig. 2) is fundamental to the system's versatility. The processing chamber includes two wafer carousels with a capacity of either 40-4" wafers, 52-3" wafers or 80-2" wafers. The carou-

Fig. 1. Veeco model 2100 MPR

Fig. 2. Veeco carousel end station

sels may be operated in two modes: (1) hybrid scanned mode, or (2) x-y scanned mode.

The hybrid mode of operation allows very rapid implanting of an entire carousel. During implantation, the carousels are rotated at ≅ 100 rpm while the ion beam is electrostatically scanned in the y axis. The resulting implantations are very reproducible (0.5 % wafer to wafer, more than $1x10^{11}$ ion/cm² run to run) and very uniform. The rotation of the wafers into and out of the beam provides very simple and effective wafer cooling. Typically, the wafer temperature rise will not exceed 120°C for doses to $5x10^{15}$ ion/cm² performed at full beam energy (540 µA at 210 keV). Production rates of 160-4" wafers/hour (320-2" wafers/hour) are possible in the hybrid mode. A quartz-oscillator beam scanner and synchronous carousel drive prohibit phase-lock patterns from being generated.

For process-development purposes, the carousels may be indexed sequentially to permit individual wafer implantation. The beam is scanned electrostatically in both the x and y directions in this mode of operation. Wafers are accurately positioned as the carousel indexes, using a stepping motor and code wheel. Each wafer position is uniquely identified and displayed, so that every wafer can be processed differently and be correctly retrieved at the end of the batch. Alternatively, several or all of the wafers can be processed identically, with the carousel indexing automatically.

The third operational mode utilizes the full line of Veeco research wafer holders. The carousels are removed and the appropriate research tool inserted into the end-station lid. Research wafer holders include a heated-/cooled model that operates from 600°C to near $LN_2$ temperatures. The x-y scanning mode is used with all research tools.

The carousels and all research tools can be provided with optional spring clips to allow processing wafer fragments, GaAs pieces, or metallurgical samples. Full wafers are easily loaded in a known, fixed position. A known wafer orientation is an aid to yield-analysis work in a broad range of semiconductor applications.

The unit also offers exceptional process flexibility. It is an intermediate-current system (540 μA $As^+$ and $P^+$, 300 μA $B^+$), yet can be controlled with minimal analyzing slit usage down to 50 nanoamperes and less. This unusually broad operating-current range is the direct result of the third-generation hot-filament ion source that is used (Fig. 2). This source is characterized by low power consumption (500 W), stable output, high brightness, optimum results from fluorinated gas feeds, long filament life (25-40 hours with As), and broad dynamic range. The use of fluorinated gas feeds yields large double-ionized beams without the mass ambiguities common with arsine and phosphine. No additonal mass filter or velocity filter is necessary to produce pure $P^{++}$ or $As^{++}$ beams.

The system employs a pre-analysis/post-acceleration design with pre-analysis voltages of 25 kV and 35 kV (front-panel selectable). The standard analyzing magnet will separate antimony at a 25 kV pre-acceleration setting. The post-acceleration voltage is variable between 0 and 175 kV, for a maximum single-ionized beam energy of 210 keV. An optional decelerator allows operation from 5 keV to 210 keV.

Several unique features directly enhance the ion implantation process yield. The use of fluorinated gas feeds permits total cryopumping of the system, including the source. Cryopumping of the source in ion implanters is possible only when the gas feeds do not produce an abundance of hydrogen. This has been the historical reason for the "hybrid" pumping of implanters. It is not physically possible to cryopump the H and $H_2$ gas load from arsine and phosphine. However, the fractions of $AsF_5$ and $PF_5$ are readily cryopumped. The system represents the culmination of over four years of research and field use of cryopumps in this application. The complete absence of diffusion-pump oil totally eliminates the possibility of yield impact from this source of contamination. The unit employs three Air Products CSW-202 pumps. Each is equipped with automatic regeneration, and isolated from the mechanical pump by a molecular sieve.

Reduction of particulate contaminants is another yield-impact-related area. The design of the carousel places all moving parts below the wafer position. Particulate contamination is minimized since all "dirt generators" are physically beneath the wafers during the entire process. The surfaces of the end station are easily cleaned when required.

The system brings together several factors which result in superior dose reproducibility. The most significant are: a unique secondary suppression system, an unusually stable ion-source design, and a standard use of the Brookhaven model-1000 D current integrator.

A unique supression system is employed, which ensures the highest doping accuracy under all process conditions. In addition to the conventional negative-electron mirror located immediately behind the aperture mask, the cylinder wall of the Faraday cup (extending from the suppressor to the target) is positively biased to repel all positive secondary ions back to the target. Negative secondaries which hit this cylinder circulate through the bias circuit, and do not go through the current integrator. This technique is essential when positive-secondary-producing processes are used, such as implanting through photoresist masks. The yield effects of this approach are especially noticeable in the low-dose range ($1 \times 10^{10}$ ions/cm$^2$ to $4 \times 10^{12}$ ions/cm$^2$).

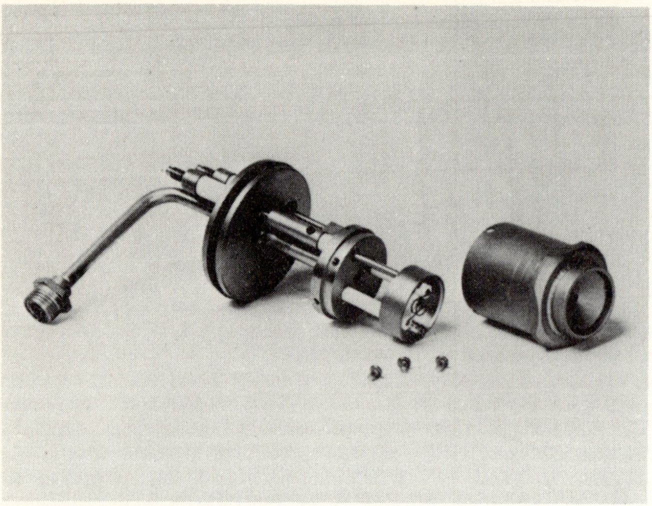

Fig. 3. Veeco hot-filament source

The advanced-design hot-filament source used (Fig. 3) is extremely stable. Short-term transient bursts are seldom observed, and have lower peak levels than those seen in Freeman-type sources. Dose inaccuracies due to bursts are further minimized by the use of the Brookhaven 1000 D current integrator. The 1000 D will accept a 10 mA pulse without saturating, making it virtually impossible to overdose a wafer due to integrator pulse-reading limitations. The 1000 D also exhibits the lowest inherent noise level available in the industry, and this contributes to the low-dose reproducibility of the 2100 MPR.

Dose uniformity yields of better than 0.75 % (1 $\sigma$ over 4" wafers) in both the hybrid and x-y scanning modes are obtained. Results substantially better than this have been achieved under certain process conditions. The beam scanner frequencies (256 Hz and 2569 Hz) are generated by a single quartz-crystal oscillator and a frequency-divider chain. The frequencies have been carefully chosen to provide optimum uniformity, both wafer to wafer and across a given wafer. In the hybrid mode, the y-axis scan is 2569 Hz, while the carousels rotate at $\cong$ 100 rpm. The beam scanner is coupled to the scan plates through a power amplifier that drives low-impedance, high-voltage output transformers. Scan frequencies and carousel-rotation speed are fixed to eliminate yield difficulties arising from operator misadjustments. Details on how the various frequencies were chosen are available in reprint form.

Total device yield is strongly influenced by reliability as well as maintenance ease in ion implanters, simply because a system cannot produce usable devices while it is being repaired. Despite an engineering of the maximum possible reliability, it is realized that service will be required at some time, and diagnostic centers have been provided to speed any repairs. When a malfunction occurs, the operator is advised that the beam gate has closed, and the general nature of the fault is shown by fifteen front-panel fault indicators. If the fault is not correctable through front-panel adjustment, a technician will then examine the diagnostic centers. Each power supply in the system (except 25/35 kV and 175 kV) is tested for voltage, current, and control. "Control" relates to the proper function of the opto-isolators used throughout the system. Data from the diagnostic centers transfer directly to flow-chart documentation. The flow-charts then lead immediately to component-level schematics, so that a repair can be quickly made.

This ion implanter represents the extension and refinement of intermediate-current technology into a high-reliability, maximum-flexibility tool. The design incorporates several advances that enhance device yield, and also includes features that minimize implanter-induced variables into the process. The system uniquely serves the needs of an organization that must use a single ion implanter for R & D, pilot-line operation, and full-level production.

## 3. A High-Production Ion Implantation System

To process 3", 4", and 5" wafers with full-contact wafer cooling, the 2100 WE series (Fig. 4) was introduced. These systems are designed to process 275 cooled wafers/hour, with typical breakage of less than 0.1 %. Veeco's three new 200-keV ion implanters deliver performance that no competitive equipment can match. At 275 wafers/hour, the Veeco 2300 WE, 2400 WE, and 2500 WE have the industry's highest throughput, with full-thermal-contact wafer cooling to accommodate high-dose implants. The Veecool™ cooled target block uses no rubber or elastomers, and is not damaged when hit by the ion beam.

The Veeco Wafermatic cassette-to-cassette wafer-transport system has the fewest moving parts of any ion implanter, which minimizes particle contamination and improves transport reliability. The industry's only

Fig. 4. Veeco 2100 WE series, cassette to cassette

double-interlocked transport logic prevents wafers from breaking in the seal assembly.

Only Veeco has a 10,000:1 dynamic range ion source to give higher beam purity, less sputtered metal contamination and better device yield, and only Veeco implanters are totally cryopumped.

The matching scan matrix of the target block allows easy conversion from one wafer size to another. A console-mounted x-y recorder can be used for initial tuning and verification of correct operation. The standard WE operates as low as 25 keV. A decelerator allows system operation down to 5 keV.

These are the only implanters with 0.5 % guaranteed reproducibility from wafer to wafer, even for doses as low as $1x10^{11}$ ions/cm², and the only 200 keV ion implanters with Veeco reliability. A reliability that's backed up by an one-year warranty covering parts, service, and service transportation.

## 4. Ion Implanter for GaAs - Device Development and Production

The Veeco model-400 MPK 400 keV ion implanter (Fig. 5) is a standard-product ion implanter, especially designed to meet the needs of production-level GaAs manufacturers. Three distinct areas of this equipment have been engineered with these special requirements in mind: (1) four standard and optional ion sources, to provide for the wide variety of ion species needed when processing III-V materials; (2) a mass-analysis system with exceptional resolution and mass range, to provide isotopically pure ion beams up to 210 amu; and (3) a production end station that handles wafers and pieces, while providing temperature control from ambient to 550°C.

This ion implanter is highly modular, and offers considerable flexibility for the user. The selection of options, or the setup of the machine,

Fig. 5. Veeco type-400 MPH implanter for GaAs

Table 1. Ion species and ion sources for 400 MPK implanter

| Species | $I_{max}$ (μA,3"scan,400 kV) | Feed Material | Ion Source |
|---|---|---|---|
| $Si^+$ | 12 | $SiF_4$ | hot-filament |
| $Si^{++}$ | 4 | $SiF_4$ | hot-filament |
| $Se^+$ | 33 | $SeF_6$ | hot-filament |
| $Se^{++}$ | 1.8 | $SeF_6$ | hot-filament |
| $Be^+$ | 8 | $BeCl_2$ | hot-filament |
| $Be^+$ | 14 | $PF_5$ | cold-cathode w/beryllium canal |
| $O^+$ | 60 | $O_2$ | hot-filament |
| $S^+$ | 40 | $H_2S$ | hot-filament |
| $H^+$ | 270 | $H_2$ | rf |
| $H^+$ | 18 | $H_2$ | hot-filament |
| $Ga^+$ | 13 | $GaCl_3$ | oven, hot-filament |
| $Mg^+$ | 31 | Mg | oven, hot-filament |

begins with the selection of the ion species. This choice determines the optimum ion source (see Table 1). A choice of four sources is offered. The standard model is delivered with an oven-style hot-filament source and a cold-cathode source. The oven-style hot-filament source operates using solid materials as well as gases and liquids. Solids are placed in the source-charge holder (approximately 0.3 cc capacity; heats as high as 600°C). As the solid vaporizes, it is ionized by the hot filament and extracted. Gases are fed directly into the source and ionized; liquids with high vapor pressures may also be used, so that they evaporate at source vacuum levels. This source has an exceptionally well-defined energy spread of 10 eV, which results in excellent mass resolution.

The cold-cathode source is extremely rugged and reliable. It can be run with a variety of gases, but is most effective for producing boron and beryllium beams. Boron is produced by $BF_3$ gas. Beryllium is produced by running the source on $PF_5$ and tuning for maximum sputter erosion of the beryllium exit canal. This is an extremely convenient means of producing this species. An optional rf ion source is available for users who wish higher currents (≅100μA) of hydrogen, or noble gases (He, Ar, Ne, Kr, Xe). The most significant application of this ion source is in proton-damage implants.

An optional hot-filament source is also available. This source offers the same performance features as the standard oven-style hot-filament source, but does not include the oven. Users limiting themselves to gaseous and liquid feed materials can utilize this ion source.

The mass-analysis capability of implanters that have not been specifically designed with GaAs requirements in mind is often inadequate to resolve the heavier ions needed for GaAs. This system has a standard MxE = 38 (mass-energy-product = 38) magnet that will analyze up to 95 amu at 400 keV. This standard magnet, with the hot-filament ion source and a $SeF_6$ gas feed, will give base-line resolution of selenium isotopes.

An optional higher-field magnet with MxE = 84 is offered. It will analyze up to 210 amu at 400 keV. The MxE = 84 magnet provides for unit resolution across its entire range of operation. Pre-acceleration post-analysis design ensures optimum resolution across the entire operating range.

The ion implanter system is equipped with a special carousel end station designed specifically for GaAs processing. Wafers are mounted on the carousel with convenient thin clips. The mounting platen is a special glass material. There are 10 platens, each capable of holding a 2-inch wafer. Four clips per platen allow the mounting of four irregular-size pieces, for a total batch capability of 40 pieces. The angle of incidence is adjustable at each individual platen, from 0° to 7° from normal. The carousel indexes sequentially to implant one wafer at a time. The ion beam is scanned in the x and y directions electrostatically.

High-intensity lamps operating in atmosphere may be used to heat the GaAs. The lamps transmit infrared radiation through a vacuum window, and also through the glass wafer-mount platens. The wavelengths of IR produced by the lamps have been selected for optimum absorption by GaAs, as well as maximum transmission through the glass platen and vaccum window. Two wafers are heated at a time, one being preheated prior to the implant, and the other during the implant, to maintain the temperature. The temperature range is from ambient to 500°C.

The end station also accepts standard unheated carousels for wafers up to 3 inches in diameter. All carousels are removable and special research wafer holders can be installed to extend the system's usefulness in research and development applications.

# The Series IIIA and IIIX Ion Implanters

D. Aitken

Applied Implant Technology Inc.
Horsham, West Sussex, England

## 1. Introduction

The demand for high-current ion implanters in the semiconductor industry has increased substantially over the last 3 - 4 years and the requirement has divided into two categories:

i. Implant machines.
ii. Pre-deposition machines.

Implant machines are high-energy (typically up to 200kV accelerating potential) machines where the implant depth is that required in the device. Pre-deposition machines are lower-energy machines (typically 40kV to 120kV accelerating potential) where the implant is used as a convenient way of introducing a uniform, precisely calibrated amount of dopant into the surface region of the wafer and the dopant can later be driven down to the required depth distribution by subsequent heat treatment. In general these machines are required for high-dose implants and therefore a high current capability is essential. The energy requirement is largely determined by the surface conditions on the wafer. If there is already an oxide layer on the wafer it is often desirable to implant through this oxide rather than have the extra process step of removing it. It is sometimes found necessary to intentionally deposit an oxide layer prior to pre-deposition implants in order to protect the surface from contamination during the implant. In the case of arsenic implants the energy requirement for implant through oxide is likely to be in the 100keV to 120keV range. Another aspect which can influence the pre-deposition implant energy is the desirability in many applications to minimise the surface concentration of the dopant.

As a consequence of the above considerations, the present trend in the pre-deposition machine requirements is towards higher energies than was originally considered necessary.

## 2. The Series III Machine

The Series III machine has been described in two previous publications [1, 2]. The Series III machine has now been superseded by two variants, IIIA and IIIX which are externally identical as shown in Fig.1. The IIIA is similar to the Series III but has a larger exit aperture in the ion source (60mm x 2mm in IIIA, 40mm x 2mm in III), and has increased pumping capacity and a larger capacity processor.

The IIIA machine, like the III, is a 200kV machine with a 40kV maximum extraction voltage and up to 160kV post-acceleration across a single gap.

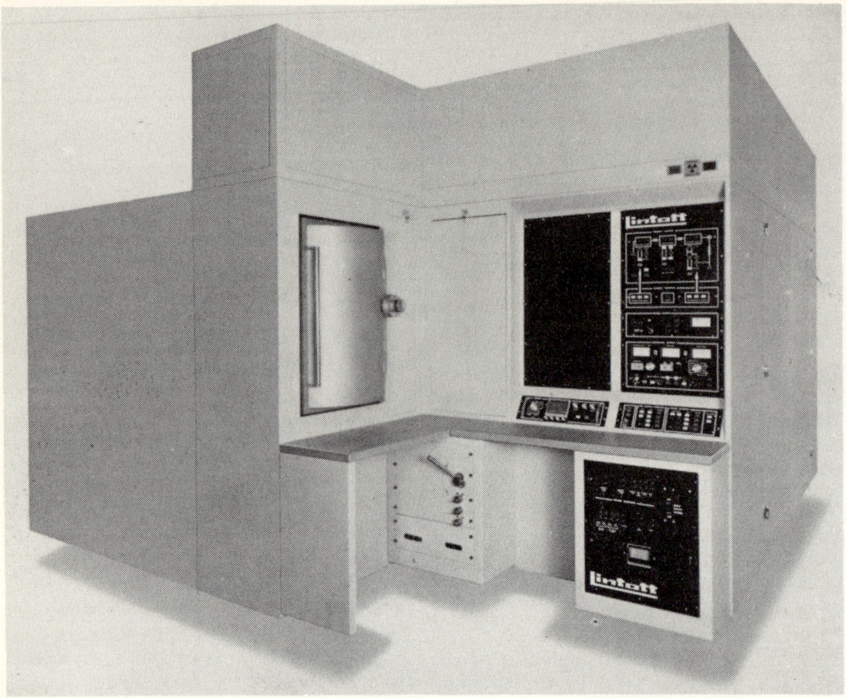


Fig.1. The Series IIIA (or X) machine

The IIIX machine is a 120kV pre-deposition variant with modified ion optics. The differences between the IIIA and IIIX are:

i. Ion source aperture size.
ii. Post-acceleration system.
iii. Position of post-acceleration cryopump.

Otherwise the machines are identical and one type can easily be converted to the other giving a highly desirable flexibility in implant capability. Fig. 2 is a schematic of the IIIX machine and the insert shows the post-acceleration module which converts the IIIX into a IIIA machine. The post-acceleration power supply is a Cockcroft-Walton type and can be converted from the low-voltage (up to 80kV) to the high-voltage requirement (up to 160kV) by increasing the number of discs in the high-voltage stack, the power capability of the supply remaining the same.

Both machines produce a large low power density beam at the target, 60 x 40mm$^2$ in IIIA, 60 x 60mm$^2$ in IIIX. This, combined with the high heat sink efficiency wafer plates now available, minimises the radiation damage problems found in machines with high power density beams and enables high power implants to be carried out onto photoresist-covered wafers.

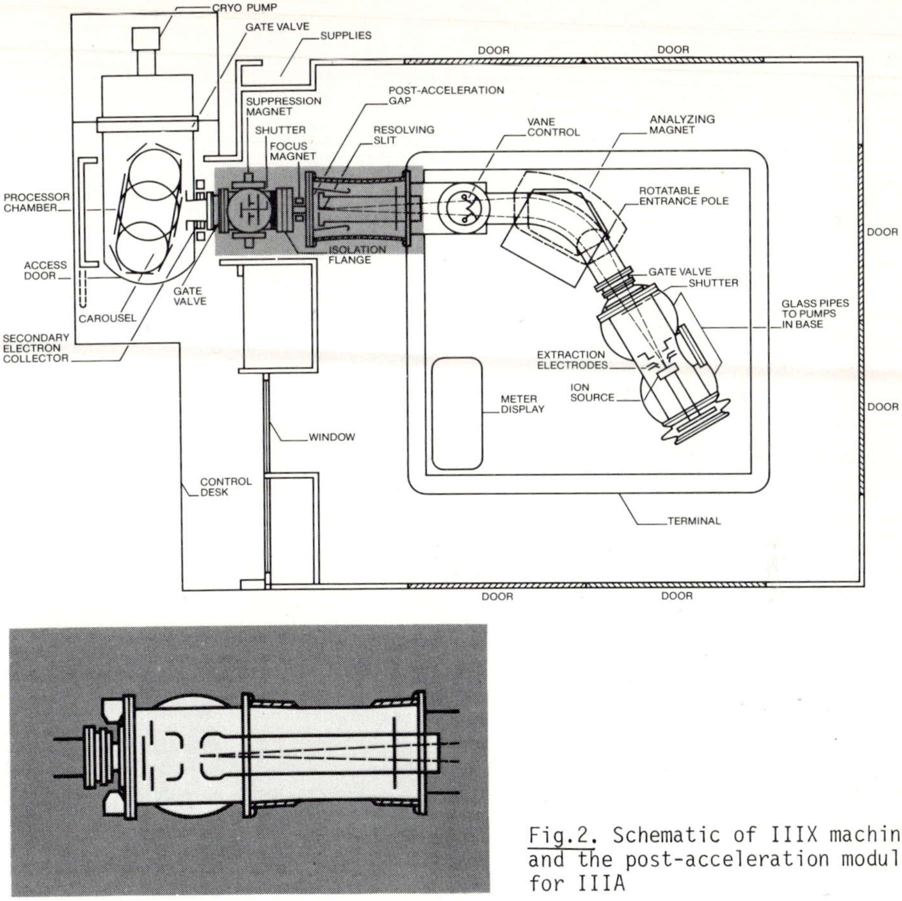


Fig.2. Schematic of IIIX machine and the post-acceleration module for IIIA

Both machines have the following important features:

1. Stabilised beam current - a vane unit controls the beam, and the system does not rely on changes in scan speed to compensate for changes in current output from the source. This is essential for the implant consistency required by state-of-the-art silicon devices.

2. The race track carousel in the processor is free from the geometric errors found in most alternative systems and the uniformity is independent of wafer size (from 2" to 150mm).

3. A wide dose range capability from $10^{10}$ to $10^{17}$ ions $cm^{-2}$.

4. High reliability Freeman source with dual 700°C vaporisers.

## 3. The Series IIIA Machine

The IIIA machine is similar to the Series III but has the following improvements:

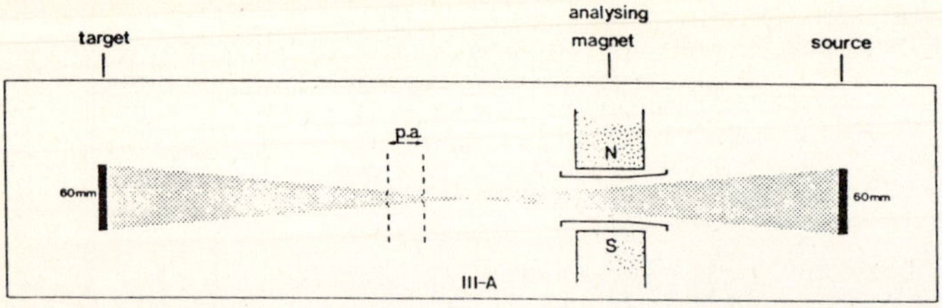


Fig. 3. Vertical plane optics for Series IIIA

Fig.4. Freeman source with external filament insulators

1. The 40 x $2mm^2$ aperture is replaced by a 60 x $2mm^2$ aperture giving increased output capability.

2. The above increase in extracted beam size is made possible by incorporating an analysis flight tube with a flared entry geometry. Fig.3 shows the geometry of the beam in the vertical plane. The beam height at the processor is 60mm.

3. The end-cap design has been improved so that the filament insulators are situated outside the arc chamber, thus greatly decreasing the rate of metallisation and significantly increasing the useful lifetime of these components (see Fig. 4).

4. The pumping speed available at the ion source region has been increased by using two diffusion pumps with three times the pumping speed of those in the original Series III design. This gives a significant benefit for ions which require a gaseous source feed material, notably boron ions from boron trifluoride.

5. The post-acceleration region is pumped by an 8" cryopump which significantly reduces the neutral content of the beam (Series III had a 100mm diffusion pump).

6. The pumping speed available at the processor has been dramatically increased by using a 20" cryopump instead of the 12" high speed diffusion used in the Series III. This gives 11,000 litres $sec^{-1}$ pumping speed for air and in excess of 30,000 litres $sec^{-1}$ for water vapour. This gives rapid pump down and consequent increased throughput as well as the important ability to cope with the rapid degassing of the wafers during the early period of a high-current implant.

7. The processor design has been improved to incorporate water cooling of the processor mechanism in order to cope with the higher power levels and the number of wafer plates and their size has been optimised for the larger wafer sizes (100 mm, 125mm and 150mm). The batch size for 100mm wafers is 40 for the standard plates and 36 for high heat sink efficiency plates.

## 4. The Series IIIX Machine

The IIIX machine is a 120kV (40kV extraction + 80kV post-acceleration) machine capable of delivering well in excess of 12.5mA of $As^+$.

The extra current capability (compared with Series III and IIIA) is achieved by using an extended extraction aperture (90 x 2mm²). This larger beam is transmitted through the same analysing magnet geometry by having a beam cross-over in the magnet flight tube as shown in Fig.5.

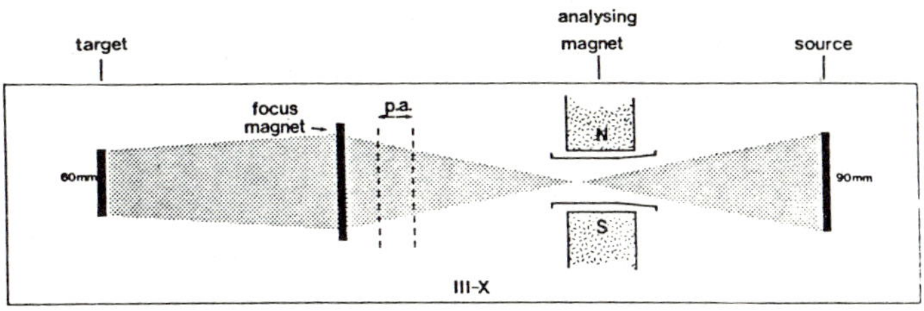


Fig. 5. Vertical plane optics for IIIX.

The beam size at the resolving slit and post-acceleration gap is consequently large (approximately 100mm height), giving a desirable low current density in this region (minimising space-charge blow-up of the beam). The beam height required in the processor is 60mm and this is achieved by having a quadrupole magnet immediately after the post-acceleration gap which focuses the beam down to the required size. This focusing is achieved automatically by using the feedback from two trim electrodes in the beam monitor which is situated between the focus magnet and the shutter. The post-acceleration electrodes are curved so that at maximum post-acceleration voltage the required focusing is achieved electrostatically. This means that the focus magnet is only required for strong focusing at low energies, and as the energy is increased the stronger is the focusing action of the post-acceleration gap. This minimises the size and power requirement of the focus magnet.

The IIIX machine shares many of the features of the IIIA:

1. Larger diffusion pumps on the ion source chamber.
2. 20" cryopump on the processor chamber, 8" cryopump for the post-acceleration region.
3. Larger-capacity processor with water-cooled mechanism.

## 5. The Electron Flood Gun

Certain types of device are prone to surface charging during high-current ion implantation,which can lead to device damage and consequent loss of yield.

An electron flood gun accessory is available (see Fig.6) which floods the surface with low-energy electrons which are then available to neutralise any positive charge buildup on the wafer surface. As the gun is completely contained within the magnetic suppression system, this electron current has no effect on the accuracy of ion dose measurement.

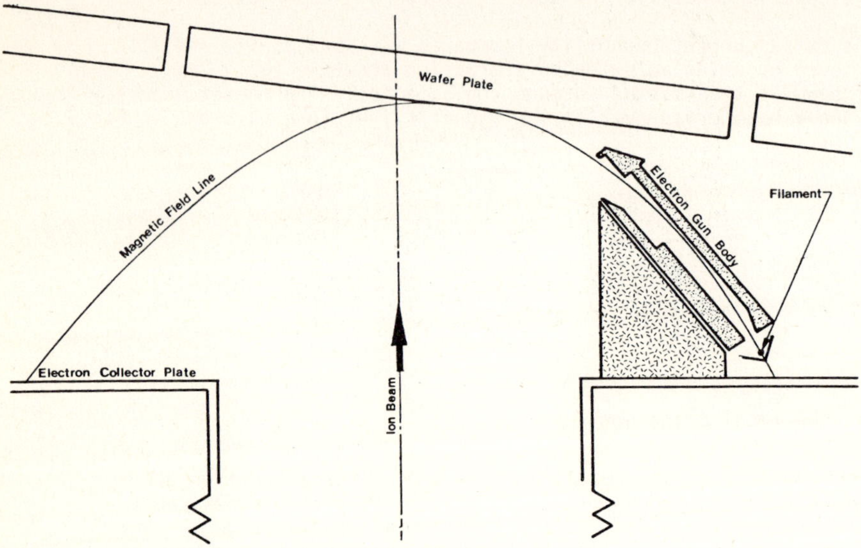


Fig. 6. Schematic of Electron Flood Gun

This system uses the magnetic suppression field to guide the beam tangentially to the surface of the wafer, and the curvature of the electron path in this magnetic field allows the system to be designed so that there is no line of sight between the gun filament (0.020" diameter tungsten wire) and the wafer. The system is generally used with an electron current equal to the ion current, but such a high current level is often not necessary.

## 6. The Process Verification System

A microprocessor-based process verification accessory is available which consists of a small video monitor screen and a keypad mounted adjacent to the implanter oscilloscope as shown in Fig.7 together with an electronic chassis mounted elsewhere.

The primary function of this unit is to receive the required implant parameters and to inhibit the start of an implant if any of these parameters are incorrect. The implant recipe is displayed alongside the actual machine parameters and any mismatch is highlighted by reverse printing the particular machine settings.

Fig.7. Oscilloscope/video monitor/keypad unit

The settings which are monitored are:

1. beam energy
2. dose per scan
3. number of scans
4. scan height
5. dopant.

The required implant parameters can be entered into the system in any of three ways:

1. The conditions can be entered directly via the keypad. Questions are displayed on the screen for each parameter and the appropriate information entered.

2. Up to 127 recipes can be stored in memory and the implant parameter recalled by entering the appropriate recipe number. The recipes are pre-programmed by means of the keypad.

3. The recipes can be downloaded from an external host computer. Communication is by RS-232C using Hewlett Packard 'SECS' protocol.

## 7. Specifications

### 7.1 IIIA

#### 7.1.1 Maximum Beam Currents into Processor

| Ion species | Beam current | Ions $cm^{-2}scan^{-1}$ |
|---|---|---|
| Boron | 2mA | $4 \times 10^{14}$ |
| Boron* | 2.5mA | $5 \times 10^{14}$ |
| Phosphorus | 6mA | $1.2 \times 10^{15}$ |
| Arsenic | 6mA | $1.2 \times 10^{15}$ |
| Argon | 3mA | $6 \times 10^{14}$ |
| Antimony (35kV) | 2mA | $4 \times 10^{14}$ |
| Boron (doubly charged) | 25μA | $2.5 \times 10^{12}$ |
| Phosphorus (doubly charged) | 400μA | $4 \times 10^{14}$ |
| Arsenic (doubly charged) | 1mA | $1 \times 10^{14}$ |

Minimum beam current for all species: 100nA ≡ $2 \times 10^{10}$ions $cm^{-2}scan^{-1}$.

*The higher boron current specification is for a standard ion source dedicated to boron

7.1.2 Machine Throughput (100mm wafers)

| Number of scans | Throughput wafers/hour | Dose in ions/cm² at 6mA |
|---|---|---|
| 1 | 300 | $1.2 \times 10^{15}$ |
| 2 | 215 | $2.4 \times 10^{15}$ |
| 3 | 168 | $3.6 \times 10^{15}$ |
| 4 | 137 | $4.8 \times 10^{15}$ |
| 5 | 116 | $6.0 \times 10^{15}$ |
| 6 | 101 | $7.2 \times 10^{15}$ |
| 7 | 89 | $8.4 \times 10^{15}$ |
| 8 | 80 | $9.6 \times 10^{15}$ |

Minimum beam current for all species: 100nA ≡ $2 \times 10^{10}$ ions $cm^{-2}scan^{-1}$

## 7.2 IIIX

7.2.1 Maximum Beam Currents into Processor:

| Ion species | Beam current | Ions $cm^{-2}scan^{-1}$ |
|---|---|---|
| Boron | 3mA | $6 \times 10^{14}$ |
| Boron* | 4mA | $8 \times 10^{14}$ |
| Phosphorus | 12.5mA | $2.5 \times 10^{15}$ |
| Arsenic | 12.5mA | $2.5 \times 10^{15}$ |
| Argon | 6mA | $1.2 \times 10^{15}$ |
| Antimony (35kV) | 6mA | $1.2 \times 10^{15}$ |
| Boron (doubly charged) | 40μA | $4 \times 10^{12}$ |
| Phosphorus (doubly charged) | 600μA | $6 \times 10^{13}$ |
| Arsenic (doubly charged) | 1.5mA | $1.5 \times 10^{14}$ |

*see footnote p. 357

7.2.2 Machine Throughput (100mm wafers)

| Number of scans | Throughput wafers/hour | Dose in ions/cm² at 12.5mA |
|---|---|---|
| 1 | 300 | $2.5 \times 10^{15}$ |
| 2 | 215 | $5 \times 20^{15}$ |
| 3 | 168 | $7.5 \times 10^{15}$ |
| 4 | 137 | $1 \times 10^{16}$ |
| 5 | 116 | $1.25 \times 10^{16}$ |
| 6 | 101 | $1.5 \times 10^{16}$ |
| 7 | 89 | $1.75 \times 10^{16}$ |
| 8 | 80 | $2 \times 10^{16}$ |

## References

1. D. Aitken, Nuc. Instr. & Meths. 139, 125 (1976)

2. D. Aitken, Rad. Effects 44, 159 (1979)

# Standard High-Voltage Power Supplies for Ion Implantation

M. Baumann

Emile Haefely & Cie AG, Lehenmattstr. 353
CH-4028 Basel, Switzerland

Ion implantation is an impressive example for the application of basic technologies to today's manufacturing processes. The development of the high-voltage power supplies needed for ion implantation has been particularly influenced by the techniques of dc accelerators. For economical reasons, it is impossible to custom-build the high-voltage power supplies for each of the different required applications. For that reason, it became a necessity to make a decision on the standardization of the power supply. Standardization means compromising and giving up the possibility of covering the entire range of users electrical data. In our standard line of high-voltage dc power supplies, special consideration is given to the following basic design features:

- The principle of high-voltage generation.
- Type of insulation (air-insulated or encapsulated).
- Rated dc voltage.
- Rated dc output power.
- Stability.

Based upon our experience in the design of high-voltage dc power supplies, we decided to use a Cockcroft-Walton circuit with an operating frequency of about 15 kHz for the high-voltage generation. The users' requirements made it clear that we had to offer both the air-insulated and the encapsulated designs. The rated voltages of our power supplies are multiples of 150 kV dc up to 750 kV dc, or multiples of 250 kV dc up to 1250 kV dc.

For the generation of the medium-frequency input power, several types of frequency converters were designed, with rated powers of 0.7 kW, 2 kW, or 5 kW.

With the new high-voltage generator and regulating system, the voltage stability is determined only by the quality (stability) of the measuring divider. For standard stability requirements up to $1 \times 10^{-3}$, the built-in divider of the system will be sufficient. For higher stability requirements up to $1 \times 10^{-5}$, an external RC voltage divider having a very low temperature coefficient is available.

The complete line of standard power supplies is represented by the series LM, LS, and TS. The series LM and LS are of the air-insulated design, whereas the series TS are of the pressurized design, using SF6 as the insulating gas. A new control unit is provided for these standard models. It has digital voltage and current displays, as well as analog indicators for coarse indication. As an example, a 250 kV supply is shown in Fig. 1.

Fig. 1. Modulator high-voltage power supply; 250 kV, 4 kW, stability $10^{-4}$

All of the components of these high-voltage dc power supplies are based upon semiconductor technology. For that reason, special consideration was given to making the power supplies and their medium-frequency supply and control systems short-circuit proof. This is an absolute necessity for ion implantation equipment where electrical breakdowns in the accelerator system cannot be avoided.

# The IONMICROPROBE A-DIDA 3000-30 for Dopant Depth Profiling and Impurity Bulk Analysis

J.L. Maul and H. Frenzel

ATOMIKA Technische Physik GmbH,
D-8000 München, Fed. Rep. of Germany

## 1. Introduction

A knowledge of implantation/diffusion depth profiles, for instance after subsequent oxidations or annealing steps, is of major interest in semiconductor device development and production. Secondary ion mass spectrometry (SIMS) is a technique for depth profiling down to the ppb level. Its potential strength also includes identification of the chemical nature of impurities in bulk material or epitaxial layers at concentration levels down to $10^{13}$ atoms/cm$^3$

The IONMICROPROBE A-DIDA 3000-30 is an uncomplicated SIMS system which makes this high potential strength of the SIMS method available in routine applications. Its uncomplicated design and underlying principle results not only in ease of operation and maintenance, but also makes the system extremely economical in terms of investment cost and downtime.

## 2. Some Features of the IONMICROPROBE A-DIDA 3000-30

In secondary ion mass spectroscopy (SIMS) the sample surface, under bombardment by "primary" ions, emits so-called "secondary" ions which carry information about the chemical composition and are mass analyzed. Depending on the conditions of primary ion bombardment, either just a small percentage of the uppermost monolayer is used for the analysis ("static" SIMS), or, at a much faster continuous removal rate, the concentration change with depth is measured ("depth profiling"), or the lateral distribution of preselected chemical elements or isotopes is imaged by raster-scanning the primary ion microbeam.

Hence, the IONMICROPROBE A-DIDA 3000-30 consists of a microbeam primary ion gun, a mass spectrometer, a vacuum system, and a data acquisition system (see Figs.1 and 2).

The IONMICROPROBE A-DIDA 3000-30 is a vertical configuration, with the primary ion gun on the top above the horizontal sample chamber. The sample transfer system with the central vacuum control is placed at the right-hand side, whereas all electronic supplies and controls are located in the cabinets on the left-hand side.

The primary ion gun is equipped with three easily interchangeable types of ion sources with a hot cathode for noble gases like Ar, a cold cathode for reactive ions like $O_2$, and heated frit for Cs. Each one is optimized in terms of ion beam quality, lifetime, and maintenance, and is designed to get the optimum trace sensitivity out of the SIMS method [1]. A special telefocus ion beam line provides, on the way down to the sample, enough space for very effective differential pumping of the ion gun, removal of impurities from the primary ion beam by mass separation, and fine focussing

Fig.1. IONMICROPROBE A-DIDA 3000-30

of the primary ion beam (the smallest beam size is a few µm). The beam is raster scanned across the sample, not only in the ion-imaging, but also in the depth-profiling mode. Raster scanning in combination with restriction of the analyzed sample area to the center of sputter crater by electronic gating [2] and removal of neutrals from the primary ion beam reduces crater edge effects so effectively that six orders of magnitude of dynamic range in implantation profile measurement are achieved [3].

The primary ion energy can be selected from 0.5 keV to 15 keV with up to some $10 mA/cm^2$ current densities, low energies for optimum depth resolution, high energies for high-speed depth profiling from tiny sample areas, and high lateral resolution in imaging.

The mass spectrometer is based on a high-transmission quadrupole mass filter with special ion optics for secondary ion collection and energy analysis in front, and a single particle-counting device at the exit. The use of a quadrupole has a number of important advantages for SIMS analysis which finally result, directly or indirectly, in low intrinsic system background and high transmission (high trace sensitivity); an absence of memory effects and problems in analyzing insulating material; fast switch positive/negative secondary ion detection from the same sample area; UHV compatibility; uncomplicated sample positioning; variation of the primary ion incidence angle the possibility of simultaneous depth profiling by SIMS and AES, and ease of operation and maintenance. All of the ion optics are made of a special patented material which together with the other design features, keeps the intrinsic system background at a very low level.

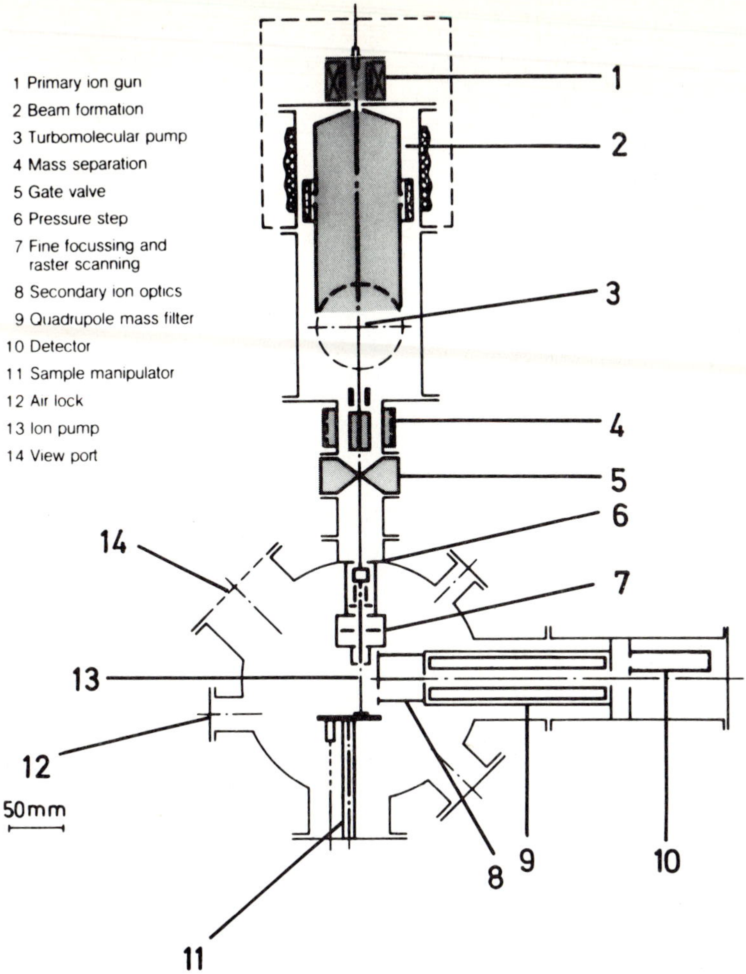


Fig.2. IONMICROPROBE A-DIDA 3000-30, schematic

The vacuum system consists of separate turbopumps for the ion gun and the sample transfer system, an ion pump with Ti-sublimation pump, and a cold wall for the analysis chamber. A unique, very effective, but uncomplicated internal bakeout system helps to keep the base pressure in routine operation in the $10^{-10}$ Torr range. It takes only 5 minutes to transfer a sample from atmospheric pressure down to $10^{-9}$ Torr operating pressure. A central vacuum control operates all valves and vacuum gauges with safety interlocks for easy push-button operation (see Fig.3). The considerable effort made to supply true state-of-the-art UHV reduces to a minimum any negative influence which residual gas bombardment of the sample might have on reproducibility, quantification and detection limits.

Data acquisition and output are controlled by a specially designed microprocessor system. The central mode controls set the system to different

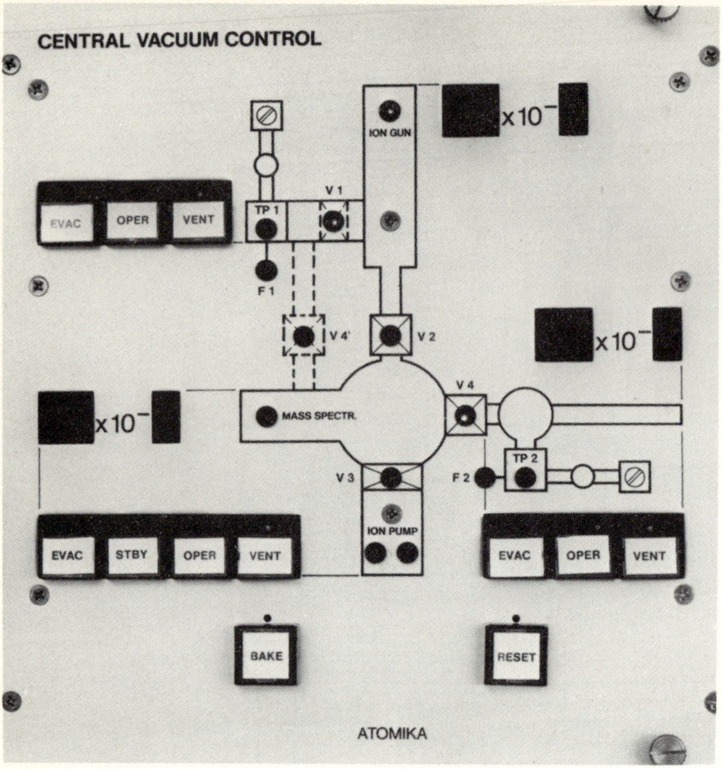


Fig.3. Central vacuum control

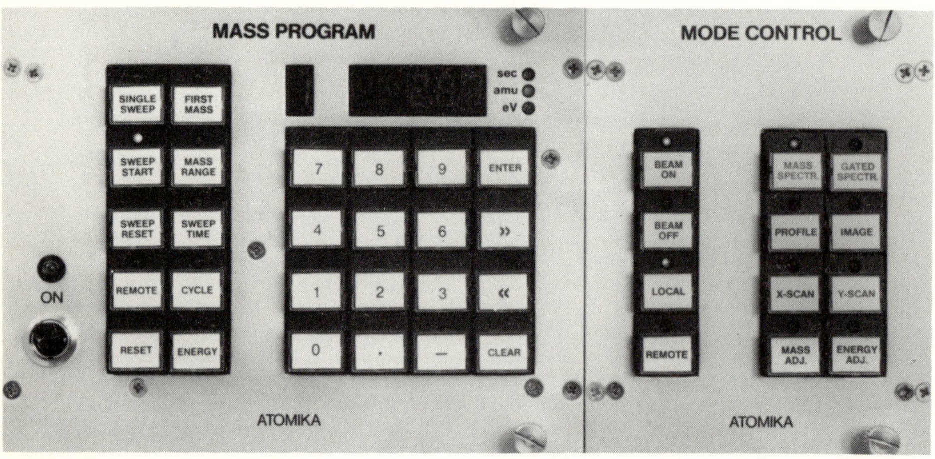


Fig.4. Microprocessor control of A-DIDA 3000-30

Fig.5. Secondary ion mass adjust. On the CRT: chosen center mass ± 1.5 amu and marker

modes of operation by single push-button commands: mass adjust, energy adjust, mass spectrum, depth profiling, imaging, etc. (see Fig.4).

The selection of secondary ion mass, for instance, and secondary ion energy — important for trace sensitivity [4] — is simply done by placing a marker on a scope which displays the mass spectrum around the desired mass (see Fig.5) and the secondary ion energy distribution of this mass. Microprocessor-controlled analog and digital output channels record the data. A PDP11-based data system is available, if more sophisticated data handling is desired.

## 3. Some IONMICROPROBE A-DIDA 3000-30 Applications

The versatility of the A-DIDA 3000-30 covers the whole range of SIMS applications from static SIMS and depth profiling to imaging, including the analysis of insulators and even powder chemicals. Its special strength, however, is high-quality depth profiling in terms of depth resolution and detection limits. Many data have already been published, so only two examples are shown

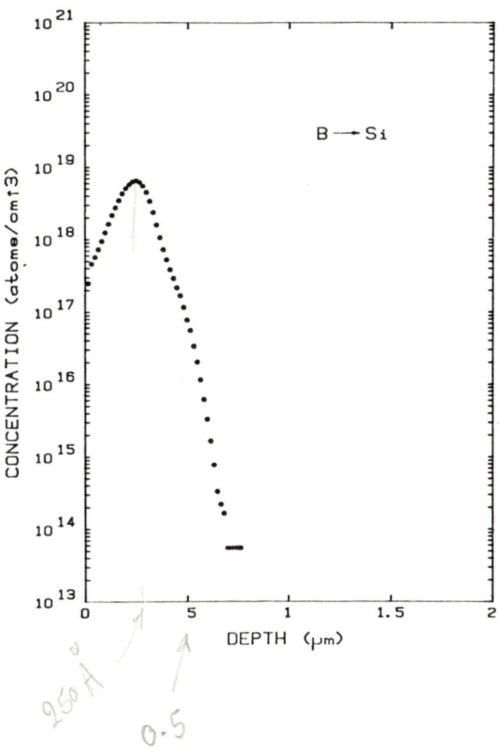


Fig.6. Depth profile of boron in silicon. Ion implantation energy 70 keV, implanted fluence $1 \times 10^{14}$ cm$^{-2}$. B concentration down to $10^{13}$ atoms/cm$^3$ range (no background subtraction) was measured

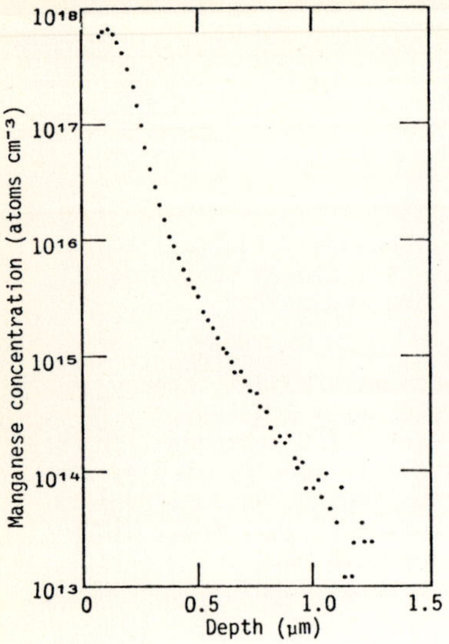


Fig.7. Depth profile of manganese in GaAs. Ion implantation energy 200 keV implanted fluence $1 \times 10^{13}$ $cm^{-2}$. Mn concentration down to $10^{13}$ atoms/$cm^3$ range (background subtracted) was measured [5]

here, one for Si and one for GaAs. For Si, a B depth profile is shown in Fig.6.; As and P profiles can be measured down to the $10^{15}$ atoms/$cm^3$ range. For GaAs, a Mn depth profile is shown in Fig.7; the $10^{13}$ atoms/$cm^3$ level is also reached for many other elements, such as Be, Mg, Cr, and Fe.

Two special features that are especially important for depth profiling may be mentioned in addition: with depth profiling, ion images of the selected masses are always displayed simultaneously, sometimes showing that a dip visible in the depth profile is just caused by sample inhomogeneity, especially at interfaces, thus avoiding misinterpretation of the measured depth profile.

In the IONMICROPROBE A-DIDA 3000-30, the sample is usually bombardes by the primary ions at normal incidence. It has been shown that in this way, quantitatively correct measurements of the segregation of dopants during subsequent oxidation of silicon can be made [6]. Such data provide important input parameters for process and device modelling.

## References

1 K. Wittmaack: Nucl. Instr. Meth. *168*, 343 (1980)
2 K. Wittmaack: J. Appl. Phys. *12*, 149 (1977)
3 K. Wittmaack: Appl. Phys. Lett. *37*, 285 (1980)
4 J.B. Clegg: Surf. Interface Anal. *2*, 91 (1980)
5 J.B. Clegg et al.: J. Appl. Phys. *52*, 1110 (1981)
6 To be published

# Index of Contributors

# Subject Index

Page numbers in *italics* refer to extended passages on the respective subjects